WIRELESS NETWORKING

The Morgan Kaufmann Series in Networking
Series Editor, David Clark, M.I.T.

For further information on these books and for a list of forthcoming titles, please visit our website at http://www.mkp.com

WIRELESS NETWORKING

Anurag Kumar

D. Manjunath

Joy Kuri

ELSEVIER

AMSTERDAM • BOSTON • HEIDELBERG • LONDON
NEW YORK • OXFORD • PARIS • SAN DIEGO
SAN FRANCISCO • SINGAPORE • SYDNEY • TOKYO

Morgan Kaufmann Publishers is an imprint of Elsevier

MORGAN KAUFMANN PUBLISHERS

Senior Acquisitions Editor	Rick Adams
Publishing Services Manager	George Morrison
Project Manager	Karthikeyan Murthy
Assistant Editor	Gregory Chalson
Design Direction	Alisa Andreola
Cover Designer	Dick Hannus
Cover Images	gettyimages.com / John Foxx
	istock.com / Jan Rysavy
Interior Printer	Sheridan Books
Cover Printer	Phoenix

Morgan Kaufmann Publishers is an imprint of Elsevier.
30 Corporate Drive, Suite 400, Burlington, MA 01803, USA

This book is printed on acid-free paper. ⊗

Library of Congress Cataloging-in-Publication Data
Kumar, Anurag.
  Wireless networking / Anurag Kumar, D. Manjunath, Joy Kuri.
       p. cm. – (The Morgan Kaufmann series in networking)
  Includes bibliographical references.
  ISBN 0-12-374254-4
1. Wireless LANs. 2. Wireless communication systems. 3. Sensor networks. I. Manjunath, D.
II. Kuri, Joy. III. Title.
  TK5105.78.K86 2008
  621.384–dc22

                                                                          2007053011

British Library Cataloguing in Publication Data
A catalogue record for this book is available from the British Library

ISBN: 978-0-12-374254-4

For information on all Morgan Kaufmann publications,
visit our website at *www.mkp.com* or *www.books.elsevier.com*

Printed and bound in the United Kingdom
Transferred to Digital Printing, 2011

Contents

Preface

Another Book on Wireless Networking?

The availability of high performance, low power, and low cost digital signal processors, and advances in digital communication techniques over the radio frequency spectrum have resulted in the widespread availability of wireless network technology for mass consumption. Several excellent books are now available that deal with the area of wireless communications, where topics of recent interest include multiple-input-multiple-output (MIMO) systems, space time coding, orthogonal frequency division multiplexing (OFDM), and multiuser detection. Wireless networks are best known in the context of first- and second-generation mobile telephony (AT&T's analog AMPS system in the first generation, and the GSM and CDMA digital systems in the second generation). There are books that provide coverage of such wireless networks, and also those that combine a comprehensive treatment of physical layer wireless communication with that of cellular networks.

In the last decade, however, there has been an explosion in the development and deployment of new wireless network technologies, and in the conceptualization of, and research in, a variety of newer ones. From the ubiquitous WiFi coffee shop and airport networks to the emerging WiMAX systems, which promise broadband wireless access to mobile users, the menu of wireless access networks promises to become so comprehensive that wired access from user devices may soon become a relic of the past. Research on wireless mesh networks (so-called ad hoc wireless networks), which started in the 1970s, is being pursued with renewed vigor due to the availability of inexpensive and interoperable mobile wireless devices. In addition, the widespread use of wireless sensor networks (in conjunction with emerging standards such as Zigbee and IEEE 802.15.4) is a clear and present possibility. Thus the variety and scale of wireless networks is unprecedented, and, in teaching courses in our institutions, we have felt the need for a comprehensive analytical treatment of wireless networking, keeping in mind the technical developments in the past, the present, and the future. This book is the outcome of our efforts to address this need.

The foremost aspect of networking, wireline or wireless, is the design of efficient protocols that work. Taking the view that the devil is in the detail, protocols with "working code" often gain widespread acceptance. With the increasing variety in networks and applications, and also in their scale, complex interactions (e.g., between devices using a particular protocol, or between protocols at the various layers) need to be understood. Although computer simulation is a useful vehicle for understanding the performance of protocols, it is not always sufficient, because, once again, the devil is in the detail. The assumptions made in deriving simulation models play an important role in the results that are obtained. If a simulation program simply encodes the standard, then running the simulation only provides a plethora of numbers, with no new insights being gained. Further, large simulation models, although possibly closer to reality, take a lot of effort to develop and debug, and are slow to execute, thus rendering them not very useful in the early stages of experimentation with algorithms. This is where analytical models become very important. First, the process of deriving such models from the standards, or from system descriptions, provides very useful insights. Second, the analytical models can be used to help verify large simulation programs, by providing exact results for subcases of the model being simulated. Third, research in analytical modeling is necessary to develop models that can be programmed into simulators, so as to increase simulation speed. Finally, the analytical approach is very important for the development of new and efficient protocols, and there is a trend toward optimization via reverse engineering of well-accepted protocols.

In addition to the variety of networks and protocols that need to be understood, there is a large body of fundamental results on wireless networks that have been developed over the last fifteen years that give important insights into optimal design and the limits of performance. Examples of such results include distributed power control in CDMA networks, optimal scheduling in wireless networks (with a variety of optimization objectives involving issues such as network stability, performance, revenue, and fairness), transmission range thresholds for connectivity in a wireless mesh network, and the transport capacity of these networks. Further, the imminence of sensor networks has generated a large class of fundamental problems in the areas of stochastic networks and distributed algorithms that are intrinsically important and interesting.

This book aims (1) to provide an analytical perspective on the design and analysis of the traditional and emerging wireless networks, and (2) to discuss the nature of, and solution methods to, the fundamental problems in wireless networking. For the sake of completeness, traditional voice telephony over GSM and CDMA wireless access networks also is covered. The approach is via various resource allocation models that are based on simple models of the underlying physical wireless communication.

About the Book and the Viewpoint

After the specification of the protocols and the verification of their correctness, we believe that networking is about resource allocation. In wireless networks, the resources are typically spectrum, time, and power. That theme pervades much of this book in our quest for models for performance analysis, for developing design insights, and also for exploring the fundamental limits. Once a problem has been analytically formulated, we draw upon a wide variety of techniques to analyze it. In this process we will use techniques drawn from, among others, probability theory, stochastic processes, constrained optimization and duality, and graph theory. We believe it is necessary to make forays into these areas in order to bring their power to bear on the problem at hand. However, we have attempted to make the book as self-contained as possible. Wherever possible, we have used only elementary concepts taught in basic courses in engineering mathematics. A brief overview of most of the advanced mathematical material that we use is provided in the appendix. Also, wherever possible we have avoided the theorem–proof approach. Instead, we have developed the theorems or results and then formally stated them.

After the introductory chapter, we begin the presentation of the main material of the book in Chapter 2 by giving an overview of the physical layer issues that are so much more important to understand wireless networks than they are for wireline networks. Wireless networks are viewed as being either *access networks* or *mesh networks*. In access networks mobile wireless nodes connect to an infrastructure node, and in mesh networks they form an independent internet and may or may not connect to an infrastructure network. Access networks are covered in Chapters 4 through 7 and mesh networks are covered in Chapters 8 through 10.

The wireless networking aspect of the book begins in Chapter 3. Like in our earlier book, *Communication Networking: An Analytical Approach,* we precede the discussion on access networks by listing the issues and setting the performance objectives of a wireless network in Chapter 3. FDM-TDMA cellular networks (of which GSM networks are a major example) are discussed in Chapter 4, with the focus on signal-to-interference ratio analysis, on channel allocation, and on the call blocking and call dropping performance. Chapter 5 is on CDMA networks where the main emphasis is on interference management via power allocation. Whereas the traffic model in Chapter 4 and in much of Chapter 5 is an arrival process of calls, each with a rate requirement, in Chapter 6, on OFDMA access networks, we consider buffered models, and discuss power allocation over time and over carriers with the objectives of stability and mean delay. In Chapter 7, we discuss the performance of distributed allocation of channel time in wireless LANs.

We begin our discussion of mesh networks in Chapter 8 by considering optimal routing and scheduling in a given mesh network. One can view this class of problems as the optimal allocation of time and space in a network. In Chapter 9 we explore fundamental limits of this time and space allocation to the flows. Chapter 10 is on the emerging area of sensor networks, a rich field of research issues including connectivity and coverage properties of stochastic networks, and distributed computation.

Some of the material in Chapter 5 and most of the material in Chapters 6 through 10 are being covered in a wireless networking textbook for the first time. We have not obtained new results for the book but we have trawled the literature to pick out the fundamental results and those that are illustrative of the issues and complexities. Wherever possible, we have simplified the models for pedagogic convenience.

Using the Book

This is a graduate text, though a final year undergraduate course could be supplemented with material from this text. Some understanding of networking concepts is assumed. A quick introduction may also be obtained from Chapter 2 of our earlier book, *Communication Networking: An Analytical Approach*. Most of the chapters are self-contained and we believe that an instructor can pick and choose the chapters. A course that needs to cover voice and data access networks (including cellular networks and wireless LANs) could be based on Chapters 4 through 7. One can say that these chapters are tied closely to real networks. Chapters 8 through 10 are of a more fundamental and abstract nature. A course with a more current research emphasis could be built around Chapters 6 through 10.

The publisher maintains a website for this book at www.mkp.com. We maintain a website for the book at ece.iisc.ernet.in/~anurag/books. These websites contain errata, additional problems, PostScript files of the figures used in the book, and other instructional material. An instructor's manual containing solutions to all the exercises and problems and some supplementary problems is also available from the authors.

Arthur Clarke had said that the communications satellite will make inevitable the United Nations of the Earth. Wireless communication and networking are making these United Nations flatter, and possibly more democratic with unbridled opportunities for all. So let's "unwire, cut the cord, and go wireless." And, while we do it, let us step back a bit and understand them from the ground up!

Acknowledgments

We are grateful to Onkar Dabeer and P.R. Kumar who reviewed the complete manuscript and provided us invaluable comments and criticisms. Saswati Sarkar was visiting us on her sabbatical when we were writing this book; her insightful comments on several chapters of the book have helped immensely in improving the accuracy of the material. Prasanna Chaporkar, N. Hemachandra, U. Jayakrishnan, Koushik Kar, Biplab Sikdar, Chandramani Kishore Singh, and Rajesh Sundaresan read various chapters of the book at our request. They found many errors and rough edges, and we remain grateful for their time and efforts. Many students helped with reviewing the problems and verifying their solutions; these include Onkar Bharadwaj, Avhishek Chatterjee, Sudeep Kamath, Pallavi Manohar, and K. Premkumar. We would like to thank Chandrika Sridhar, our ever-helpful lab secretary in network_labs@ece.iisc.ernet.in, who typed parts of an initial draft from the lecture notes of the first author, and also prepared all the problems and solutions, both from handwritten manuscripts.

This book has been developed out of the two-part survey article "A Tutorial Survey of Topics in Wireless Networking," by Anurag Kumar and D. Manjunath, published in *Sādhanā*, Indian Academy of Sciences Proceedings in Engineering Sciences, Vol. 32, No. 6, December 2007. We are grateful to the publishers of *Sādhanā* for permitting us to use several extracts and figures from our survey article.

Parts of an early draft of this book have been used by Ed Knightly (Rice University), and Utpal Mukherji (Indian Institute of Science). We hope that the finished version will meet their expectations for the courses they teach.

Book writing grants have been provided by the Centre for Continuing Education of the Indian Institute of Science to the first author and by the Curriculum Development Program of the Indian Institute of Technology, Bombay, to the second author.

Finally, we are grateful to our families for bearing patiently our absence from the regular call of duty at home, in the evenings, holidays, and weekends during the several months over which this book was developed.

Anurag Kumar   D. Manjunath   Joy Kuri
I.I.Sc., Bangalore   I.I.T., Bombay   I.I.Sc., Bangalore

CHAPTER 1

Introduction

The idea of sending information over radio waves (i.e., *wireless communication*) is over a hundred years old. When several devices with radio transceivers share a portion of the radio spectrum to send information to each other, we say that we have a *wireless communication network*, or simply a wireless network.

In this chapter we begin by developing a three-layered view of wireless networks. We delineate the subject matter of this book—that is, wireless networking—as dealing with the problem of resource allocation when several devices share a portion of the RF spectrum allocated to them. Next, we provide a taxonomy of current wireless networks. The material in the book is organized along this taxonomy. Then, in this chapter, we identify the common basic technical elements that underlie any wireless network as being (1) physical wireless communication; (2) neighbor discovery, association, and topology formation; and (3) transmission scheduling.

Finally, we provide an overview of the contents of the remaining nine chapters of the book.

1.1 Networking as Resource Allocation

Following our viewpoint in [89] we view wireline and wireless communication networks in terms of the three-layered model shown in Figure 1.1. Networks carry the flows of information between distributed applications such as telephony, teleconferencing, media-sharing, World Wide Web access, e-commerce, and so on. The points at which distributed information applications connect to the generators and absorbers of information flows can be viewed as sources and sinks of traffic (see Figure 1.1). Examples of traffic sources are microphones in telephony devices, video cameras, and data, voice, or video files (stored on a computer disk) that are being transmitted to another location. Examples of traffic sinks are telephony loudspeakers, television monitors, or computer storage devices.

As shown in Figure 1.1, the sources and sinks of information and the distributed applications connect to the communication network via common *information services*. The information services layer comprises all the hardware and software required to facilitate the necessary transport services, and to attach the sources or sinks to the wireless network; for example, voice coding, packet buffering and playout, and voice decoding, for packet telephony; or similar

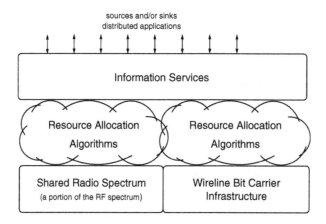

Figure 1.1   A conceptual view of distributed applications utilizing wireline and wireless networks. Wireless networking is concerned with algorithms for resource allocation between devices sharing a portion of the radio spectrum. On the other hand, in wireline networks the resource allocation algorithms are concerned with sharing the fixed resources of a bit transport infrastructure.

facilities for video telephony or for streaming video playout; or mail preparation and forwarding software for electronic mail; or a browser for the World Wide Web.

We turn now to the bottom layer in Figure 1.1. In wireline networks the information to be transported between the endpoints of applications is carried over a static bit-carrier infrastructure. These networks typically comprise high-quality digital transmission systems over copper or optical media. Once such links are properly designed and configured, they can be viewed as "bit pipes," each with a certain bit rate, and usually a very small bit error rate. The bit carrier infrastructure can be dynamically reconfigured on the basis of traffic demands, and such actions are a part of the cloud labeled "resource allocation algorithms" in the figure.

The left side of the bottom layer in Figure 1.1 corresponds to wireless networks. Typically, each wireless network system is constrained to operate in some portion of the RF spectrum. For example, a cellular telephony system may be assigned 5 MHz of spectrum in the 900 MHz band. Information bits are transported between devices in the wireless network by means of some physical wireless communication technique (i.e., a PHY layer technique, in terms of the ISO-OSI model) operating in the portion of the RF spectrum that is assigned to the network. It is well known, however, that unguided RF communication between mobile wireless devices poses challenging problems. Unlike wireline communication, or even point-to-point, high-power microwave links between dish antennas mounted on tall towers, digital wireless communication between mobile devices has to deal with a variety of time-varying channel impairments such as obstructions by steel and concrete buildings, absorption in partition walls or in foliage, and interference between copies of a signal that traverse multiple paths. In order to combat these problems, it is imperative that in a mobile or ad hoc wireless network the PHY layer should be adaptable. In fact, in some systems multiple modulation schemes are available, and each of these may have variable parameters such as the error control codes, and the transmitter powers. Hence, unlike a wired communication network, where we can view networking as being concerned with the problems of resource sharing over a static bit carrier infrastructure, in wireless networking, the resource allocation mechanisms would include these adaptations of the PHY layer. Thus, in Figure 1.1 we have actually "absorbed" the physical wireless communication mechanisms into the resource allocation layer. Hence, we can define our view of *wireless networking* as being concerned with all the mechanisms, procedures, or algorithms for efficient sharing of a portion of the radio spectrum so that all the instances of communication between the various devices obtain their desired *quality of service (QoS)*.

1.2 A Taxonomy of Current Practice

In this book, instead of pursuing an abstract, technology agnostic approach, we will develop an understanding of the various wireless networking techniques in the context of certain classes of wireless networks as they exist today. Thus we begin our treatment by taking a look at a taxonomy of the current practice of wireless networks. Figure 1.2 provides such a taxonomy. Several commonly used terms of the technology will arise as we discuss this taxonomy. These will be highlighted by the *italic* font, and their meanings will be clear from the context. Of course, the attendant engineering issues will be dealt with at length in the remainder of the book.

Fixed wireless networks include line-of-sight microwave links, which until recently were very popular for long distance transmission. Such networks basically comprise point-to-point line-of-sight digital radio links. When such links are set up, with properly aligned high gain antennas on tall masts, the links can be viewed as point-to-point bit pipes, albeit with a higher bit error rate than wired links. Thus in such fixed wireless networks no essentially new issues arise than in a network of wired links.

On the other hand the second and third categories shown in the first level of the taxonomy (i.e., access networks and ad hoc networks) involve *multiple access* where, in the same geographical region, several devices share a radio spectrum to communicate among themselves (see Figure 1.3). Currently, the most important role of wireless communications technology is in mobile access to wired networks. We can further classify such access networks into two categories: one in which resource allocation is more or less static (akin to circuit multiplexing), and the other in which the traffic is statistically multiplexed, either in a centralized manner or by distributed mechanisms.

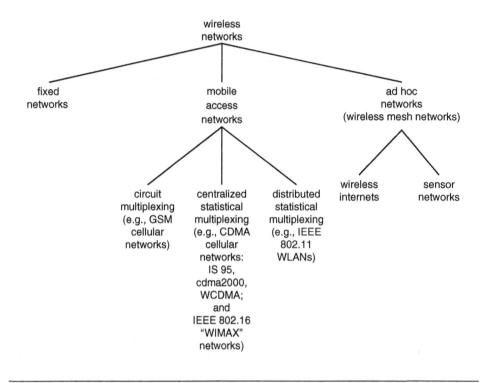

Figure 1.2   A taxonomy of wireless networks.

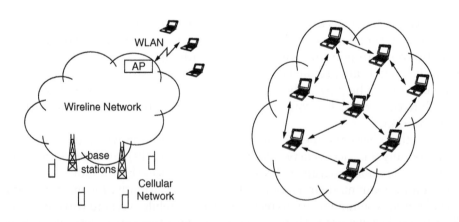

Figure 1.3   The left panel shows some access networks (a cellular telephony network, and a wireless local area network (WLAN), where the access is via an AP (access point)), and the right panel shows a mesh wireless network of portable computers.

Cellular wireless networks were introduced in the early 1980s as a technology for providing access to the wired phone network to mobile users. The network coverage area is partitioned into regions (with diameters ranging from hundreds of meters to a few kilometers) called *cells*, hence the term "cellular." In each cell there is a *base station* (BS), which is connected to the wired network, and through which the mobile devices in the cell communicate over a one hop wireless link. The cellular systems that have the most widespread deployment are the ones that share the available spectrum using *frequency division multiplexed time division multiple access* (FDM-TDMA) technology. Among such systems by far the most commercially successful has been the GSM system, developed by a European consortium. The available spectrum is first partitioned into a contiguous *up-link* band and another contiguous *down-link* band. Each of these bands is statically or dynamically partitioned into reuse subbands, with each cell being allocated such a subband (this is the FDM aspect). The partitioning of the up-link and down-link bands is done in a paired manner so that each cell is actually assigned a pair of subbands. Each subband is further partitioned into channels or carriers (also an FDM aspect), each of which is digitally modulated and then slotted in such a way that a channel can carry up to a fixed number of calls (e.g., 8 calls) in a TDM fashion. Each arriving call request in a cell is then assigned a slot in one of the carriers in that cell; of course, a pair of slots is assigned in paired up-link and down-link channels in that cell. Thus, since each call is assigned dedicated resources, the system is said to be *circuit multiplexed*, just like the wireline phone network. These are *narrowband systems* (i.e., users' bit streams occupy frequency bands just sufficient to carry them), and the radio links operate at a high signal-to-interference-plus-noise-ratio (SINR), and hence careful *frequency planning* (i.e., partitioning of the spectrum into reuse subbands, and allocation of the subbands to the cells) is needed to avoid *cochannel interference*. The need for allocation of frequency bands over the network coverage area (perhaps even dynamic allocation over a slow timescale), and the grant and release of individual channels as individual calls arrive and complete, requires the control of such systems to be highly centralized. Note that *call admission control*, that is, call blocking, is a natural requirement in an FDM-TDMA system, since the resources are partitioned and each connection is assigned one resource unit.

Another cellular technology that has developed over the past 10 to 15 years is the one based on *code division multiple access* (CDMA). In these networks, the entire available spectrum is reused in every cell. These are *broadband* systems, which means that each user's bit stream (a few kilobits per second) occupies the entire available radio spectrum (a few megahertz). This is done by *spreading* each user's signal over the entire spectrum by multiplying it by a *pseudorandom sequence*, which is allocated to the user. This makes each user's signal appear like noise to other users. The knowledge of the spreading sequences permits the receivers to separate the users' signals, by means of *correlation receivers*. Although no frequency planning is required for CDMA systems, the performance is *interference limited* as every transmitted signal is potentially an interferer for every other signal. Thus at any point of time there is an allocation of powers to all the transmitters sharing the spectrum, such that their desired receivers can decode their transmissions, in the presence of all the cross interferences. These desired power levels need to be set depending on the locations of the users, and the consequent channel conditions between the users and the base stations, and need to be dynamically controlled as users move about and channel conditions change. Hence tight control of transmitter power levels is necessary. Further, of course, the allocation of spreading codes, and management of movement between cells needs to done. We note that, unlike the FDM-TDMA system described earlier, there is no dedicated allocation of resources (frequency and time-slot) to each call. Indeed, during periods when a call is inactive no radio resources are utilized, and the interference to other calls is reduced. Thus, we can say that the traffic is *statistically multiplexed*. If there are several calls in the system, each needing certain *quality of service (QoS)* (bit rate, maximum bit error rate), then the number of calls in the system needs to be controlled so that the probability of QoS violation of the calls is kept small. This requires call admission control, which is an essential mechanism in CDMA systems, in order that QoS objectives can be achieved. Evidently, these are all centrally coordinated activities, and hence even CDMA cellular systems depend on central intelligence that resides in the *base station controllers* (BSCs).

Until recently, cellular networks were driven primarily by the needs of circuit multiplexed voice telephony; on demand, a mobile phone user is provided a wireless digital communication channel on which is carried compressed telephone quality (though not "toll" quality) speech. Earlier, we have described two technologies for *second generation (2G)* cellular wireless telephony. Recently, with the growing need for mobile Internet access, there have been efforts to provide packetized data access on these networks as well. In the FDM-TDMA systems, low bit rate data can be carried on the digital channel assigned to a user. As is always the case in circuit multiplexed networks, flexibility in the allocation of bandwidth is limited to assigning multiple channels to each user. Such an approach is followed in the GSM-GPRS (General Packet Radio Service) system, where, by combining multiple TDM slots on an FDM carrier, shared packet switched access is provided to mobile users. A further evolution is the EDGE (Enhanced Data rates for GSM Evolution) system, where, in addition to combining TDM slots, higher order modulation schemes, with adaptive modulation, are utilized to obtain shared packet switched links with speeds up to 474 Kbps. These two systems often are viewed, respectively, as 2.5G and 2.75G evolutions of the GSM system. These are data evolutions of an intrinsically circuit switched system that was developed for mobile telephony. On the other hand there is considerable flexibility in CDMA systems where there is no dedicated allocation of resources (spectrum or power). In fact, both voice and data can be carried in the packet mode, with the user bit rate, the amount of spreading, and the allocated power changing on a packet-by-packet basis. This is the approach taken for the *third generation (3G)* cellular systems, which are based entirely on CDMA technology, and are meant to carry *multimedia traffic* (i.e., store and forward data, packetized telephony, interactive video, and streaming video). The most widely adopted standard for 3G systems is WCDMA (wideband CDMA), which was created by the 3G Partnership Project (3GPP), a consortium of standardization organizations from the United States, Europe, China, Japan, and Korea.

Cellular networks were developed with the primary objective of providing wireless access for mobile users. With the growth of the Internet as the de facto network for information dissemination, access to the Internet has become an increasingly important requirement in most countries. In large congested cities, and in developing countries without a good wireline infrastructure, *fixed wireless access* to the Internet is seen as a significant market. It is with such an application in mind that the IEEE 802.16 standards were developed, and are known in the industry as WiMAX. The major technical advance in WiMAX is in the adoption of several high performance physical layer (PHY) technologies to provide several tens of Mbps between a base station (BS) and fixed subscriber stations (SS) over distances of several kilometers. The PHY technologies that have been utilized are *orthogonal frequency division multiple access (OFDMA)* and multiple antennas at the transmitters and the receivers. The latter are commonly referred to as *MIMO (multiple-input-multiple-output)* systems. In an OFDMA system, several subchannels are statically defined in the system bandwidth, and these subchannels are digitally modulated. In order to permit up-link and down-link transmissions, time is divided into frames and each frame is further partitioned into an up-link and a down-link part (this is called *time division duplexing (TDD)*). The BS allocates time on the various subchannels to various down-link flows in the down-link part of the frame and, based on SS requests, in the up-link part of the frame. This kind of TDD MAC structure has been used in several earlier systems; for example, satellite networks involving very small aperture satellite terminals (VSATs), and even in wireline systems such as those used for the transmission of digital data over cable television networks. WiMAX specifications now have been extended to include broadband access to mobile users.

We now discuss the third class of networks in the mobile access category in the first level of the taxonomy shown in Figure 1.2—distributed packet scheduling. Cellular networks have emerged from centrally managed point-to-point radio links, but another class of wireless networks has emerged from the idea of *random access,* whose prototypical example is the Aloha network. Spurred by advances in digital communication over radio channels, random access networks can now support bit rates close to desktop wired Ethernet access. Hence random access wireless networks are now rapidly proliferating as the technology of choice for wireless Internet access with limited mobility. The most important standards for such applications are the ones in the IEEE 802.11 series. Networks based on this standard now support physical transmission speeds from a few Mbps (over 100s of meters) up to 100 Mbps (over a few meters). The spectrum is shared in a *statistical TDMA* fashion (as opposed to slotted TDMA, as discussed, earlier, in the context of first generation FDM-TDMA systems). Nodes contend for the channel, and possibly collide. In the event of a collision, the colliding nodes *back off* for independently sampled random time durations, and then reattempt. When a node is able to acquire the channel, it can send at the highest of the standard bit rates that can be decoded, given the channel condition between it and its receiver. This technology is predominantly deployed for creating *wireless local area networks* (WLANs) in campuses and enterprise buildings, thus basically providing a one hop untethered access to a building's Ethernet network. In the latest enhancements to the IEEE 802.11 standards, MIMO-OFDM physical layer technologies are being employed in order to obtain up to 100 Mbps transmission speeds in indoor environments.

With the widespread deployment of IEEE 802.11 WLANs in buildings, and even public spaces (such as shopping malls and airports), an emerging possibility is that of carrying interactive voice and streaming video traffic over these networks. The emerging concept of fourth-generation wireless access networks envisions mobile devices that can support multiple technologies for physical digital radio communication, along with the resource management algorithms that would permit a device to seamlessly move between 3G cellular networks, IEEE 802.16 access networks and IEEE 802.11 WLANs, while supporting a variety of packet mode services, each with its own QoS requirements.

With reference to the taxonomy in Figure 1.2, we now turn to the category labeled "ad hoc networks" or "wireless mesh networks." Wireless access networks provide mobile devices with one-hop wireless access to a wired network. Thus, in such networks, in the path between two user devices there is only one or at most two wireless links. On the other hand a wireless ad hoc network comprises several devices arbitrarily located in a space (e.g., a line segment, or a two-dimensional field). Each device is equipped with a radio transceiver, all of which typically share the same radio frequency band. In this situation, the problem is to communicate between the various devices. Nodes need to discover neighbors in order to form a topology, good paths need to be found, and then some form of time scheduling of transmissions needs to be employed in order to send packets between the devices. Packets going from one node to another may need to be forwarded by other nodes. Thus, these are *multihop* wireless packet radio networks, and they have been studied as such over several years. Interest in such networks has again been revived in the context of *multihop wireless internets* and *wireless sensor networks*. We discuss these briefly in the following two paragraphs.

In some situations it becomes necessary for several mobile devices (such as portable computers) to organize themselves into a multihop wireless packet network. Such a situation could arise in the aftermath of a major natural disaster such as an earthquake, when emergency management teams need to coordinate their activities and all the wired infrastructure has been damaged. Notice that the kind of communication that such a network would be required to support would be similar to what is carried by regular public networks; that is, point-to-point store and forward traffic such as electronic mails and file transfers, and low bit rate voice and video communication. Thus, we can call such a network a *multihop wireless internet.* In general, such a network could attach at some point to the wired Internet.

Whereas multihop wireless internets have the service objective of supporting instances of point-to-point communication, an ad hoc wireless sensor network has a global objective. The nodes in such a network are miniature devices, each of which carries a microprocessor (with an energy efficient operating system); one or more sensors (e.g., light, acoustic, or chemical sensors); a low power, low bit rate digital radio transceiver; and a small battery. Each sensor monitors its environment and the objective of the network is to deliver some global information or an inference about the environment to an operator who could be located at the periphery of the network, or could be remotely connected to the sensor network. An example is the deployment of such a network in the border areas of a country to monitor intrusions. Another example is to equip a large building with a sensor network comprising devices with strain sensors in order to monitor the building's structural integrity after an earthquake. Yet another example is the use of such sensor networks in monitoring and control systems such as those for the environment of an office building or hotel, or a large chemical factory.

1.3 Technical Elements

In the previous section we provided an overview of the current practice of wireless networks. We organized our presentation around a taxonomy of wireless networks shown in Figure 1.2. Although the technologies that we discussed may appear to be disparate, there are certain common technical elements that constitute these wireless networks. The efficient realization of these elements constitutes the area of wireless networking.

The following is an enumeration and preliminary discussion of the technical elements.

1. *Transport of the users' bits over the shared radio spectrum.* There is, of course, no communication network unless bits can be transported between users. Digital communication over mobile wireless links has evolved rapidly over the past two decades. Several approaches are now available, with various tradeoffs and areas of applicability. Even in a given system, the digital communication mechanisms can be adaptive. First, for a given digital modulation scheme the parameters can be adapted (e.g., the transmit power, or the amount of error protection), and, second, sophisticated physical layers actually permit the modulation itself to be changed even at the packet or burst timescale (e.g., if the channel quality improves during a call then a higher order modulation can be used, thus helping in store and forward applications that can utilize such time varying capacity). This adaptivity is very useful in the mobile access situation where the channels and interference levels are rapidly changing.

2. *Neighbor discovery, association and topology formation, routing.* Except in the case of fixed wireless networks, we typically do not "force" the formation of specific links in a wireless network. For example, in an access network each mobile device could be in the vicinity of more than one BS or *access point* (AP). To simplify our writing, we will refer to a BS or an AP as an access device. It is a nontrivial issue as to which access device a mobile device connects through. First, each mobile needs to determine which access devices are in its vicinity, and through which it can potentially communicate. Then each mobile should associate with an access device such that certain overall communication objectives are satisfied. For example, if a mobile is in the vicinity of two BSs and needs certain quality of service, then its assignment to only a particular one of the two BSs may result in satisfaction of the new requirement, and all the existing ones.

In the case of an access network the problem of routing is trivial; a mobile associates with a BS and all its packets need to be routed through that BS. On the other hand, in an ad hoc network, after the associations are made and a topology is determined, good routes need to be determined. A mobile would have several neighbors in the discovered topology. In order to send a packet to a destination, an appropriate neighbor would need to be chosen, and this neighbor would further need to forward the packet toward the destination. The choice of the route would depend on factors such as the bit rate achievable on the hops of the route, the number of hops on the route, the congestion along the route, and the residual battery energies in devices along the route.

We note that association and topology formation is a procedure whose timescale will depend on how rapidly the relative locations of the network nodes is changing. However, one would typically not expect to associate and reassociate a mobile device, form a new topology, or recalculate routing at the packet timescale.

If mobility is low, for example in wireless LANs and static sensor networks, one could consider each fixed association, topology, and routing, and compute the performance measures at the user level. Note that this step requires a scheduling mechanism, discussed as the next element. Then that association, topology, and routing would be chosen that optimizes, in some sense, the performance measures. In the formulation of such a problem, first we need to identify one or more performance objectives (e.g., the sum of the user utilities for the transfer rates they get). Then we need to specify whether we seek a cooperative optimum (e.g., the network operator might seek the global objective of maximizing revenue) or a noncooperative equilibrium. The latter might model the more practical situation, since users would tend to act selfishly, attempting to maximize their performance while reducing their costs. Finally, whatever the solution of the problem, we need an algorithm (centralized or distributed) to compute it online.

If the mobility is high, however, the association problem would need to be dynamically solved as the devices move around. Such a problem may be relatively simple in a wireless access network, and, indeed, necessary since cellular networks are supposed to handle high mobility users. On the other hand such a problem would be hard for a general mesh network; highly mobile wireless mesh networks, however, are not expected to be "high performance" networks.

3. *Transmission scheduling.* Given an association, a topology, and the routes, and the various possibilities of adaptation at the physical layer, the problem is to schedule transmissions between the various devices so that the users' QoS objectives are met. In its most general form, the schedule dynamically needs to determine which transceivers should transmit, how much they should transmit, and which physical layer (including its parameters, e.g., transmit power) should be used between each transceiver pair. Such a scheduler would be said to be *cross-layer* if it took into account state information at multiple layers; for example, channel state information, as well as higher layer state information, such as link buffer queue lengths. Note that a scheduling mechanism will determine the schedulable region for the network; that is, the set of user flow rates of each type that can be carried so that each flow's QoS is met.

In general, these three technical elements are interdependent and the most general approach would be to jointly optimize them. For example, in a mobile Internet access network the mobile devices are associated with base stations. The channel qualities between the base stations and the mobile devices determine the bit rates that can be sustained, the transmission powers required, and transmission schedule required to achieve the desired QoS for the various connections. Thus, the overall problem involves a joint optimization of the association, the physical layer parameters, and the transmission schedule.

In addition to the preceding elements that provide the basic communication functionality, some wireless networks require other functional elements that could be key to the networks' overall utility. The following are two important ones, which are of special relevance to ad hoc wireless sensor networks.

- *Location determination.* In an ad hoc wireless sensor network the nodes make measurements on their environment, and then these measurements are used to carry out some global computation. Often, in this process it becomes necessary to determine from which location a measurement came. Sensor network nodes may be too small (in terms of size and available energy) to carry a GPS (global positioning system) receiver. Some applications may require the nodes to be placed indoors, where GPS signals may not penetrate. Hence GPS-free techniques for location determination become important. Even in cellular networks, there is a requirement in some countries that, if needed, a mobile device should be geographically locatable. Such a feature can be used to locate someone who is stranded in an emergency situation and is unaware of the exact location.

- *Distributed computation.* This issue is specific to wireless sensor networks. It may be necessary to compute some function of the values measured by sensors (e.g., the maximum or the average). Such a computation may involve some statistical signal processing functions such as data compression, detection, or estimation. Since these networks operate with very simple digital radios and processors, and have only small amounts of battery energy, the design of efficient self-organizing wireless ad hoc networks and distributed computation schemes on them is an important emerging area. In such networks there is communication delay and also data loss; hence existing algorithms may need to be redesigned to be robust to information delay and loss.

1.4   Summary and Our Way Forward

We began with a discussion of our view of networking as resource allocation. Figure 1.1 summarizes our view. This was followed by a taxonomy of current wireless practice in Section 1.2. Next, the common technical elements that underlie the apparently disparate technologies were abstracted and discussed in Section 1.3.

Before we can proceed to the core topic of this book—resource allocation to meet specified QoS objectives—we will need to understand basic models of, and notions associated with, the wireless channel. Along with this, the important techniques employed in digital communication will be covered in Chapter 2. These concepts will be like the building blocks in terms of which our resource allocation problems will be posed, and answers sought. Essentially, in Chapter 2, our discussion will be confined to the so-called PHY layer.

However, before commencing our study of resource allocation problems, we will pause and take a look at the applications that usually are carried on communication networks. Our objectives will be to understand the characteristics of the bit streams or the packet streams generated by various applications (the top layer of Figure 1.1), as well as the performance requirements the streams demand. This will be the topic of Chapter 3.

Beginning with Chapter 4, we will consider, one by one, the different wireless networks shown at the second level of our taxonomy (Figure 1.2). In each case, the emphasis will be on posing and solving resource allocation problems specific to that type of network. In Chapter 4, narrowband cellular systems will be studied. Power, bandwidth, and time are the resources here, and the principal objective is to maintain the signal-to-interference ratio (SIR) at an adequately high level. Our discussion will give rise to several important concepts, including frequency reuse, sectorization, spectrum efficiency, handover blocking, and channel reservation.

Continuing with cellular access networks, we will focus on CDMA systems in Chapter 5. The distinguishing feature here is that of universal frequency reuse. As before, the main theme is to assign power so as to ensure that the signal-to-interference-plus-noise ratio (SINR) is adequately high. We will see how the notions of other-cell interference, power control, and hard as well as soft handover arise in this context.

In Chapter 6, we will turn to OFDMA-TDMA systems, where power, frequency, and time constitute the basic resources to be allocated. Unlike FDM-TDMA and CDMA systems, where to each flow a fixed bit rate is assigned, in OFDMA-TDMA systems, the resources are assigned dynamically over time, depending on time varying user requirements and channel conditions. Generally speaking, the objective is to maximize the aggregate bit capacity of a time-varying channel, subject to a constraint on the average power. The important notion of the water-filling power allocation will emerge from our discussions.

In Chapter 7, the focus shifts to random-access systems and, in particular, IEEE 802.11 WLANs. The principal resource here is channel time, and distributed control of access to the channel is of interest. In a system of n colocated WLAN nodes, what is the saturation throughput that each can achieve? We will analyze this important question. Various issues pertaining to the transport of voice and data traffic over WLANs will also be discussed.

Continuing with our discussion of the various networks according to our taxonomy, we will study multihop wireless mesh networks in Chapters 8 and 9. In Chapter 8, we assume that a wireless mesh network is given. On this network, we will address the fundamental question of optimal routing and link scheduling of packet flows for a given set of source-destination pairs. Again, the basic resources here are bandwidth, time, and power, and it is of interest to know which nodes should get access to the bandwidth at what times so as to achieve the objective of maximizing throughput. Our analysis will lead to the notions of optimal scheduling and routing. We first consider open loop flows. The flow rates may be given or they may be unknown. For the latter case, the important maximum weight scheduling is described in detail. We also consider routing and scheduling for elastic flows so as to maximize a network utility function.

In Chapter 9, we will address some fundamental questions that arise in the context of wireless mesh networks. First, we ask, what is the minimum power level that nodes can use while ensuring that the network of nodes remains connected? After a suitable definition of the network capacity we also obtain the capacity of arbitrary and random networks. Although asymptotic analyses provide interesting insights, wherever possible, we also consider finite networks.

Finally, in Chapter 10, we will turn to wireless sensor networks. Apart from power and bandwidth, each sensor itself can be considered as a resource now.

A variety of new problems arise; for example, if sensors are deployed in a random manner over a given area, how many of them are required so that every point in the area is sensed by not less than k sensors? As mentioned before in Section 1.3, wireless sensor networks often have special needs; for example, localization and distributed computation. Resource allocation problems for meeting such objectives will also be discussed.

CHAPTER 2

Wireless Communication: Concepts, Techniques, Models

W
e recall from Figure 1.1 in Chapter 1 that, when studying wireless networks, we will not take the links as given bit carriers but will be concerned with the sharing of the wireless spectrum resource as well. The strictly layered approach would view the wireless physical layer as providing a bit carrier service to the link layer. The link layer just offers packets to the physical layer, which does the best it can. If on the other hand, there is interaction between the layers and the link layer can be aware of the time varying quality of the wireless communication, then it could prioritize, schedule, defer, or discard packets in order to attempt to meet the QoS requirements of the various flows. It is therefore important to obtain an understanding of how digital radio communication is performed, and the issues, constraints, and trade-offs that are involved. The material in this chapter is well established and is available in great detail and in much more generality in many books on digital communications. An excellent up-to-date coverage of this topic is provided in [131] and [43]. Readers familiar with digital wireless communication can skip this chapter with no loss of continuity.

Overview

Our approach to modeling, analyzing, and designing resource allocation in wireless networks will be based on simple models of the techniques that are used for carrying bit streams over wireless channels. Because of their place in the seven-layer OSI model, these are also called *physical layer* techniques or, as an abbreviation, *PHY* techniques. In this chapter we will provide these models, and show how they arise.

In Section 2.1 we will study, in some detail, the simplest binary modulation over a very simple radio channel in which the only phenomenon that corrupts the user's data is *additive noise*. We will see that the receiver can make errors when attempting to extract the transmitted bits from the noisy received signal, and we will relate the *bit error rate (BER)* to the received *signal-to-noise ratio (SNR)*. We will see how higher bit rates can be obtained by using higher order *constellations* into which blocks of user bits can be mapped. We will briefly discuss how adding redundant bits at the transmitter, or *channel coding*, can be used to reduce the BER at the expense of a reduction in the user level bit rate. Then, in Section 2.1.4, we will understand other ways in which propagation over a radio channel can corrupt the user's data: these are *path loss, shadowing,* and *multipath fading.* The latter two are stochastic phenomena, and we will see how they are modeled. Section 2.1 will close with an understanding of how random fading causes a deterioration in the BER achievable for a given SNR.

In Section 2.2 we will explain the idea of *channel capacity,* and we will provide Shannon's formula for the capacity of an additive white Gaussian channel. The idea of the *ergodic capacity* of a fading channel will also be introduced.

In Section 2.3 we will study how *diversity* can mitigate the effect of a fading channel. Diversity can be obtained in various ways, one of them being by the use of multiple receive antennas. We will then see that multiple transmit and receive antennas (i.e., *MIMO antenna systems*) can also provide a capacity gain by making the channel look like several independent parallel channels.

Recent mobile wireless access networks have relied heavily on the techniques of *code division multiple access (CDMA),* and also, more recently, *orthogonal frequency division multiple access (OFDMA).* In these systems, the resources (e.g., bandwidth and time) are not statically partitioned over the users. Instead, the available spectrum is shared dynamically between the users, with the resource allocation being dynamically adjusted as the user demands and channel conditions vary over time. We study CDMA and OFDMA in Section 2.4.1 and in Section 2.4.2, respectively.

2.1 Digital Communication over Radio Channels

The primary resource that is shared in a wireless network is the *radio spectrum.* We will limit ourselves to the situation in which the communicating nodes share a radio spectrum of *bandwidth*[1] W, centered at the carrier frequency f_c (see Figure 2.1).

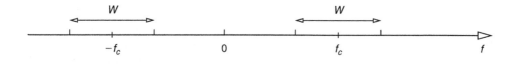

Figure 2.1 The nodes in a wireless network share a portion of the radio spectrum.

[1]The term *bandwidth* has varied and confusing usage in the wireless networking literature. The RF spectrum in which a system operates has a bandwidth. When a digital modulation scheme is used over this spectrum then a certain bit rate is provided; often this aggregate bit rate may also be referred to as bandwidth, and we may speak of users sharing the bandwidth. This latter usage is unambiguous in the wire-line context. In multiaccess wireless networks, however, users would be sharing the same RF spectrum bandwidth, but would be using different modulation schemes and thus obtaining different (and time varying) bit rates, rendering the use of a phrase such as "bandwidth assigned to a user" very inappropriate.

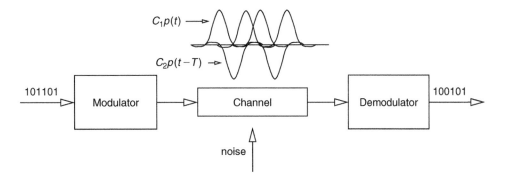

Figure 2.2 A sequence of pulses is modulated with the bits to be transmitted. The basic pulse is $p(t)$. Notice that the bit sequence 101101 is transmitted as $+\sqrt{E_s}\ p(t)$, $-\sqrt{E_s}\ p\ (t-T)$, $+\sqrt{E_s}\ p\ (t-2T)$,..., $+\sqrt{E_s}\ p\ (t-5T)$. There is an error in the third bit, so that, after detection, the received sequence is 100101.

It is assumed that $f_c \gg W$; for example, $f_c = 2.4\,\text{GHz}$ and $W = 5\,\text{MHz}$. All communication between any pair of nodes in the network can utilize this entire spectrum.

2.1.1 Simple Binary Modulation and Detection

As shown in Figure 2.2, digital communication is achieved over the given radio spectrum by *modulating* a sequence of *pulses* by the given bit pattern. The pulse, $p(t)$ (also called the *baseband pulse*), is chosen so that when translated to the carrier f_c its spectrum fits into the given radio spectrum; that is, in this case, the spectrum of the baseband pulse will occupy the frequencies $\left(-\frac{W}{2}, +\frac{W}{2}\right)$. Taking $T = \frac{1}{W}$, it is possible to define a pulse $p(t)$, that is bandlimited[2] to $\left(-\frac{W}{2}, +\frac{W}{2}\right)$, and is such that $p(t - kT), k \in \{\ldots, -3, -2, -1, 0, 1, 2, 3, \ldots\}$, constitute an orthonormal set, that is, $\int_{-\infty}^{+\infty} p(t)p(t - kT)\mathrm{d}t = 0$ for $k \neq 0$. Further, $\int_{-\infty}^{+\infty} p^2(t)\mathrm{d}t = 1$, that is, the energy of the pulse is 1. The pulses are repeated every T seconds.

In the situation depicted in Figure 2.2, the modulation is very simple: each pulse in the pulse train is multiplied by $+\sqrt{E_s}$ if the bit to be transmitted is 1, and by $-\sqrt{E_s}$ if the bit to be sent is 0. Notice that the energy of the modulated pulse becomes E_s. It is said that the modulator maps bits into channel *symbols*. Thus, in this example, the symbol set is $\{-\sqrt{E_s}, +\sqrt{E_s}\}$. In general, there could be more than just two possible symbols; for example, four symbols would permit two incoming bits to be mapped into each channel symbol. Continuing our simple

---

[2]Mathematically, a pulse, $p(t)$, that is bandlimited (e.g., to $(-\frac{W}{2}, +\frac{W}{2})$) occupies infinite time. Practically, a pulse that is chosen for a digital modulation scheme has negligible energy beyond a small multiple of T on either side of its main lobe.

example, let C_k denote the *symbol* into which the k-th bit is mapped. When the pulses are repeated every T seconds, the modulated pulse stream can be written as

$$X(t) = \sum_{k=-\infty}^{\infty} C_k \, p(t - kT) \tag{2.1}$$

Given this continuous time signal, and recalling the orthonormality of the various shifts of $p(t)$ by kT, it is easy to see that the following operation recovers the information carrying sequence C_k.

$$C_k = \int_{-\infty}^{+\infty} X(t)p(t - kT)\mathrm{d}t$$

The *baseband signal* $X(t)$ is then translated to the radio spectrum shown in Figure 2.1 by multiplying it with a sinusoid at the carrier frequency. The resulting signal is

$$S(t) = \sqrt{2} \sum_{k=-\infty}^{\infty} C_k \, p(t - kT) \cos(2\pi f_c t) \tag{2.2}$$

The multiplication by $\sqrt{2}$ is to make the energy in the modulated symbols equal[3] to E_s. Thus, the *symbol energy* in the transmitted signal is E_s Joules/symbol, and since the symbol rate is $\frac{1}{T}$ symbols/second, the transmitted signal power is therefore $\frac{E_s}{T}$ Watts. In Figure 2.2 we do not show the translation of the signal by the carrier. It is as if the channel has been shifted to the baseband.

As shown in Figure 2.2, as the modulated signal passes through the channel, and is processed in the front-end of the receiver, it is corrupted by noise. This is taken to be zero mean *additive white Gaussian noise (AWGN)*, which means that noise just adds to the signal and is a Gaussian random process with a power spectrum that is constant over the passband of the channel (hence the term "white," since all frequencies ("colours") have the same power). The signal occupies a band of W Hz around the carrier frequency f_c ($\frac{W}{2}$ Hz below and $\frac{W}{2}$ Hz above $\pm f_c$; see Figure 2.1). Hence, we need only be concerned with noise that occupies this band. Such *bandpass white Gaussian noise*, with a power spectral density of $\frac{N_0}{2}$, is mathematically represented as (see [113])

$$N(t) = U(t) \cos(2\pi f_c t) \tag{2.3}$$

---

[3]To see why we have chosen the symbols C_k to be $\pm\sqrt{E_s}$, and the reason for the factor $\sqrt{2}$, notice that the energy in each transmitted pulse is $2 \int_{-\infty}^{+\infty} (C_k)^2 p^2(t) \cos^2(2\pi f_c t) \, \mathrm{d}t$ which can be shown to be equal to E_s.

where the process $U(t)$ is a zero mean white Gaussian process with power spectral density N_0, bandlimited to $\left(-\frac{W}{2}, +\frac{W}{2}\right)$. We can view the noise process $U(t)$ as a baseband noise process that is translated to the carrier frequency and placed in the passband of the channel.

It can now be shown (see this chapter's Appendix) that the previously described modulation scheme, and the additive white Gaussian noise model, along with receiver processing, results in the following *symbol-by-symbol channel model* that relates the source symbol sequence C_k and the predetection statistic Y_k, from which the source symbol sequence has to be inferred.

$$Y_k = C_k + Z_k \qquad (2.4)$$

where Z_k is a sequence of i.i.d. zero mean Gaussian random variables with variance $\frac{N_0}{2}$.

Figure 2.3 depicts the probability density of Y_k under the two possible values of C_k. These are both Gaussian densities with variance $\frac{N_0}{2}$. The detector concludes that the bit sent was 0 if the value of Y_k is smaller than the threshold and 1 if the value of Y_k is more than the threshold. An error occurs if 1 is sent and Y_k falls below the threshold, and vice versa. When the source produces 0s and 1s with equal probabilities then the threshold is midway between the means of the two densities, that is, the threshold is 0. The probability of error if a 0 was sent is then given by:

$$\Pr(Y_k > 0 \mid 0 \text{ was sent}) = Q\left(\sqrt{\frac{2E_s}{N_0}}\right)$$

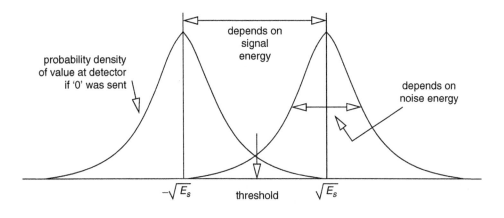

Figure 2.3 The probability densities of the statistic Y_k under the two possible symbols.

where $Q(\tau) := \int_\tau^\infty \frac{1}{\sqrt{2\pi}} e^{-\frac{x^2}{2}} \, dx$. This can be seen to be the same as the probability of error if a 1 was sent. Hence the probability of error of the binary modulation scheme that we have described, under AWGN, is given by

$$P_{error-AWGN} = Q\left(\sqrt{\frac{2E_s}{N_0}}\right) \tag{2.5}$$

Note that in this example, since each symbol is used to send one bit, the error rate obtained is also the bit error rate (BER). In Problem 2.1 we find that $P_{error-AWGN}$ decreases exponentially with $\frac{E_s}{N_0}$. In particular, for BERs of 10^{-3} and 10^{-6} the $\frac{E_s}{N_0}$ values required are approximately 7 dB and 10.5 dB, respectively. We note that if 1500 byte packets have to be transmitted over a wireless link, then in order to obtain a packet error probability of 0.01, we need BER $\leq 10^{-6}$.

We see that the probability of correct detection depends on $\frac{E_s}{N_0}$, which is the ratio of the symbol energy to the noise power spectral density. Increasing the symbol energy increases the separation between the two Gaussian probability densities in Figure 2.3, and hence, for given noise variance, reduces the probability of Y_k falsely crossing the threshold. Similarly, decreasing the noise reduces the width of the two Gaussian probability densities, thus also reducing the error probability for a given signal energy.

2.1.2 Getting Higher Bit Rates

In the simple example in Section 2.1.1, since each pulse is modulated by one of two possible symbols, and the symbol rate is $\frac{1}{T}$, the bit rate is therefore $\frac{1}{T}$ bps. One of the goals in designing a digital communication system over a radio spectrum is to use this spectrum to carry as high a bit rate as possible. With the binary modulation example in mind there are two possibilities for increasing the bit rate.

1. Increase the symbol rate; that is, decrease T.

2. Increase the number of possible symbols, from 2 to $M > 2$.

Then, in general, the bit rate will be given by $\frac{\log_2 M}{T}$. There are, however, limits on both these possibilities.

Note that if the pulse bandwidth is limited to $\frac{W}{2}$, the channel bandwidth, then the pulse duration will not be time limited, and in fact the received signal in a symbol interval will be the sum of the pulse in that interval and parts of pulses in neighboring intervals. The pulses therefore have to be appropriately designed to take care of this effect. This leads to the so-called *Nyquist criterion*, which limits the pulse rate to no more than W (i.e., $\frac{1}{T} \leq W$).

Before we proceed, it is useful to make an observation. We saw in Section 2.1.1 that the probability of error for that binary signaling system depended on the ratio $\frac{E_s}{N_0}$. If the signaling rate is $\frac{1}{T}$, then the average power in the transmitted signal is $E_s \times \frac{1}{T}$. The noise power in the channel bandwidth is $W N_0$. Hence the *signal power to noise power ratio (SNR)* is given by $\frac{E_s}{TWN_0}$. If, in addition, the symbol rate is such that $T \times W = 1$, then the SNR is just $\frac{E_s}{N_0}$. Thus we see that for this example the probability of error depends on the SNR. This is sometimes called the *predetection SNR*, as it is the SNR before the receiver attempts to decide which symbol was sent.

Let us now consider the other alternative for increasing the bit rate; that is, increasing the number of possible symbols that can modulate the pulses. Figure 2.4(a) shows the binary symbol set that we have already discussed. This is called binary *pulse amplitude modulation (PAM)*, or 2-PAM. An example of the simplest possibility is shown in Figure 2.4(b); this is called 4-PAM. Since each of the 2-bit patterns 00, 01, 10, 11 can be mapped to one of the symbols, this scheme can transmit 2 bits per symbol. However, in order to achieve a particular probability of error with a given noise power, the distance between the symbols has to be retained as in the binary case; to see this consider Figure 2.3, add a Gaussian density for each new symbol added, and then consider the probability of error between neighboring symbols. This means that the symbol energy when transmitting the left-most and right-most symbols in Figure 2.4(b) will be 3^2 times larger than that for the other two symbols. This in turn implies a larger average signal power, and hence a larger SNR (assuming the same noise power) for achieving the same probability of error. Yet another alternative is shown in Figure 2.5(a) where we have *two-dimensional symbols*. Each symbol can be written in the form $ce^{j\theta}$, with $c = 1$ and $\theta \in \left\{0, \frac{\pi}{2}, \pi, \frac{3\pi}{2}\right\}$. This symbol set is called QPSK (*quadrature phase shift*

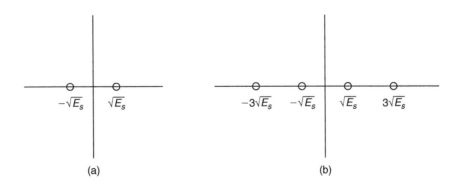

(a)   (b)

Figure 2.4   Some symbol sets: (a) binary antipodal, (b) 4-level amplitude modulation.

(a)   (b)

Figure 2.5 (a) A complex symbol set with 4 symbols; (b) the symbol set with noise added.

keying) since all the symbols have the same amplitude but they have different phases. Now, instead of the form in (2.2), the transmitted signal takes the general form

$$S(t) = \sqrt{2} \sum_{k=-\infty}^{\infty} C_k \cos(\Theta_k) p(t - kT) \cos(2\pi f_c t)$$

$$- \sqrt{2} \sum_{k=-\infty}^{\infty} C_k \sin(\Theta_k) p(t - kT) \sin(2\pi f_c t) \qquad (2.6)$$

Here, the sequence (C_k, Θ_k) depends on the modulating bits. Thus, basically, the x-coordinate (i.e., $C_k \cos(\Theta_k)$) of the symbol modulates the carrier $\cos(2\pi f_c t)$ and the y-coordinate (i.e., $C_k \sin(\Theta_k)$) of the symbol modulates $-\sin(2\pi f_c t)$, which is also called the *quadrature* carrier (since it is $\frac{\pi}{2}$ out of phase with the *in-phase* carrier). The bandpass additive noise $N(t)$ has the general form

$$N(t) = U(t) \cos(2\pi f_c t) - V(t) \sin(2\pi f_c t)$$

where $U(t)$ and $V(t)$ are independent zero mean Gaussian processes with power spectral density N_0, bandlimited to $\left(-\frac{W}{2}, \frac{W}{2}\right)$. We can interpret $U(t)$ and $V(t)$ as the in-phase and quadrature noise processes, respectively.

In fact, we notice that the QPSK signal shown in (2.6) is the superposition of two orthogonal 2-PAM signals; the in-phase and quadrature signals are both 2-PAM signals. After down conversion (multiplying the signal by $\sqrt{2} \cos(2\pi f_c t)$ and also by $-\sqrt{2} \sin(2\pi f_c t)$ and filtering out the high frequency terms), and multiplication and integration with the pulse $p(t)$, we will obtain the following pair of statistics:

$$Y_k^{(i)} = C_k \cos(\Theta_k) + Z_k^{(i)}$$
$$Y_k^{(q)} = C_k \sin(\Theta_k) + Z_k^{(q)}$$

where (i) and (q) denote the in-phase and quadrature components. The sequences $Z_k^{(i)}$ and $Z_k^{(q)}$ are independent, and each is a sequence of i.i.d. zero mean Gaussian random variables with variance $\frac{N_0}{2}$. We can write this more compactly by using complex numbers. Define $Y_k = Y_k^{(i)} + jY_k^{(q)}$ and $Z_k^{(i)} + jZ_k^{(q)}$. Then, we can write

$$Y_k = X_k + Z_k \tag{2.7}$$

where $X_k = C_k e^{j\Theta_k}$ is the k-th channel symbol. We say that the sequence of complex random variables Z_k are *circularly symmetric complex Gaussian*.

In Figure 2.5(b) we show the received symbols after corruption by noise; the noise now has the two-dimensional Gaussian density that is circularly symmetric about each symbol. Notice from the geometry in Figure 2.5(a) that, by utilizing both dimensions, for a given probability of error, a *smaller symbol spacing* can be used than for the symbol set in Figure 2.4(b), and hence a given BER can be achieved with less average power. Thus, we have noisy observations of the two coordinates of the transmitted complex symbol, from which the transmitted symbol has to be detected. Since in each symbol only one of the phases is used (and the other is 0, owing to the simple QPSK symbol set), the average signal power is that of a 2-PAM signal.

As is evident from Figure 2.5 many more symbol sets are possible. If the amplitude as well as phase of the symbols can vary then it is called QAM (*quadrature amplitude modulation*), whereas if only the phase can vary then it is called a PSK symbol set. Symbol sets are also called *constellations*. The probability of error of all the digital modulation and demodulation schemes based on the basic ideas discussed earlier can be expressed as a function of the SNR at the receiver.

2.1.3 Channel Coding

In a given situation, owing to physical limitations it may not be possible to increase the SNR so as to achieve the desired BER. The application being transported on the wireless link may require a lower BER in order to achieve reasonable performance. For example, if the link is used to transport packets and the packet length is L bits, then a BER of ϵ yields a packet error rate of $1 - (1 - \epsilon)^L$. We will see in Section 3.4.3 that a high packet error rate can seriously affect the performance of TCP transfers. Hence, this may place a minimum BER requirement on the link.

For a given digital modulation scheme, the BER as seen by the data source can be reduced by *channel coding*. The simplest viewpoint is shown in Figures 2.6 and 2.7. The channel with the given modulation scheme is viewed as an error prone binary channel. Blocks of the incoming bits of length K are coded into *codewords* of length $N(> K)$, thus introducing redundancy. If the code length and the codes are judiciously chosen, even after the channel introduces errors, an errored codeword can be expected to stay close to the original codeword. In Figure 2.7 we show source bit strings of length K being mapped into code blocks of length N. Since the number of possible code strings (2^N) is larger than

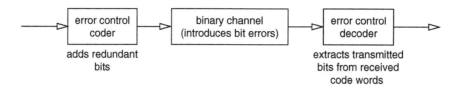

Figure 2.6 Channel coding: adding redundant bits to protect against channel errors.

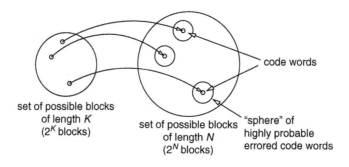

Figure 2.7 A channel code maps source bit strings into longer code bit strings (or codewords); decoding involves identifying the codeword nearest to the received bit string.

the number of possible source strings (2^K), the code words can be chosen so that there is sufficient spacing between them. Now even if the channel causes errors, the errored codewords will occupy spheres of high probability around the transmitted codewords. Hence, by using *nearest codeword decoding*, the transmitted codeword, and hence the original source string, can be inferred with a small residual error probability. The trade-off is that the information bit rate of the communication link becomes $\frac{K}{N}$, which is less than 1 information bit per code bit. This is called the rate of the code, denoted by R.

One trivial way of improving error performance is to increase N, because this results in the codewords being spaced farther apart; but this reduces the information rate. It is possible, however, to increase K with N, keeping the information rate R constant, while reducing the bit error rate to arbitrarily small values. Shannon's noisy channel coding theorem states that there is a number C, called the *channel capacity*, such that if $R < C$, then, as the block length increases, an arbitrarily small bit error rate can be achieved (of course, at the cost of a large block coding delay). If we attempt to use $R > C$, then the bit error rate cannot be reduced to 0.

Recall our analysis of the two-level modulation carried out earlier in this chapter. We recall that for bit error rates of 10^{-3} and 10^{-6} the $\frac{E_s}{N_0}$ values required were approximately 7 dB and 10.5 dB, respectively. As an example, with a high quality rate $\frac{1}{2}$ code, the required $\frac{E_s}{N_0}$ can be reduced by 2 dB for 10^{-3} and by 5 dB for 10^{-6}. Of course, the user bit rate drops to $\frac{1}{2}$ bit per symbol. This reduction in $\frac{E_s}{N_0}$ is called *coding gain*.

A coder is followed by a digital modulation scheme that maps code bits into channel symbols. As discussed in Section 2.1.1, the modulator maps a certain number of code bits (e.g., 2 in 4-QPSK) into each channel symbol. Thus the capacity of the overall system (coder—modulator—channel—demodulator—decoder) can be expressed in terms of bits per symbol. At this point, it is obvious that in order to achieve this capacity *the receiver must know the channel coding and modulation scheme that the transmitter is using*. Shannon also provided the fundamental relationship between the channel capacity (C) and the signal-to-noise ratio for an *additive white Gaussian noise channel*. We will introduce this relationship later in this chapter. First we need to study models for signal power attenuation between the transmitter and the receiver.

2.1.4 Delay, Path Loss, Shadowing, and Fading

In the previous discussion we assumed that the transmitted signal was contaminated by only additive white Gaussian noise. This yielded the simple model shown in (2.4). However, in practical channels, signals undergo attenuation and delay. In wireless channels, because of propagation over multiple paths, and mobility of the scatters or of the communicating devices, the attenuation can vary with time and the relative location between the transmitter and the receiver. We have seen that the BER performance of a digital communication system depends on the *received* SNR. Hence, we are interested in the received signal power after the signal has passed through the channel.

Radio waves are scattered by the objects on which they impinge. Hence, unless a very narrow antenna beam is used, the receiver's antenna receives the transmitted signal along several paths. There is often a direct or *line-of-sight* path, and there are several paths along which the signal reaches the receiver after one or more reflections from various objects. Energy is lost in reflections, and is absorbed by media through which the signal passes (partitions and walls). Hence the received signal is a sum of attenuated and delayed versions of the original signal.

Delay Spread and Intersymbol Interference

Superposition of the delayed signals from the various paths can cause a symbol from one path to overlap with a neighboring symbol from another path. Let us examine this issue first. These are electromagnetic signals and hence they travel at the speed of light; let us take the propagation time to be roughly 0.33 μsec per 100 meters. Hence, this is the kind of delay that can be expected if the various path lengths differ by no more than 100 m.

If the symbol time is several μseconds (e.g., 100,000 symbols per second) then there will not be significant overlap between the neighboring symbols, and we can assume that the symbols are still separately discernible, except that each is multiplied by a complex "attenuation." If this happens then the channel is said to have *flat fading*. We will understand the term "flat" when we interpret this phenomenon in the frequency domain. Then, motivated by (2.7), we can write the k-th received symbol after down conversion as

$$Y_k = G_k X_k + I_k + Z_k \qquad (2.8)$$

where the various new terms are understood as follows.

1. G_k is the random attenuation of the k-th symbol. $G_k, k \geq 1$, is a complex valued random process. Thus, a transmitted symbol is not only attenuated, but can also be *rotated*. Note that the symbol energy is multiplied by $|G_k|^2$. Let us write $H_k = |G_k|^2$; H_k, $k \geq 1$, is a random process, and we need to characterize it in order to understand the effect of the channel on the received signal power, and hence the SNR. We note that the H_k are also called *channel gains*.

2. I_k is a complex random variable that models the interference (from other transmissions in the same or nearby spectrum[4]). We recall that Z_k is a sequence of complex random variables that models the additive noise (for example, the thermal noise in the electronic circuitry of the receiver) and is taken to be a white Gaussian random process. A commonly used simplification is to use the same model even for the interference process, with the noise and interference processes being modeled as being independent. The BER then becomes a function of the *signal to interference plus noise ratio (SINR)*.

For a transmitter receiver pair, the difference between the smallest signal delay and the largest signal delay is called the *delay spread*, T_d. For example, if the path lengths differ by no more than 100s of meters then the delay spread would be in 100s of nanoseconds. When the delay spread is not very small compared to the symbol time then the superposition of the signals received over the variously delayed paths at the receiver results in *intersymbol interference (ISI)*. We then obtain the following linear model:

$$Y_k = \sum_{j=0}^{J_d-1} G_k(j) X_{k-j} + I_k + Z_k \qquad (2.9)$$

For every k, $G_k(j), 0 \leq j \leq J_d - 1$, are complex random variables that model the way the channel attenuates and phase shifts the transmitted symbols. $G_k(j)$ models

---

[4]Note that we are taking the simplified approach of treating other users' signals as interference. More generally, it is technically feasible to extract multiple users' symbols even though they are superimposed. This is called *multiuser detection*.

the influence that the input j symbols in the past has on the channel output at k. Thus, in general, a channel has memory; in the model, the memory extends over J_d symbols. The memory arises as a consequence of there existing several paths from the transmitter to the receiver, with the different paths having different delays. The notation shows that the channel gain at the k-th symbol could be a function of the symbol index k; this models the fact that fading is a time-varying phenomenon. As the devices involved in the communication move around, the radio channel between them also keeps changing.

The delay spread, T_d, has been explained previously as a time domain concept. It can also be viewed in the frequency domain as follows. The symbols X_k are carried over the RF spectrum by first multiplying them with a (baseband) pulse of bandwidth approximately W (e.g., 200 KHz), and then upconverting the resulting signal to the carrier frequency (e.g., 900 MHz) (recall (2.2)). The delay spread in the channel (i.e., T_d) can be such that superposition of variously delayed versions of some frequency components in the baseband pulse can cancel out. In such a case, some of the frequency components in the pulses can get selectively attenuated, resulting in the corruption of the symbols they carry; this is called *frequency selective fading*. On the other hand, if $T_d << \frac{1}{W}$, then the pulse would be passed through with only an overall attenuation; we recall that this situation was called flat fading. The reciprocal of the delay spread is called the *coherence bandwidth*, W_c. Thus, if $W_c >> W$ then all the frequencies fade together and we have flat fading.

The assumption of flat fading is reasonable for a *narrowband system*, where the available radio spectrum is channelized and each bit stream occupies one channel. Then the symbol duration becomes larger than the delay spread, and the model of (2.8) is applicable. This will be the channel model that we will use when analyzing FDM-TDMA cellular systems in Chapter 4.

On the other hand, consider the situation in which W_c is small compared to the system bandwidth (T_d is large compared to $\frac{1}{W}$); that is, the channel is frequency selective. Then, in relation to the model in (2.9), and recalling that the intersymbol interval is $\frac{1}{W}$, we observe that frequency selectivity corresponds to the channel memory extending over more than 1 symbol, and hence to the existence of ISI. Thus, when high bit rates are carried over wideband channels (i.e., large W) then techniques have to be used to combat ISI, or to avoid it altogether. We will encounter CDMA and OFDMA later in this chapter, as two *wideband systems* that actually exploit delay spread or frequency selectivity to achieve *diversity* (a concept explained in Section 2.3).

In some systems, we can combat ISI by passing the received signal through a *channel equalizer*, which can compensate for the various channel delays, making the overall system (i.e., the channel followed by the equalizer) appear like a fixed delay channel. In a mobile wireless situation, owing to mobility, the paths that a signal takes between a transmitter and a receiver may keep changing; hence a channel equalizer needs to be adaptive. In some systems the problem of signals arriving over multiple paths is turned into an advantage. If the paths can be *resolved*, and if they fade independently, then their signals can be combined to reduce the probability of error, for a given received signal-to-noise ratio. Such a receiver is said to exploit *multipath diversity*.

A Characterization of the Power Attenuation Process

It follows from the linear model with flat fading, shown in Equation 2.8, that the received sequence, $Y_k, k \geq 1$, is also a complex valued random process. The problem for the receiver, on receiving the sequence of complex numbers $Y_k, k \geq 1$, is to carry out a *detection* of which symbols $X_k, k \geq 1$, were sent and hence which user bits were sent. This problem is particularly challenging in mobile wireless systems since the channel is randomly changing with time. The analysis and design of modulation schemes often is based on the analysis of received signal power to noise power ratios. Hence, it is important to have an effective but simple model of the channel power attenuation process, H_k.

The process $\{H_k\}$ is characterized by writing it in terms of three multiplicative components, that is,

$$H_k = \left(\left(\frac{d_k}{d_0} \right)^{-\eta} \cdot S_k \cdot R_k^2 \right) \qquad (2.10)$$

Let us write the marginal terms of the stationary random processes in this expression by dropping the symbol index k. We will now discuss each of these terms.

The term $\left(\frac{d}{d_0} \right)^{-\eta}$ is the path loss factor. Here, d is the distance between the transmitter and the receiver when the k-th symbol is being received, d_0 is the "far field" *reference distance* beyond which this model is applicable, and η is the path loss exponent, which is typically in the range 2 to 5. The value of d_0 relates to the antenna dimensions and the propagation environment. For distances less than d_0, a different path loss exponent may be used, or, when d_0 is very small, we may assume no path loss.

If the attenuation is measured at various points at a distance d from the transmitter, then the attenuation will be found to be random, owing to variations in the terrain, and in the media through which the signal may have passed. Empirical studies have shown that this randomness is captured well if the second factor S, in (2.10), has the form $10^{-\frac{\xi}{10}}$, with ξ being a Gaussian random variable with mean 0 and variance σ^2. This is called the *shadowing* component of the attenuation, and, since $\log_{10}$ of this term has a Gaussian (or normal) distribution, it is called *log-normal shadowing*. It is often convenient to express values of power and power ratios in the *decibel (dB)* unit which is obtained by taking $10 \log_{10}$ of the value. Hence the shadowing attenuation in signal power is $10 \log_{10} S = -\xi$ dB, which is zero mean Gaussian with variance σ^2. A typical value of σ is 8 dB. Considering two standard deviations above and below the mean, this value means that, with a high probability, shadowing can result in a variation of channel gain of $40 \left(\approx 10^{\frac{2 \times 8}{10}} \right)$ times to $0.025 \left(\approx 10^{\frac{-2 \times 8}{10}} \right)$ times the mean path loss.

Shadow fading is spatially varying, and hence if there is relative movement between the transmitter and the receiver then shadow fading will vary. The correlation in the shadow fading in dB between two points separated by a distance D is given by $\sigma^2 e^{-\frac{D}{D_0}}$, where D_0 is a parameter that depends on the terrain. Some measurements have given $D_0 = 500$ m for suburban terrains, and $D_0 = 50$ m for urban terrains. Hence if the distance is varying by a few meters per second (note that 36 Kmph = 10 meters/second) then the shadowing will vary over seconds, which means that the variations will occur over hundreds of thousands of symbols.

We now turn to the third factor, R^2, in the expression for attenuation in (2.10). Typical carrier frequencies used in mobile wireless networks are 900 MHz, 1.8 GHz (e.g., these two frequency bands are used in cellular wireless telephony systems), or 2.4 GHz (e.g., used in IEEE 802.11 wireless LAN systems). Hence, the carrier wave periods are a few picoseconds. Thus, when the transmitted signal arrives over several paths then very small differences in the path lengths (a few centimeters) can cause large differences in the phases of the carriers that are being superimposed. Thus, although these time delays may not result in ISI, the superposition of the delayed carriers results in constructive and destructive carrier interference, leading to variations in signal strength. This phenomenon is called *multipath fading*. This is a random attenuation that has strong autocorrelation over a time duration called the *coherence time*, T_c; that is, the attenuations at two time instants separated by more than the coherence time are weakly correlated. The coherence time is related to the *Doppler frequency*, f_d, which is related to the carrier frequency, f_c, the speed of movement, v, and the speed of light, c, by $f_d = f_c \frac{v}{c}$. Roughly, the coherence time is the inverse of the Doppler frequency. For example, if the carrier frequency is 900 MHz, and $v = 20$ meters/sec, then $f_d = 60$ Hz, leading to a coherence time of 10s of milliseconds. In the indoor office or home environment, the Doppler frequency could be just a few Hz (e.g., 3 Hz), with coherence times of 100s of milliseconds. The marginal distribution of R^2 depends on whether all the signals arriving at the receiver are scattered signals, or if there is a *line-of-sight* signal as well. In the former case, assuming uniformly distributed arrival of the signal from all directions, the distribution of R^2 is exponential with mean $\mathsf{E}(R^2)$, that is,

$$f_{R^2}(x) = \frac{1}{\mathsf{E}(R^2)} e^{(-x/\mathsf{E}(R^2))}$$

The distribution of the amplitude attenuation (i.e., R) is Rayleigh; hence this is also called *Rayleigh fading*. On the other hand if there is a line-of-sight component so that a fraction $\frac{K}{K+1}$ of the signal arrives directly, and the remaining signal arrives uniformly over all directions, then

$$f_{R^2}(x) = \frac{K+1}{E(R^2)} e^{\left(-K - \frac{(K+1)x}{E(R^2)}\right)} I_0 \left(2 \sqrt{\left(\frac{K(K+1)x}{E(R^2)}\right)}\right)$$

where

$$I_0(x) = \frac{1}{2\pi} \int_0^{2\pi} e^{-x \cos(\theta)} d\theta$$

This is called the Ricean distribution.

With this characterization of the attenuation in the received signal power we can now write the received SNR (denoted by Ψ_{rcv}) in terms of the ratio of the transmitted signal power to the received noise power (denoted by Ψ_{xmt}). We have

$$\Psi_{rcv} = \Psi_{xmt} \cdot H$$

$$= \Psi_{xmt} \cdot \left(\frac{d}{d_0}\right)^{-\eta} \cdot 10^{\frac{-\xi}{10}} \cdot R^2 \tag{2.11}$$

Then, in dB, we can write the received SNR as

$$(\Psi_{rcv})_{dB} = (\Psi_{xmt})_{dB} + 10 \log_{10} H$$

$$= (\Psi_{xmt})_{dB} - 10\eta \log_{10} \left(\frac{d}{d_0}\right) - \xi + 10 \log_{10} R^2 \tag{2.12}$$

BER with Fading

We now turn to the calculation of the performance of the wireless link in the presence of fading. We have seen that, although the transmitter may send at a fixed power, in the presence of fading, the received power, and hence the received SNR, is time varying. The rate of variation of the SNR depends on the mobility of the receiver. A receiver that moves short distances over the duration of a "conversation" (e.g., a voice call, or a file transfer) would sample the distribution of the Rayleigh fading but would see roughly constant values of path loss and shadowing. On the other hand a receiver that makes large movements during a call duration would see variations in all the three attenuation factors during the call. Let us consider the former situation. In this case the *in-call* performance depends on the value of path loss and shadow fading sampled by the call, and on the distribution of Rayleigh fading, but the performance *across calls* depends on the variation in path loss and shadowing as well. We would like the performance not to fall below some value. For example, there could be a desired upper bound on BER; exceedance of this bound would be termed an *outage*. Let us examine this point in the context of the binary modulation scheme discussed in Section 2.1.1. The BER for this modulation scheme was given by (2.5):

$$P_{error-AWGN}(\Psi_{rcv}) = Q\left(\sqrt{2\Psi_{rcv}}\right)$$

If the path loss and shadowing factors *during a call* are fixed, then we can calculate the *in-call, BER averaged over the fading,* as follows:

$$\int_0^\infty P_{error-AWGN}\left(\left(\frac{d}{d_0}\right)^{-\eta} \cdot 10^{\frac{-\xi}{10}} \cdot \gamma \cdot \Psi_{xmt}\right) f_{R^2}(\gamma)d\gamma$$

where, as mentioned earlier, for Rayleigh fading, $f_{R^2}(\cdot)$ is the exponential probability density with mean $E(R^2)$.

Let us write the SNR during the call, with the fading averaged out, as

$$\overline{\Psi}_{rcv} := \left(\frac{d}{d_0}\right)^{-\eta} \cdot 10^{\frac{-\xi}{10}} \cdot \left(E\left(R^2\right)\Psi_{xmt}\right)$$

In many cases it can be shown that the preceding integral expression for in-call BER can be simplified to the form

$$P_{error-fading}\left(\overline{\Psi}_{rcv}\right)$$

for some function $P_{error-fading}$. For example, for the binary modulation scheme discussed earlier, it can be shown that $P_{error-fading}\left(\overline{\Psi}_{rcv}\right) = \frac{1}{2}\left(1 - \sqrt{\frac{\overline{\Psi}_{rcv}}{1+\overline{\Psi}_{rcv}}}\right)$, which for large $\overline{\Psi}_{rcv}$ can be observed to *decrease reciprocally with SNR* (i.e., as $\frac{1}{\overline{\Psi}_{rcv}}$), *rather than exponentially,* as for unfaded AWGN (see Problem 2.1; see also Problem 2.4).

During a call, we can write the average SNR (with the averaging being over the fading), $\overline{\Psi}_{rcv}$, in dB as

$$\left(\overline{\Psi}_{rcv}\right)_{dB} = \left(\Psi_{xmt}E\left(R^2\right)\right)_{dB} - 10\eta \log_{10}\left(\frac{d}{d_0}\right) - \xi$$

The term $\left(\Psi_{xmt}E(R^2)\right)_{dB}$ is the *Rayleigh faded SNR "referred to"* d_0. We see that the received SNR, in dB, at a distance d from the transmitter is Gaussian with mean $\left(\Psi_{xmt}E(R^2)\right)_{dB} - 10\eta \log_{10}\left(\frac{d}{d_0}\right)$ and variance σ^2. In order to achieve a certain BER, say, ϵ, the received SNR will be required to be above a threshold, say, β; that is,

$$\overline{\Psi}_{rcv} > \beta \Rightarrow P_{error-fading}\left(\overline{\Psi}_{rcv}\right) < \epsilon$$

Violation of this requirement would be called an outage, the probability of which we would like to limit to P_{outage}. We note that, since we assumed that during a call the path loss and shadowing are fixed, P_{outage} is the outage probability across calls; that is, the fraction of calls that experience a BER larger than ϵ. The BER and outage requirement can then be expressed in the following form:

$$\Pr((\overline{\Psi}_{rcv})_{dB} < (\beta)_{dB}) < P_{outage}$$

Equivalently,

$$\Pr\left(\left(\Psi_{xmt}\mathsf{E}\left(R^2\right)\right)_{dB} - 10\eta\log_{10}\left(\frac{d}{d_0}\right) - \xi < (\beta)_{dB}\right) < P_{outage}$$

Let us look at an example. Given that $\frac{d}{d_0} = 10$, $\eta = 3$, the shadowing standard deviation $\sigma = 8$ dB, the received SNR threshold is $\beta = 10$ dB, and $P_{outage} = 0.01$, the requirement just displayed is satisfied if

$$\left(\Psi_{xmt}\mathsf{E}\left(R^2\right)\right)_{dB} - 30 - 2.3 \times 8 = 10$$

where the factor 2.3 is obtained from a table of the Gaussian distribution. This yields

$$\left(\Psi_{xmt}\mathsf{E}\left(R^2\right)\right)_{dB} = 58.4 \ dB$$

2.2 Channel Capacity
2.2.1 Channel Capacity without Fading

Consider the following simple version of the general linear model that was shown in (2.9)

$$Y_k = X_k + Z_k \tag{2.13}$$

where we notice that we have removed the model of ISI, the multiplicative fading term, and also the additive interference term, leaving just a model in which the output random variable at symbol k is the input symbol X_k with an additive noise term Z_k. Thus, there is no attenuation of the transmitted symbol, but there is perturbation by additive noise. When the X_k are taken from a one-dimensional constellation (as in the beginning of Section 2.1), then the model for the random process Z_k, $k \geq 1$, is that these are i.i.d. Gaussian random variables with mean 0 and variance σ^2 (see Equation 2.4). This is called an *additive white Gaussian noise (AWGN)* channel. The information bits are mapped to the channel symbols

X_k, which are corrupted by additive noise. The observations Y_k have to be used to infer which symbols were transmitted.

Suppose that the input symbols have the following *power constraint*:

$$\lim_{n\to\infty} \frac{1}{n}\sum_{k=1}^{n} |x_k|^2 \leq \overline{P} \tag{2.14}$$

that is, the average energy per symbol is bounded by $\overline{P}$ Joules/symbol. This is a practical constraint as power amplifiers operate well only in certain limited power ranges. Also, microwave radiations can be harmful to the body; hence there are safety regulations on how much power can be radiated by radio transmitters. Further, when several systems coexist then intersystem interference needs to be managed. Hence, some form of power constraint usually is required in wireless communication systems.

If the input symbols are allowed to be only real numbers, then Shannon's celebrated Noisy Channel Capacity Theorem states that the maximum rate at which information can be transmitted over this AWGN channel, in bits/symbol, is given by

$$C = \frac{1}{2}\log_2\left(1 + \frac{\overline{P}_{rcv}}{\sigma^2}\right) \text{ bits/symbol} \tag{2.15}$$

where, $\overline{P}_{rcv}$ is the *received* signal power per symbol, and $\frac{\overline{P}_{rcv}}{\sigma^2}$ is the received signal-to-noise power ratio. Evidently, here, in the no fading case, we have $\overline{P}_{rcv} = \overline{P}$. What this result means is that this rate can be achieved with the bit error rate going to zero as channel coding is done over longer and longer blocks, with the block length going to ∞.

In Section 2.1.1, we derived the symbol-by-symbol channel model by starting with a continuous time model for a modulation scheme that used only real valued symbols. Let us now apply this formula to derive the capacity of that system. We saw that the additive noise sequence has variance $\frac{N_0}{2}$. If the power constraint on the transmitted signal (i.e., $S(t)$ in (2.2)) is P Watts, then the power constraint per symbol is $\overline{P} = PT = \frac{P}{W}$ Joules/symbol. Since we are assuming no channel loss, using (2.15), we obtain the capacity

$$C = \frac{1}{2}\log_2\left(1 + \frac{2P}{N_0 W}\right) \text{ bits/symbol} \tag{2.16}$$

If, in (2.13), the input symbols are complex numbers, then the additive noise is modeled as a sequence of complex valued random variables, which is taken to be a sequence of i.i.d. zero mean, *circularly symmetric Gaussian* random variables with variance σ^2 (recall (2.7)). This means that the real and the imaginary parts are independent sequences of zero mean i.i.d. Gaussian random variables with the same variance, $\frac{\sigma^2}{2}$. The capacity formula then takes the simple form

$$C = \log_2\left(1 + \frac{\overline{P}_{rcv}}{\sigma^2}\right) \text{ bits/symbol} \qquad (2.17)$$

where $\overline{P}_{rcv}$ is the average received power per symbol. Without channel loss $\overline{P}_{rcv} = \overline{P}$.

Now let us apply this to the modulation with complex symbols that led to the channel model in (2.7). There Z_k are i.i.d. zero mean circularly symmetric Gaussian with variance with the real and imaginary parts have variance $\frac{N_0}{2}$. Then, without channel loss, and a power constraint P on the transmitted continuous time signal, the constraint on the average received energy per symbol is $\frac{P}{W}$, yielding the channel capacity

$$C = \log_2\left(1 + \frac{P}{N_0 W}\right) \text{ bits/symbol} \qquad (2.18)$$

It is instructive to compare the expressions (2.16) and (2.18); see Problem 2.5.

We note that these capacity expressions gave the answer in bits per symbol. Often, in analysis it is better to work with natural logarithms. With this in mind we can rewrite (2.17) as

$$C = \ln\left(1 + \frac{\overline{P}_{rcv}}{\sigma^2}\right) \text{ nats/symbol}$$

Since $\ln x = \log_2 x \times \ln 2$, the capacity in nats per symbol is obtained by multiplying the capacity in bits per symbol by $\ln 2 \approx 0.693$.

If the symbol rate is $\frac{1}{T}$ then, for the AWGN channel with complex symbols, Shannon's formula yields the bit rate

$$\frac{1}{T} \log_2\left(1 + \frac{P_{rcv}}{N_0 W}\right) \text{ bits/second}$$

where P_{rcv} is average power in the received signal. For the system bandwidth W, the bit rate, therefore, is limited to

$$W \log_2\left(1 + \frac{P_{rcv}}{N_0 W}\right) \text{ bits/second} \qquad (2.19)$$

An important measure of performance of a digital modulation scheme is $\frac{C}{W}$ bits/Hz; that is, the number of information bits that can be carried per Hertz of system bandwidth. Let us write $P_{rcv} = E_b \times C$, where we can call E_b the received energy per bit. (2.19) can then be written as

$$\frac{C}{W} = \log_2\left(1 + \frac{E_b}{N_0} \frac{C}{W}\right)$$

from which we obtain

$$\frac{E_b}{N_0} = \frac{2^{\frac{C}{W}} - 1}{\frac{C}{W}}$$

The quantity on the left is the *ratio of the received energy per bit to the power spectral density of the additive noise*, and is called the *signal-to-noise ratio per bit*. We conclude that, in order to achieve $\frac{C}{W}$ bits/Hz, we require an $\frac{E_b}{N_0}$ of at least $\frac{2^{\frac{C}{W}} - 1}{\frac{C}{W}}$. For example, for $\frac{C}{W} = 1$ bit/Hz (a typical number for a FDM-TDMA system such as GSM), the minimum value of $\frac{E_b}{N_0} = 1$ or 0 dB. Practical modulation and coding schemes need larger values of $\frac{E_b}{N_0}$, as seen in the examples earlier in this chapter.

2.2.2 Channel Capacity with Fading

How does a time varying channel attenuation affect the Shannon capacity formula? If the channel attenuation is h, and the noise is AWGN, then, for transmitted power P_{xmt}, the channel capacity is given by (2.19):

$$W \log_2 \left(1 + \frac{hP_{xmt}}{N_0 W}\right)$$

Suppose that the transmitter is unaware of the extent of the channel fading, and uses a fixed power and a fixed modulation and coding scheme. Suppose also that the fading level varies slowly. Then, for a given level of fading, the receiver must know h in order for the communication to achieve the Shannon capacity. To see this, let us look at Figure 2.4(b). If the channel's power attenuation is h, the received symbols are multiplied by $\sqrt{h}$. This results in the symbols being "squeezed" together or spread apart. Obviously, the detection thresholds will need to depend on the level of fading.

Suppose that H_k is a stationary and ergodic process. It can then be shown that if the transmitter cannot adapt its coding and modulation, but the receiver can exactly track the fading, then the channel capacity with fading is given by

$$C_{fading-CSIR} = \int W \log_2 \left(1 + \frac{hP_{xmt}}{WN_0}\right) g_H(h)dh \qquad (2.20)$$

where $g_H(\cdot)$ is the marginal density of the channel attenuation process H_k. For example, $g_H(h)$ is exponential for Rayleigh fading (see Section 2.1.4). The acronym *CSIR* stands for *channel state (or side) information at the receiver*. Thus the transmitter can encode at any *fixed* rate $R < C_{fading-CSIR}$, and for large enough code blocks the error rate can be made arbitrarily small, provided the receiver can track the channel. It is important to bear in mind that this is an ideal result; to achieve it, the channel fades will have to be averaged over and this will result in large coding delays.

In Problem 2.6 we see that $C_{fading-CSIR} \leq W \log_2 \left(1 + \frac{E(H)P_{xmt}}{WN_0}\right)$, that is, the capacity with fading is less than that with no fading with the same average SNR. With fading, there will be times when the SNR is higher than the average and times when the SNR will be lower than the average. Yet this result shows that the resulting channel capacity is less than that without fading, as long as the same average SNR is maintained.

2.3 Diversity and Parallel Channels: MIMO

We emphasise that we are discussing direct point-to-point communication between a transmitter and a receiver. We have already seen that the signal from the transmitter can reach the receiver over multiple paths. Since it can be expected that random fading along these paths will be independent, combining the signals from these paths in some manner might lead to better performance than working with the aggregate signal. Such *diversity* can be obtained in various ways. If the receiver has multiple antennas (see Figure 2.8), and if the antennas are spaced sufficiently far apart (at least half the carrier wavelength) then, for the same transmitted signal,

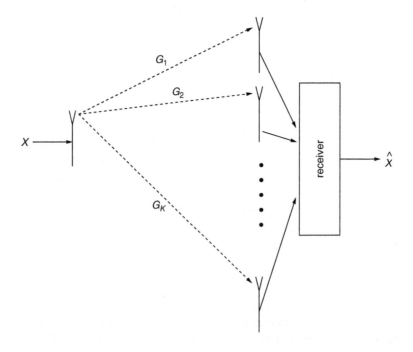

Figure 2.8 A single-input-multiple-output (SIMO) system comprising one transmit antenna and *K* receive antennas.

the signals received at the different antennas fade approximately independently.[5] To see how such independently faded copies can be exploited, let us consider the following model for the signal received along each path.

$$Y_k = G_k X + Z_k$$

where $k, 1 \leq k \leq K$, indexes the diversity "paths," and X is the transmitted (complex) symbol. The $Z_k, 1 \leq k \leq K$, are zero mean, i.i.d. circularly symmetric normal random variables, each with variance σ^2. Recalling the notation $H_k = |G_k|^2$, let us write $G_k = \sqrt{H_k} e^{j\theta_k}$, that is, on the k-th path, the transmitted symbol X is scaled by $\sqrt{H_k}$ and rotated by θ_k.

Assuming that the receiver knows the values of $\theta_k, 1 \leq k \leq K$, it can be shown that the optimum strategy is to form a linear combination of the K received signals by using complex weights $\mu_k e^{-j\theta_k}$, to obtain

$$Y = \sum_{k=1}^{K} \mu_k e^{-j\theta_k} Y_k$$

$$= \left(\sum_{k=1}^{K} \mu_k \sqrt{H_k} \right) X + \sum_{k=1}^{K} \mu_k e^{-j\theta_k} Z_k$$

Note that rotation by θ_k does not destroy the circular symmetry of the noise, Z_k. Let the transmitted power be $\bar{P}$, that is, $\mathsf{E}(|X|^2) = \bar{P}$. If the symbol detection is based on the statistic Y, then the performance of this receiver algorithm will be based on the received SNR

$$\Psi_{rcv} = \frac{\left(\sum_{k=1}^{K} \mu_k \sqrt{H_k} \right)^2 \bar{P}}{\left(\sum_{k=1}^{K} \mu_k^2 \right) \sigma^2}$$

Now, by the Cauchy-Schwartz inequality, we have

$$\left(\sum_{k=1}^{K} \mu_k \sqrt{H_k} \right)^2 \leq \sum_{k=1}^{K} \mu_k^2 \sum_{k=1}^{K} H_k$$

---

[5] To understand the relationship between antenna spacing and low correlation between received signals, let us recall the concept of coherence time. Multipath fading observed by a mobile has low correlation between time instants separated by T_c, which is roughly the reciprocal of $f_d = \frac{f_c}{c} v$, where f_c is the carrier frequency, c is the speed of light, and v is the speed of the mobile. Equivalently, $f_d = \frac{v}{\lambda_c}$, where λ_c is the wavelength of the carrier. It follows that fade correlations are weak over a distance equal to the carrier wavelength. A precise analysis of the phenomenon actually shows that the correlations are weak over distances as little as half the wavelength. Note that $\lambda_c = 30$ cm for $f_c = 1$ GHz, and $\lambda_c = 6$ cm for $f_c = 5$ GHz.

with equality when $\mu_k = a\sqrt{H_k}$ for some a (i.e., when the vector $(\mu_1, \mu_2, \ldots, \mu_K)$ is a multiple of the vector $(\sqrt{H_1}, \sqrt{H_2}, \ldots, \sqrt{H_K})$). Choosing the weights μ_k, $1 \le k \le K$, in this way maximizes the predetection SNR, yielding

$$\Psi_{rcv} = \left(\sum_{k=1}^{K} H_k \right) \Psi_{xmt}$$

where, as before, $\Psi_{xmt} = \frac{P}{\sigma^2}$ is the transmit SNR. We now wish to study the bit error probability for this approach. Suppose that the bit error probability with AWGN decreases exponentially with the received SNR (see Problem 2.1). Then, the average bit error rate is proportional to

$$\mathsf{E}\left(e^{-\Psi_{rcv}} \right) = \mathsf{E}\left(e^{-\left(\sum_{k=1}^{K} H_k \right)\Psi_{xmt}} \right)$$

Recall our discussion in Section 2.1.4, and hence, write $H_k = \pi\Phi_k$ where π is the path loss and shadowing factor from the transmitter to the receiver (taken to be a constant over the time scale to which this analysis applies), and $\Phi_k, 1 \le k \le K$, represent Rayleigh fading over the various paths. This yields

$$\mathsf{E}\left(e^{-\Psi_{rcv}} \right) = \mathsf{E}\left(e^{-\left(\sum_{k=1}^{K} \pi\Phi_k \right)\Psi_{xmt}} \right)$$

Assuming that the fading at the different antennas are independent and identically distributed, we take the $\Phi_k, 1 \le k \le K$, to be i.i.d. exponentially distributed with mean, say, ϕ. We then have

$$\mathsf{E}\left(e^{-\Psi_{rcv}} \right) = \left(\mathsf{E}\left(e^{-\Phi_1 \pi \Psi_{xmt}} \right) \right)^K$$

$$= \left(\frac{1}{1 + \phi\pi\Psi_{xmt}} \right)^K$$

$$\approx \left(\overline{\Psi}_{rcv} \right)^{-K}$$

where the approximation holds for large average received SNR $\overline{\Psi}_{rcv} = \phi\pi\Psi_{xmt}$. Recall that for Rayleigh fading the probability of error decreased only as the reciprocal of $\overline{\Psi}_{rcv}$. Thus, by combining the received signals over multiple paths, the bit error probability performance has been substantially improved. From the form for the decay of the bit error probability with $\overline{\Psi}_{rcv}$, we say that we have a *diversity gain* of K.

The transmitter could also just repeat the signal over time, and if the repetitions are spaced apart by more than the coherence time (see Section 2.1.4) then the received signals fade independently. It turns out that commonly used channel codes provide a better chance of successful decoding if the channel error process is uncorrelated over the code symbols. We saw earlier that the channel fade process, G_k, is correlated over periods called the channel coherence time, which depends on the speed of movement of the mobile device. *Interleaving* is a way to obtain an uncorrelated fade process from a correlated one. Basically the transmitter does not send successive symbols of a codeword over contiguous channel symbols, but successive symbols are separated out so that they see uncorrelated fading. In between, other codewords are interleaved. We say that interleaving exploits *time diversity*, that is, the fact that channel times separated by more than the coherence time fade independently. Observe that interleaving introduces *interleaving delay*, which adds to the link delay, and hence to the end-to-end delay over the wireless network. Also, interleaving fails if the fading is very slow, for example if the relative motion stops, and the transmitter-receiver pair are caught in a bad fade.

In the discussion earlier in this section, we considered the case in which multiple independently faded copies of a transmitted symbol arrive at the receiver. By appropriate combining of these received symbols, the probability of error is reduced. Suppose that the channel is such that the transmitter can, in parallel, transmit several symbols, each of which is then independently faded and received. Then the available power $\overline{P}$ can be distributed over the parallel channels to obtain a higher bit rate than if all the power was used on a single channel; see Problem 2.7, and, for more details, Chapter 6.

Physically, parallel channels between a transmitter-receiver pair can arise if the system bandwidth is partitioned into several orthogonal channels (e.g., by partitioning in frequency and time), and then several of these channels are simultaneously available for communication between the transmitter-receiver pair. Even for narrow-band systems, multiple parallel channels can arise if the transmitter and receiver use multiple antennas (see Figure 2.9). As before, let the system bandwidth be W Hz. Suppose that there are N transmit antennas and M receive antennas. Let $G_{k,i,j}$ $(1 \leq i \leq M, 1 \leq j \leq N)$ denote the channel gain between the transmit antenna j and the receive antenna i, at the k-th symbol. As we know, these channel gains will capture the path loss, the shadowing, and the multipath fading, and will be modeled as complex valued random variables. Let $\mathbf{G}_k$ denote the $M \times N$ channel gain matrix at symbol k. Let $\mathbf{X}_k = (X_{k,1}, X_{k,2}, \ldots, X_{k,N})^\mathsf{T}$ denote the input symbols into the N transmit antennas at the k-th symbol time. These too are complex valued, as would be the case for general two-dimensional constellations. Let $\mathbf{Y}_k = (Y_{k,1}, Y_{k,2}, \ldots, Y_{k,M})^\mathsf{T}$ denote the corresponding complex valued channel outputs. Hence, we can write

$$\mathbf{Y}_k = \mathbf{G}_k \mathbf{X}_k + \mathbf{Z}_k$$

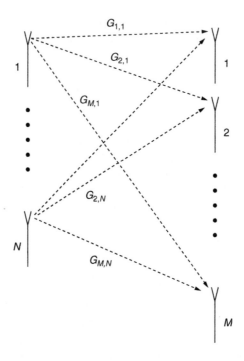

Figure 2.9 A multiple-input-multiple-output (MIMO) system comprising _N_ transmit antennas and _M_ receive antennas.

where $\mathbf{Z}_k$ is the $M \times 1$ additive noise process. The components of $\mathbf{Z}_k$ are zero mean i.i.d. circularly symmetric Gaussian random variables, each with variance σ^2; also the $\mathbf{Z}_k$ sequence is i.i.d. over k. This also means that $Z_{k,i}, 1 \leq i \leq M$, are complex with their real and imaginary parts being zero mean independent Gaussian random variables, each with variance $\frac{\sigma^2}{2}$.

There is a total transmit power constraint of $\overline{P}$:

$$\lim_{n \to \infty} \frac{1}{n} \sum_{k=1}^{n} \sum_{j=1}^{N} |\mathbf{X}_{k,j}|^2 < \overline{P}$$

Define, as before, $\Psi_{xmt} = \frac{\overline{P}}{\sigma^2}$. Let us also assume i.i.d. Rayleigh fading between each transmit-receive antenna pair. Then $H_{k,i,j} = |G_{k,i,j}|^2$ are i.i.d. exponentially distributed with a common mean over the antennas, say, ϕ. If the distance between the transmit antennas and the receive antennas is large, then the path losses between the antenna pairs would be the same. Let us denote this common path loss by π, as in the diversity analysis shown earlier. We "pull" the average path loss, and the mean of the Rayleigh fading out of the channel gain matrix, leaving the mean power gain of the elements in the channel gain matrix to be 1. Then the received SNR, averaged over Rayleigh fading, is (as before) written as

$$\overline{\Psi}_{rcv} = \phi\pi\Psi_{xmt}$$

The transmitter does not know the channel gains, and it can be shown that the best strategy is for the transmitter to split its power equally over the N transmit antennas. Then, given a sample of the gain matrix, say, $\mathbf{G}$, it can be shown that the capacity of this channel is given by

$$C = W\log_2\left(\det\left(\mathbf{I}_M + \frac{\overline{\Psi}_{rcv}}{N}\mathbf{G}\cdot\mathbf{G}^\dagger\right)\right) \text{ bits/second} \qquad (2.21)$$

where $\det(\cdot)$ denotes the "matrix determinant," $\mathbf{I}_M$ denotes the $M \times M$ identity matrix, and $\mathbf{G}^\dagger$ denotes "conjugate-transpose." Now $\mathbf{G}\cdot\mathbf{G}^\dagger$ is an $M \times M$ Hermitian matrix (i.e., its conjugate-transpose is the same as itself). The theory of matrices provides the following facts:

1. The eigenvalues of $\mathbf{G}\cdot\mathbf{G}^\dagger$ are real and nonnegative.

2. The number of positive eigenvalues is no more than $\min\{M, N\}$.

Let us index the eigenvalues in decreasing order of magnitude and denote them by $\lambda_1 \geq \lambda_2 \geq \ldots \geq \lambda_{\min\{M,N\}}$. Then using the fact that the determinant of a square matrix is equal to the product of its eigenvalues, and that the eigenvalues of $\mathbf{I}_M + \frac{\overline{\Psi}_{rcv}}{N}\mathbf{G}\cdot\mathbf{G}^\dagger$ are of the form $1 + \lambda_j\frac{\overline{\Psi}_{rcv}}{N}$, we obtain the following simplification:

$$C = W\sum_{j=1}^{\min\{M,N\}}\log_2\left(1 + \lambda_j\frac{\overline{\Psi}_{rcv}}{N}\right) \text{ bits/second} \qquad (2.22)$$

We see that, under the assumptions we have made, the multiple transmit antenna and multiple receive antenna system (also called a *multiple-input-multiple-output (MIMO)* system) is equivalent to several parallel channels. Note that, for different realizations of the channel gain matrix, the gains of the parallel channels (the eigenvalues $\lambda_j, 1 \leq j \leq \min\{M, N\}$) will be different. In effect, we have parallel channels with random gains.

Let us consider the situation in which $M = N$, and all the eigenvalues are equal, say, λ. Then

$$C = WM\log_2\left(1 + \frac{\lambda\overline{\Psi}_{rcv}}{M}\right) \text{ bits/second}$$

We see that for a single transmit and receive antenna system the capacity (i.e., $W \log_2(1 + \overline{\Psi}_{rcv})$) scales as $\log \overline{\Psi}_{rcv}$ for large $\overline{\Psi}_{rcv}$, whereas for an $M \times M$ MIMO system (with equal eigenvalues) the capacity scales as $M \log \overline{\Psi}_{rcv}$. This is called *multiplexing gain*.

Thus, we find that a multiple antenna system can be used to obtain diversity gain (as explained above for one transmit antenna and M receive antennas), or can be used to increase the channel capacity by the creation of parallel spatial channels between the transmit and receive antenna groups. For an N transmit antenna, and M receive antenna system, the diversity gain is bounded by $M \times N$, whereas the multiplexing gain is limited to $\min\{M, N\}$.

We note that the above discussion assumed that the channel gains are unknown at the transmitter. If channel gain estimates could be provided to the transmitter, then it could judiciously choose the transmitted symbols and their powers so that the better of the parallel spatial channels are assigned the larger transmit powers. We will study such optimal power allocation problems in the OFDMA context in Chapter 6.

2.4   Wideband Systems

Unlike the narrow-band digital modulation used in FDM-TDMA systems, in CDMA and OFDMA the available spectrum is not partitioned, but all of it is dynamically shared among all the users. The simplest viewpoint is to think of CDMA in the time domain and OFDMA in the frequency domain. In a wideband system, a user's symbol rate is much smaller than the symbol rate that the channel can carry (i.e., $\frac{1}{W}$).

2.4.1   CDMA

In CDMA a user's symbol, which is of duration L channel symbols (also called *chips*), is multiplied by a *spreading code* of length L chips. This is called *direct sequence spread spectrum (DSSS)*, since this multiplication by the high rate spreading code results in the signal spectrum being spread out to cover the system bandwidth. If the user's bit rate is R and the chip rate is $R_c \, (> R)$, then $L = \frac{R_c}{R} \, (> 1)$ and is called the *spreading factor*. The spreading codes take values in the set $\{-1, +1\}^L$ and are chosen so that each code is approximately orthogonal to all the time shifts of the other codes, and also to its own time shifts. Then the spread symbols are transmitted. All the signals interfere because they occupy the same radio bandwidth. We provide a simple analysis of such a system, with reference to the depiction in Figure 2.10.

There are M users. The symbol duration is $\frac{1}{R}$, during which there are L chips. Denote the chip time by $\tau_c = \frac{1}{R_c}$. We can write the transmitted signal from User 1 (see Figure 2.10) as

$$x_1 \sum_{j=0}^{L-1} S_{1,j} \, p(t - j\tau_c)$$

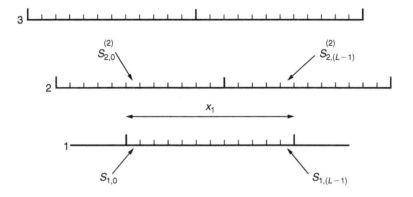

Figure 2.10 Depiction of the superposition of CDMA symbols. The transmissions of three users are shown. The tall ticks denote symbol boundaries and the short ticks denote chip boundaries. A symbol of User 1 that has the value $x_1 \in \{+\sqrt{E_1}, -\sqrt{E_1}\}$ has been shown. It has been spread by the code $S_{1,j}$, $0 \le j \le L - 1$. Interfering symbols of the other users are also shown. The interfering users are assumed to be chip synchronous but their symbols are randomly offset from that of the symbols of User 1.

where x_1 is the user's information carrying symbol, $S_{1,j}, 0 \le j \le L - 1$, is User 1's spreading code, and $p(t)$ is the baseband pulse that is bandlimited to $\left(-\frac{W}{2}, \frac{W}{2}\right)$, and has the property

$$\int_{-\infty}^{\infty} p^2(u)\, du = 1$$

Let $x_i \in \{+\sqrt{E_i}, -\sqrt{E_i}\}$, where E_i corresponds to the transmit power used by User i. Let $\sqrt{h_{i,1}}$ denote the magnitude of the channel attenuation from the transmitter of User i to the receiver of User 1. For simplicity, let us work at the baseband, and then we can write the received signal at the receiver of User 1, over the duration of one symbol, $0 \le t \le \frac{1}{R}$, as

$$y(t) = \sum_{j=0}^{L-1} \sqrt{h_{1,1}}\, x_1\, S_{1,j}\, p(t - j\tau_c) + \sum_{i=2}^{M} \sum_{j=0}^{L-1} \sqrt{h_{i,1}}\, x_{i,j}\, S_{i,j}^{(i)}\, p(t - j\tau_c) + N(t)$$

where $x_{i,j}$ denotes the value of the symbol of User i that interferes with User 1 at the j-th chip in User 1's symbol (see Figure 2.10), $S_{i,j}^{(i)}$ denotes that a shifted version (denoted by the superscript (i)) of the spreading code of User i interferes with the chips of User 1, and $N(t)$ is additive white Gaussian noise with power spectral density N_0, bandlimited to $\left(-\frac{W}{2}, \frac{W}{2}\right)$. The receiver of User 1 now performs the following operation:

$$\int_{-\infty}^{+\infty} y(u) \sum_{j=0}^{L-1} S_{1,j}\, p(u - j\tau_c)\, du$$

yielding the following statistic,[6] based on which the transmitted symbol from User 1 has to be detected:

$$\sqrt{h_{1,1}}\, x_1\, L + \sum_{i=2}^{M} \sum_{j=0}^{L-1} \sqrt{h_{i,1}}\, x_{i,j}\, S_{i,j}^{(i)} S_{1,j} + Z$$

where Z is zero mean Gaussian with variance $N_0 L$ (to see how this is obtained, see the derivation in the appendix of this chapter). The spreading codes are pseudo-random sequences taking values in $\{-1, +1\}$, and hence we model $x_{i,j}\, S_{i,j}^{(i)} S_{1,j}, 0 \leq j \leq L - 1$, as i.i.d. random variables taking values in $\{+\sqrt{E_i}, -\sqrt{E_i}\}$, each with equal probability. Thus we obtain the following symbol-by-symbol model for the CDMA channel:

$$Y_k = L\sqrt{h_{1,1}} X_k + I_k + Z_k \tag{2.23}$$

where I_k is the interference, Z_k is additive noise (which is an i.i.d. Gaussian sequence with zero mean and variance $N_0 L$), and we have assumed that the channel gains are not varying with time. Since the interference is the sum of contributions from many independent random variables, we model it also as having a Gaussian distribution. Note, from this calculation, that I_k has zero mean, and variance

$$\sum_{i=2}^{M} L\, h_{i,1}\, E_i$$

Hence, the detection performance will depend on (see Section 2.1.1)

$$\sqrt{\frac{L^2\, h_{1,1}\, E_1}{\sum_{i=2}^{M} L\, h_{i,1}\, E_i + N_0 L}} = \sqrt{\frac{L\, h_{1,1}\, P_1}{\sum_{i=2}^{M} h_{i,1}\, P_i + N_0 W}}$$

---

[6]The integration over $(-\infty, +\infty)$ will actually cover neighboring symbols as well. But, because the pulses $p(t - j\tau_c)$ are orthogonal, the terms that we display are all that we will get.

where we have taken $R_c = W$, and $P_i = E_i \times R_c$ as the transmit power of User i (the power is the energy per chip times the chip rate). Thus the detection performance depends on the ratio

$$\frac{L\, h_{1,1}\, P_1}{\sum_{i=2}^{M} h_{i,1}\, P_i + N_0 W} \tag{2.24}$$

Now we can see why L is also called the *processing gain*. The effective predetection signal-to-interference-plus-noise ratio (SINR) for a user is the received SINR (i.e., $\frac{h_{1,1}\, P_1}{\sum_{i=2}^{M} h_{i,1}\, P_i + N_0 W}$) multiplied by the processing gain L.

Recalling the notation T_d for the delay spread of the channel, let us write $L_d = \frac{T_d}{\tau_c}$: L_d is the number of chip-times that correspond to the delay spread. Now consider the signal arriving over paths that have delays that are multiple of the chip-times. If the receiver can lock into any of these paths, then the transmitted symbol can be decoded as described earlier. If the paths fade independently, however, then we can exploit *multipath diversity* in much the same way as explained in Section 2.3. Because of the orthogonality property mentioned earlier, at the receiver, multiplication of the received signal by various shifts of the spreading code, and appropriate linear combination of the results, yields a detection statistic that is the sum of several faded copies of the user symbol. Since these shifts correspond to as many paths from the transmitter to the receiver, this is called *multipath resolution*. In the context of CDMA systems this is achieved by the *Rake receiver*. We note that this is exactly the same procedure as explained for receive antenna diversity in Section 2.3. Thus the Rake receiver permits a desired bit error rate to be achieved with a smaller SINR. Advanced receiver techniques such as *interference cancellation* now also are employed in CDMA systems.

Scheduling transmissions in a CDMA system involves a decision as to the spreading codes and the power levels to be allocated to the users. These determine the rate at which a particular bit flow can be transmitted. Of course, this decision will have to be made jointly for all users, since the decision for one user impacts every other user. We turn to such resource allocation problems in Chapter 5.

2.4.2 OFDMA

We begin by recalling some notation. The system bandwidth is denoted by W, and the delay spread by T_d. In a wideband system we are dealing with a situation in which $T_d >> \frac{1}{W}$, so that intersymbol interference has to be dealt with if we directly do digital modulation at a symbol rate of $\frac{1}{W}$. For example, we may have $W = 5$ MHz, and $T_d = 5$ μsec.

OFDMA is based on OFDM (orthogonal frequency division multiplexing) (see [43]), which can be viewed as statically partitioning the available spectrum into several (e.g., 128 or 512) *subchannels*, each of bandwidth B, such that $B << \frac{1}{T_d}$. Thus, a flat fading model can be used for each subchannel.

If there are n subchannels, then we say that the OFDM *block length* is n. The user bit stream is mapped into successive blocks of n channel symbols that are then transmitted in parallel over the n carriers, so as to occupy the *block time* $T (>> T_d)$; see Figure 2.11. The term *orthogonal* in OFDM refers to the fact that the center frequencies of the subchannels are separated by the reciprocal of the OFDM block time, T (see Figure 2.12). This makes the carriers approximately orthogonal over the block time. The subchannels can then be over-lapping (i.e., $B > \frac{1}{T}$), while the orthogonality between the subcarriers facilitates demodulation at the receiver.

Let $X_{j,k}, 1 \leq j \leq n$, denote the j-th symbol in the k-th OFDM block (see Figure 2.11). The batch of n symbols, which are transmitted in parallel, is also called an OFDM *symbol*. Then the earlier discussion suggests that the predetection channel output can be written as

$$Y_{j,k} = G_{j,k} X_{j,k} + Z_{j,k} \qquad (2.25)$$

where $j, 1 \leq j \leq n$, indexes the subcarrier and $k \geq 1$ indexes the successive OFDM symbols. $G_{j,k}$ denotes the fading on the j-th subcarrier during the k-th OFDM symbol. $Z_{j,k}$ denotes an additive noise sequence, which is taken to be i.i.d. zero mean, Gaussian.

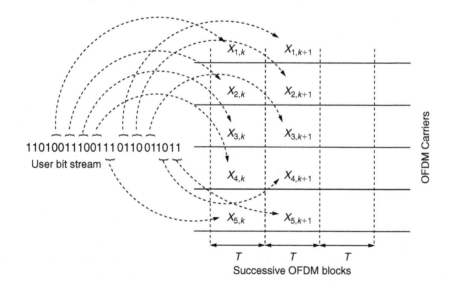

Figure 2.11 Depiction of the mapping of user bits into OFDM symbols. Here there are five OFDM carriers. Serially arriving user bits are split into pairs that are mapped successively into five parallel channel symbols ($X_{1,k}$, $X_{2,k}$,..., $X_{5,k}$), $k \geq 1$ (for example, the 4-QPSK constellation could be used). These five channel symbols comprise an OFDM block, which is transmitted over the block time T.

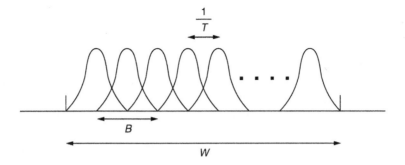

Figure 2.12 In OFDMA, the system bandwidth, _W_, is partitioned into overlapping subchannels, each of bandwidth _B_, with their center frequencies spaced apart by $\frac{1}{T}$, where _T_ is the OFDMA symbol duration.

Let us see how this model can be justified. By the orthogonality requirement, the carrier spacing is the reciprocal of the OFDM block time, $\frac{1}{T}$. Then the number of carriers, n, is related to the system bandwidth, W, by

$$\frac{1}{T} \times n = W$$

As an example, consider $T = 100$ μsec, so that the carrier spacing is 10 KHz and, for $W = 5$ MHz, $n = 500$. Suppose the channel delay spread, T_d, is such that

$$\frac{1}{W} \times n >> T_d$$

even though $\frac{1}{W} < T_d$. Then, combining the previous two equations we find that

$$T >> T_d$$

that is, the delay spread is much smaller than the OFDM symbol duration. We see that this is true in our numerical example, where $T = 100$ μsec and $T_d = 5$ μsec. Thus, the model in (2.25) is justified if the condition $\frac{1}{W} \times n >> T_d$ holds. We see that a frequency selective channel (for which $\frac{1}{W} < T_d$) gets converted to n parallel channels, each of which is frequency nonselective.

If it is further true that $T \times N << T_c$, the channel coherence time, for some $N \geq 1$, then over N OFDM blocks (which could constitute an *OFDM frame*) the channel can be taken to be constant. For example, taking $N = 50$ OFDM symbols, for the numerical values chosen earlier, $N \times T = 5$ ms. Then, for a channel coherence time of 10 ms, we can take the channel to be constant over an OFDM frame.

Continuing our running numerical example, suppose that the 64 QAM constellation is used to modulate the user's bits. Then, the previously described system carries one 64 QAM symbol per carrier, per OFDM symbol time, or OFDM block time, T. Thus, ideally, we can send 6 bits per carrier over time T, or we obtain a bit rate of $6 \times \frac{1}{T} \times n$ bits per second, or, with the preceding numerical values, $6 \times 10^4 \times 500 = 30$ Mbps, which is the same as if we could directly use a symbol rate of $W = 5$ Msps (mega symbols per second), with 6 bits being carried per symbol. The latter is, however, hard to achieve due to the difficulty of managing ISI over a time varying fading channel. Thus, it would seem that OFDM avoids ISI and provides the ideal bit rate. However, this discussion has ignored several practical overheads such as guard spaces between neighbouring subchannels, a guard time to eliminate any intersymbol interference, channel estimation overheads, and framing overheads.

It can be shown that fading is uncorrelated between subcarriers that are spaced apart by more than the coherence bandwidth, W_c Hz, which, we recall, is related roughly reciprocally to the delay spread, T_d. Hence, just as time diversity is exploited in TDM systems, frequency diversity can be exploited in OFDM systems: successive symbols of a user's codeword can occupy independently fading subcarriers.

It is easy to see how this concept can be used to share the flows from multiple users over a single OFDM link. Depending on the rate requirement of each user, a certain number of subchannels can be dynamically allotted to each of the users. Scheduling transmissions over an OFDMA link involves a decision as to how many subchannels to assign to a user, and what constellations, channel coding, and power levels to use from time to time, depending on the channel conditions and user rate requirements. Of course, the decisions for various users are interrelated. We note that, unlike static allocation on FDM-TDMA systems, the resource allocation decisions in OFDMA can vary from frame to frame, depending on channel conditions and traffic demands. We provide some OFDMA resource allocation formulations and their solutions in Chapter 6.

2.5 Additional Reading

In this chapter we have provided a tutorial overview of digital communication over wireless channels. The main objective was to provide an understanding of several models that we will use in the remainder of the book. We will require these physical layer models in the analysis, optimization, and control of the flow of traffic from various applications over wireless networks. The subject of this chapter has been extensively covered in several excellent textbooks; two well-established texts are the ones by Proakis [113], and by Lee and Messerschmitt [92]. The area of mobile multiuser communication over wireless channels has made rapid progress in recent years. It is important to gain an understanding of the mobile wireless channel, and the analysis of digital communication schemes over mobile wireless channels. We have provided only the basics; extensive coverage is provided by Stuber [123] and by Rappaport [116]. Modern topics, such as OFDMA and MIMO communication are covered in two recent textbooks, one by Tse and Viswanath [131], and the other by Goldsmith [43].

Appendix
Derivation of Equation 2.4

The signal received at the front-end of the receiver is given by

$$Y(t) = S(t) + N(t) \tag{2.26}$$

At the receiver, the signal is translated back to the baseband by multiplying the received signal by $\sqrt{2} \cos(2\pi f_c t)$, which yields

$$\sum_{k=-\infty}^{\infty} C_k \, p(t - kT) \, \left(\cos(4\pi f_c t) + 1\right) + U(t) \frac{\left(\cos(4\pi f_c t) + 1\right)}{\sqrt{2}}$$

The receiver filters this signal to the frequency interval $\left(-\frac{W}{2}, +\frac{W}{2}\right)$. Hence, the high frequency terms are filtered out (since $f_c \gg W$), and we are left with

$$\sum_{k=-\infty}^{\infty} C_k \, p(t - kT) + \frac{1}{\sqrt{2}} U(t) \tag{2.27}$$

The noise $\frac{1}{\sqrt{2}} U(t)$ is white Gaussian with power spectral density $\frac{N_0}{2}$ Watts/Hz. Since the signal is now bandlimited to the interval $\left(-\frac{W}{2}, +\frac{W}{2}\right)$, the average noise power is $W \times \frac{N_0}{2} = \frac{WN_0}{2}$ Watts; this means that

$$\lim_{t \to \infty} \frac{1}{t} \int_0^t \left(\frac{U(x)}{\sqrt{2}}\right)^2 \, dx = \frac{WN_0}{2}$$

where the integrand on the left is the power dissipation if the noise was put across a 1 ohm resistor; the integration yields energy over $(0, t)$, and the division by time yields the average power.

The receiver also needs to *synchronize* to the pulse boundaries. Once this is done the demodulator then needs to look at each received pulse and determine which symbol it is carrying. This step is called *detection*. Let us now see how the k-th symbol is detected, that is, how it is determined whether $C_k = +\sqrt{E_s}$, or $C_k = -\sqrt{E_s}$. The received signal is multiplied by the pulse $p(t - kT)$ and integrated over $(-\infty, +\infty)$, the pulse $p(t)$ being assumed to be known at the receiver.[7] Since $\int_{-\infty}^{+\infty} p^2(t) \, dt = 1$, and the shifted pulses are orthogonal, this yields

$$C_k + \int_{-\infty}^{+\infty} \frac{U(t)}{\sqrt{2}} p(t - kT) \, dt$$

Now $U(t)$ is a zero mean Gaussian process; hence, using the fact that a linear combination of Gaussian random variables is again Gaussian, we conclude that

---

[7]Since the pulses are practically time limited to some small multiple of T, such an integration can be performed by storing the received signal for some multiple of T, before starting the integration.

$\int_{-\infty}^{+\infty} \frac{U(t)}{\sqrt{2}} p(t - kT) \, dt$ is a zero mean Gaussian random variable, which we denote by Z_k. Thus, $E(Z_k) = 0$, and the variance of Z_k is obtained as

$$E\left(\left(\int_{-\infty}^{+\infty} \frac{U(t)}{\sqrt{2}} p(t - kT) \, dt\right)^2\right)$$

$$= \frac{1}{2} E\left(\int_{-\infty}^{+\infty} \int_{-\infty}^{+\infty} U(t)p(t - kT)U(x)p(x - kT) \, dt \, dx\right)$$

Since $U(t)$ is a white Gaussian noise process, with power spectral density N_0, bandlimited to $\left(-\frac{W}{2}, +\frac{W}{2}\right)$, it can be shown that the covariance function of $U(t)$ is given by

$$E(U(t)U(x)) = N_0 \frac{\sin \pi W(x - t)}{\pi(x - t)}$$

It then follows that

$$\frac{1}{2} E\left(\int_{-\infty}^{+\infty} \int_{-\infty}^{+\infty} U(t)p(t - kT)U(x)p(x - kT) \, dt \, dx\right)$$

$$= \frac{N_0}{2} \int_{-\infty}^{+\infty} \left(\int_{-\infty}^{+\infty} \frac{\sin \pi W(x - t)}{\pi(x - t)} p(t - kT) \, dt\right) p(x - kT) \, dx$$

However, $\frac{\sin \pi Wx}{\pi x}$ is just the transfer function of an ideal low pass filter with pass band $\left(-\frac{W}{2}, +\frac{W}{2}\right)$. Since the pulse $p(t)$ is bandlimited to this same range of frequencies, we have

$$\int_{-\infty}^{+\infty} \frac{\sin \pi W(x - t)}{\pi(x - t)} p(t - kT) \, dt = p(x - kT)$$

We therefore conclude that

$$E\left(Z_k^2\right) = E\left(\left(\int_{-\infty}^{+\infty} \frac{U(t)}{\sqrt{2}} p(t - kT) \, dt\right)^2\right) = \frac{N_0}{2}$$

In a similar manner it can be shown that $E(Z_k Z_l) = 0$, for $k \neq l$. Hence, since they are jointly Gaussian, Z_k and Z_l are independent for $k \neq l$.

Thus, we find that we have the symbol-by-symbol channel model

$$Y_k = C_k + Z_k$$

where Z_k is a sequence of i.i.d. zero mean Gaussian random variables with variance $\frac{N_0}{2}$. ∎

Problems

2.1 Show that $P_{error-AWGN}$ decreases exponentially with $\frac{E_s}{N_0}$. (Hint: for large x, $Q(x) \approx \frac{1}{x\sqrt{2\pi}} e^{-\frac{x^2}{2}}$.)

2.2 Consider a mobile radio environment in which we model only path loss and Rayleigh fading. The path loss exponent is η. The transmit power, averaged over Rayleigh fading, at the reference distance d_0 from a transmitter is P.

a. Write down an expression for the random received power $P_{rcv}(d)$ at a receiver at a distance $d = ad_0$, and obtain the distribution of $P_{rcv}(d)$.

b. Two cochannel transmitters (indexed 1 and 2) are simultaneously transmitting at distances $d_1 = a_1 d_0$ and $d_2 = a_2 d_0$ from the receiver. A transmission can be decoded if its signal to interference ratio exceeds γ. Ignoring the receiver noise, obtain the probability that the transmission from Transmitter 1 is decoded, treating the signal from Transmitter 2 as interference. This is called the *capture probability* (of Transmitter 1 over Transmitter 2).

c. Determine β such that if $a_2 > (1+\beta)a_1$ then the probability of transmission 1 being decoded is greater than $1-\epsilon$ ($\epsilon > 0$ is very small).

2.3 Consider the binary modulation scheme analyzed in Section 2.1.1. Obtain the bit error rates for various SNR values $\gamma = 12$ dB, 11 dB, 10 dB, and 9 dB. In each case, calculate the probability of packet error for 1500 byte packets. Hence compare the plots in Figure 2.9 with the AWGN plot in Figure 2.12. Hint: Use the approximation $Q(x) \approx \frac{1}{x\sqrt{2\pi}} e^{-\frac{x^2}{2}}$.

2.4 For the same situation as Problem 2.3 consider Rayleigh fading. For average (Rayleigh-faded) SNRs $\gamma = 12$ dB, 24 dB, and 36 dB, obtain the fraction of time that the SNR is less than 9 dB. Hence explain why a very large SNR is required in Figure 2.12 to obtain a high throughput.

2.5 By using the concavity of $\log(1 + x)$, show that the capacity in (2.16) is less than that in (2.18). What practical insight do we get from this?

2.6 Use Jensen's inequality to show that $C_{fading-CSIR} \leq W \log_2 \left(1 + \frac{E(H)P_{xmt}}{WN_0}\right)$.

2.7 Consider two AWGN channels with the same (power) fading h, and noise power σ^2. We have an amount of power P to assign. If the power P_i is assigned to Channel i, the capacity achieved is $\ln\left(1 + \frac{hP_i}{\sigma^2}\right)$. Is it better to put all the power into one channel or to split the power over the two channels? What is the optimal power assignment, assuming that the transmitter *knows* that the two channels have the same power gain?

CHAPTER 3

Application Models and Performance Issues

I n Chapter 2 we provided an understanding of the issues and techniques involved in carrying bit streams over wireless channels. The resources required to carry a bit stream depend on the characteristics of the stream (e.g., the average rate, peak rate, and rate variability), and the performance required by the application generating the bit stream (e.g., an interactive voice call requires end-to-end delay bounds, but can tolerate some data loss). In this chapter we will discuss the major types of applications that telecommunication networks are used for, and the performance issues related to these applications, particularly when wireless networks are involved.

Overview

We begin by providing a "big-picture" view of telecommunication networks as they exist today, showing all the elements, including the phone network, the Internet, and various wireless access networks. Then we outline various application scenarios that can arise in these interconnected networks. We classify network applications into three types according to the traffic they generate for the network to carry, namely, *elastic traffic*, *real-time stream traffic*, and *store-and-forward stream traffic*. Taking interactive telephony as the principal example of real-time stream traffic, we point out that the traffic offered to the network can be either *constant bit rate (CBR)* or *variable bit rate (VBR)*. We provide the *quality of service (QoS)* objectives for each case. The predominant use of the Internet is for applications such as e-mail and web browsing, which generate elastic traffic. Such applications are also increasingly important for wireless access networks, as users begin to use their handheld devices for Internet access. The remainder of the chapter provides an understanding of this important type of traffic. We show the need for *feedback control* of elastic traffic sources. In the Internet such feedback control is exercised by the *Transmission Control Protocol (TCP)*, which uses an adaptive window mechanism for managing the rates of elastic traffic sources. The latter part of this chapter will be devoted to a discussion of the performance of TCP over wireless links.

3.1 Network Architectures and Application Scenarios

In Figure 3.1 we provide a simplified view of telecommunication networks as they exist today. The *public switched telephone network (PSTN)* has carried telephone calls for nearly a century. In this network, the calls are multiplexed onto the links by using circuit switching. Packet switched public data networks have evolved from the early X.25 networks to the present ubiquitous Internet. Cellular networks have provided mobile access since the early 1980s, and already have evolved through three generations of commercial deployment. As discussed in Chapter 1, a variety of resource allocation techniques are employed in cellular networks; these techniques will be the subject of Chapters 4, 5, and 6. In campuses and enterprises, mobile devices such as laptops and personal digital assistants (PDAs) obtain access to Internet services via wireless local area networks (WLANs). Figure 3.1 also shows some emerging technologies, namely, *wireless metropolitan area networks (WMANs)*, and ad hoc multihop wireless mesh networks.

Figure 3.1 also shows certain devices, generically called *gateways (GW)*, interconnecting the PSTN, the Internet, and cellular networks. Note that we show only *bearer* gateways that are in the path of the actual application traffic. Signaling gateways are also needed because signaling protocols are different in networks

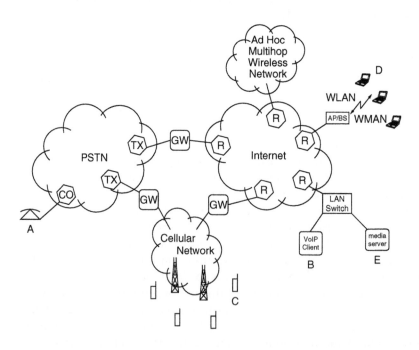

Figure 3.1 A simplified view of the public switched telephone network (PSTN) and the Internet, and how they connect with each other and various wireless networking technologies. GW denotes a gateway; there are gateways for signaling and call control, and also for transferring traffic across network boundaries. Other abbreviations are: CO – Central office, TX – Trunk exchange; R – Router.

that have evolved independently. For example, a signaling protocol called *Session Initiation Protocol (SIP)* is used to set up voice calls in the Internet, whereas the *Signaling System No. 7 (SS #7)* is used in the PSTN.

In this setting we can identify several different instances of point-to-point communication, each of which gives rise to certain resource allocation issues. A common instance is that of a voice call between a fixed line telephone instrument on the PSTN (e.g., A in Figure 3.1) and a cellular phone (e.g., C). We will often use the more generic term *mobile station (MS)* for a mobile device such as a cell phone or a PDA; in some technologies the terms *station (STA)* or *subscriber station (SS)* are used instead of MS. One of the functions of the gateway, GW, between the PSTN and the cellular network is to convert between the *constant bit rate (CBR)* flow of voice bytes in the PSTN to a lower bit rate voice coding scheme over the resource limited cellular wireless network. In cellular networks, typically, voice is handled as a CBR stream of a lower bit rate than that in the PSTN. In an FDM-TDMA system, such as GSM, there are channels, each of which can carry one call at a fixed rate. An accepted call is assigned to a channel for the entire duration of the call. Thus, this is essentially a circuit multiplexing system. A call for which no free channel is available is *blocked* and the main performance measure for the system is the *blocking probability*, defined as follows. Let $A(t)$ be the number of call arrivals until time t, and $B(t)$ be the number of calls blocked in the same time; then the blocking probability P_b is given by

$$P_b = \lim_{t \to \infty} \frac{B(t)}{A(t)}$$

whenever the limit exists. In a CDMA cellular system, the voice connection is handled by assigning to it the required coded rate. However, unlike typical FDM-TDMA systems, a variety of rates can be assigned. Each accepted call needs to be assigned a transmit power level, and whether or not a call is accepted depends on the rate requirements of the other calls that have already been accepted, and the resulting interference levels once the new call is accepted. Again, the performance measure of interest is the probability of call blocking.

Another possible type of connection is a voice call between the PSTN instrument A and a *voice over IP (VoIP)* endpoint, say, B. The GW between the PSTN and the Internet then would convert between the CBR flow of voice bytes in the PSTN to an asynchronous flow of voice packets in the Internet. Since the packets flow asynchronously in the packet network, and can be queued in buffers in the network routers, certain new issues arise, which will be discussed in Section 3.3. Similarly, there could be a voice call between either A or B and the endpoint D, which accesses the Internet via a WLAN or a WMAN. In such cases, the packet multiple access mechanism over the wireless access network affects the performance of the voice call.

Yet another scenario is that of the MS C being used to browse the contents of the web server E which is attached to the Internet. We will see later, in Section 3.4, that this kind of application is quite different from voice, as there is no *intrinsic* rate at which the data should be transferred from the web server to the mobile phone. Feedback-based rate control algorithms are employed to ensure some sort of rate fairness between such connections, and efficient utilization of network resources. In the Internet, such control is exercised by TCP in conjunction with implicit or explicit congestion feedback from the network. In addition, a cellular access network such as a CDMA cellular network, or an OFDMA network, would have its own rate control algorithms. Unlike these centrally controlled cellular systems, a random access WLAN (see Chapter 7) does not have an explicit rate allocation mechanism. Hence, when a device such as D is engaged in web transfers from the Internet, it is of interest to determine what the throughput is, and what kind of fairness is achieved.

Finally, the MS C or the device D could be displaying a video that is stored in the server E. Now, once the video starts playing, the network should provide this connection the average rate required to transport the video stream. Variability in the rate at which the network transports the video can be compensated by buffering a sufficient amount of the video in the playout device. Such buffering must be done in such a way that the playout does not starve (i.e., the buffer empties out), nor does the buffer overflow.

Figure 3.1 also shows an ad hoc multihop wireless mesh network attached to the Internet. Although multihop mobile wireless networks have been studied for more than three decades (in the early years under the name *packet radio networks*), even today the deployments of such networks are still experimental. It is one of the more active research areas in wireless networking, and we will provide a research oriented discussion in Chapters 8, 9, and 10. The IEEE 802.16 suite of protocols (popularly known as WiMax) now contains the definition of a mesh networking standard. Under this standard, nodes that cannot directly access a WiMax base station (BS) can form a static mesh network that is connected at some point to the WiMax BS. We can think of these as managed mesh networks. Such networks will be expected to carry all Internet services, respecting the QoS objectives that we will describe. On the other hand, ad hoc wireless mesh networks, such as community networks formed from WiFi access points in homes, cannot be expected to provide any consistent QoS to the applications they carry. We might expect that these would be used primarily for nonreal-time store-and-forward applications, such as e-mail and web browsing.

3.2 Types of Traffic and QoS Requirements

Based on the discussion of the various scenarios in Section 3.1, we can infer that applications generate one of the types of traffic in the following list. Some example applications that generate each type of traffic are also listed.

- Elastic traffic; e.g., WWW browsing, FTP file transfers, and electronic mail

- Real-time stream traffic; e.g., packet voice telephony
- Store-and-forward stream traffic; e.g., streaming movies or music over the Internet.

In the remainder of this section we discuss the characteristics of these traffic types, and also their quality of service (QoS) requirements.

Elastic Traffic

Consider a data file, residing on the disk of the server E (shown in Figure 3.1) that needs to be transferred to the disk of a portable computer attached to the Internet (e.g., the laptop D, which connects via a WLAN), or to the memory of the cell phone C. Although the human (or some machine application) that wishes to achieve this file transfer would like to have the transfer completed in, say, a second or two, the source of data itself does not demand any specific transfer rate. If the data transfer does not lose data, no matter how fast or slow it is (but as long as the rate is positive), the file will sooner or later get transferred to the destination device. We say that, from the point of view of the network, this source of traffic is *elastic*. Many store-and-forward services (with the exception of media streaming services) are elastic; e.g., file transfer, WWW download, electronic mail (e-mail). In this list, the first two are distinguished by the fact that they are *nondeferable* (i.e., the network should initiate the transfer immediately), whereas e-mail is *deferable*.

We observe that elastic traffic does not have an intrinsic temporal behavior, and can be transported at arbitrary transfer rates. Thus the following are the QoS requirements of elastic traffic.

- Transfer delay and delay variability can be tolerated. An elastic transfer can be performed over a wide range of transfer rates, and the rate can even vary over the duration of the transfer.

- The application cannot tolerate data loss. This does not mean, however, that the network cannot lose any data. Packets can be lost in the network (owing to uncorrectable transmission errors or buffer overflows) provided that the lost packets are recovered by an automatic retransmission procedure. Thus effectively the application would see a lossless transport service. Since elastic sources do not require delay guarantees, the delay involved in recovering lost packets can be tolerated.

In practice, of course, users will not tolerate arbitrarily poor throughput, high throughput variability, and large delays. Hence a network carrying elastic traffic will need to manage its resource-sharing mechanisms in a way such that some minimum level of throughput is provided. Further, some sort of *fairness* must also be ensured between the ongoing elastic transfers.

Elastic traffic can also be carried over circuit multiplexed networks (e.g., the PSTN or GSM cellular networks), or over networks that allocate a fixed rate to the elastic connection (e.g., a second generation CDMA cellular network). In this case, *shaping* of the traffic so as to match the allocated rate should be carried out by the source. Obvious examples would be Internet access over a dial-up line in the PSTN, or a fixed rate connection over a cellular access network being used for Internet access.

Real-Time Stream Traffic

Consider digitized speech emanating from an end-device involved in interactive telephony. This could be a periodic stream of bytes or packets, or, if *silence suppression* is employed then it could be an on-off stream of bytes or packets. Obviously, this source of traffic has an *intrinsic temporal behavior*, and this pattern needs to be preserved for faithful reproduction of the speech at the receiver. The network will introduce delay: fixed propagation delay, and, in packet networks, queuing delay that can vary from packet to packet (see Figure 3.2). Playout delay introduced at the receiver (to mitigate the effect of random packet delay variation) will be larger the more variable the packet delay. Hence, the network cannot serve such a source at arbitrary rates, as it could in the case of elastic traffic. In fact, depending on the adaptability of such a real-time stream source, the network may need to reserve bandwidth and buffers in order to provide an adequate transport service to the source. Applications such as real time interactive speech or video telephony are examples of real-time stream sources.

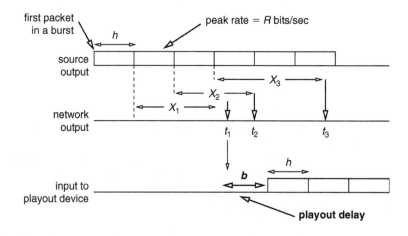

Figure 3.2   A sequence of packets from a voice talk-spurt being transported across a packet network, and then being played out at the receiver after a playout delay. Each source packet contains *h* seconds of voice, the packet delays are $X_1, X_2, \ldots,$ the packets arrive at the receiver at times $t_1, t_2, \ldots,$ and the playout delay is *b*. Notice that immediate playout at time t_1 would have resulted in the talk-spurt being broken.

The following are the typical QoS requirements of real-time stream sources.

- Delay (average and variation) needs to be controlled. Real-time interactive traffic such as that from packet telephony would require tight control of source-to-sink delay; for example, for wide area packet telephony the delay may need to be controlled to less than 200 ms with a probability more than 0.99. Packets that do not conform to the delay bound are considered to be lost.

- There is tolerance to data loss. Note that, from the point of view of the receiver, packets can be lost for two reasons: (1) buffer overflows, or unrecovered link losses in the network, or (2) late arrivals at the receiver. Owing to the high levels of redundancy in speech and images, a certain amount of data loss is imperceptible. As an example, for packet voice in which each packet carries 20 ms of speech, and the receiver does lost-packet interpolation, 5 to 10% of the packets can be lost without significant degradation of the speech quality [81], [54]. Because of the delay constraints, the acceptable data loss target cannot be achieved by first losing and then recovering the lost packets; in other words, stream traffic expects the *intrinsic loss rate* from the packet transport service to be bounded.

Store-and-Forward Stream Traffic

We can distinguish what we have just described as real-time stream traffic from the kind of traffic that is generated by applications such as streaming audio and video. Such applications basically involve a *one-way transfer* of an audio or video file stored on the disk of a media server. Consider a video stored in a server being played over a network. For example, the computer D or the handheld device C in Figure 3.1 may be used to watch a movie stored in the Server E. In order for the received video to be useful, the playout device should be continuously "fed" with video frames so that it is able to reproduce a smooth video output. This can be achieved by providing a guaranteed rate to the transfer, as would be done, for example, in a CDMA cellular system in the context of the device C. Alternatively, owing to the fact that the transfer is one way, a more economical way is to treat the transfer as elastic, and buffer the video frames as they are received. This would be the approach taken when the video is transferred over the random access WLAN to the computer D. Playout is initiated only after a sufficient number of video frames has been buffered so that a smooth video playout can be achieved in spite of a variable transfer rate across the contention-based WLAN. Note that the same description holds for streaming audio.

Thus, the problem of transporting streaming audio or video becomes just another case of transferring elastic traffic, with appropriate receiver adaptation. Note, however, that the elasticity here is constrained since the *average rate* at which the network transports the video bit stream must match the rate at which the video has been coded. Simple interactivity, such as the ability to rewind, can also be supported by the receiver storing frames that have already been played out. This, of course, puts a burden on the amount of storage that the playout device needs to have. An alternative is to trade off sophistication at the receiver with the possibility of interactivity across the network; that is, the press of the rewind button stops the video playout, frames stored in the playout device are used to create a rewind effect, and meanwhile additional past frames are fetched from the server. But this would need some delay and throughput guarantees from the network, requiring a service model somewhere in between the elastic and the real-time stream model that we have described earlier. We conclude that the QoS requirements of a store-and-forward stream transfer would be the following:

- The average transfer rate provided in the network should match (in fact, should be greater than) the average rate at which the stored media has been encoded.

- The transfer rate variability should not be too large.

Thus store-and-forward stream traffic is like stream traffic since it has an intrinsic average rate at which it must be transported, but it does not have strict delay bounds, and hence the network can provide it a time varying transfer rate. In fact, TCP can be used to transport store-and-forward streaming media, provided the average TCP throughput does not drop below the average coded rate of the media. The added benefit of TCP is that it recovers lost packets.

Closed and Open Loop Traffic:
It is appropriate to refer to real-time stream traffic as *open loop* as it has an intrinsic temporal behavior. Typically, the rate of flow on a connection is determined by the application, and these sources are not controlled by the network. In some systems, a limited amount of controllability is possible, by the sink alerting the source of poor playout quality, to which the source can respond by using a lower bit rate coder. On the other hand, closed loop controls invariably are used when transporting elastic traffic, and, hence, such traffic can be called *closed loop*. By means of implicit feedback (packet loss) or explicit feedback (control bits in packet headers) the source of the traffic is made to continually adjust its rate of emitting data.

3.3 Real-Time Stream Sessions: Delay Guarantees
In this section we will discuss traffic modeling and QoS issues for real-time stream sessions in the context of voice telephony.

3.3.1 CBR Speech
Consider a voice call between a pair of endpoints in Figure 3.1. For example, the PSTN phone A and the cellular phone C, or between B and D, or between C

and B. In each end device, electrical signals from a microphone are digitized and coded by a *speech codec*. A typical approach is to sample the analog signal from the microphone at 8000 samples per second, to quantize the resulting continuous amplitude samples into 256 predetermined levels, and then encode each of these levels into 8 bits (one byte). The output of such a speech coder is called PCM (Pulse Code Modulation) coded speech (ITU's G.711 standard). The PCM encoder, thus, yields a CBR source that produces 1 byte every 125 μseconds. A PCM source can be compressed to yield CBR sources at various rates. For example, ITU's G.729 vocoder takes PCM as the input and produces 10 bytes of coded speech every 10 ms, thus yielding a coded bit rate of 1 KBps (kilobytes per second). However, this speech coder has a coding delay of 15 ms and a decoding delay of 7.5 ms. An important measure of the performance of network telephony is the *Mouth-to-Ear (MtoE)* delay—the delay between a sound being produced at the source device and this being heard at the other end. Thus, if the G.729 speech coder is employed, then there is a minimum MtoE delay of 22.5 ms.

In order to carry a CBR voice source, it is necessary for the network to use a service rate greater than or equal to the voice bit rate. Further, if the source is allocated exactly the constant bit rate then there will not be any queuing. Hence, for CBR sources it is sufficient to allocate the CBR rate. Consider a voice call between the PSTN phone A and the cell phone C. If the gateway GW converts PCM speech arriving over the PSTN to CBR speech at rate R, then the cellular network can just allocate resources so that the voice call is provided a service rate of R. This is typically what is done in an FDM-TDMA cellular system (such as GSM), or in a CDMA cellular system. We will discuss resource allocation issues in these two types of systems in Chapters 4 and 5, respectively.

3.3.2 VBR Speech

In speech generated by interactive telephony, there are low energy periods that correspond to silences while the speaker listens, or to gaps between words, sentences, and utterances. The coder output corresponding to these *inactive periods* can be discarded or encoded at a lower rate. This yields a *variable bit rate (VBR)* coded speech. The VBR speech can be handled as a variable rate byte stream, or can be *packetized* for transport over a packet network. One approach is to take a certain number of bytes from the source (e.g., 160 bytes or 20 ms of speech from a PCM source) and generate a packet from these. It may happen that a talkspurt finishes before 160 bytes have been collected; in such a case a short packet is generated. The packetizer must wait to accumulate a packet; thus bytes that arrive early in the packet have to wait for those that arrive later, until the packet is formed. This results in a *packetization delay*, which can, obviously, be reduced by using shorter packets. Packets cannot be very short either, as there could be a significant amount of header overhead in each packet (e.g., in the Internet there would be at least 12 bytes for RTP (Real-time Transport Protocol), 8 bytes for UDP (User Datagram Protocol), and 20 bytes for IP). If the coder output during speech inactivity periods is discarded, then the output of a packetizer will comprise bursts of packets (during which packets are generated at a constant rate) and periods during which no packets are generated.

Note that although the inactive periods do not have speech information in them, the duration of the gaps is indeed information that needs to be conveyed to the receiver. One of the difficulties in the transport of packetized VBR speech is in the retention of such timing information. Since packets are transmitted only during active periods, the inactive periods can only be approximately replicated at the receiver. It has been found that the resulting errors are not noticeable if the inactive periods are long. Thus the voice activity detection function (after the speech encoder) does not discard bytes from short inactive periods.

Consider again the voice conversation between the PSTN phone A and the cell phone C in Figure 3.1, with VBR speech being used in the cellular network; that is, the voice arrives over the PSTN as a CBR flow, but is encoded into a VBR flow at the gateway, GW. Suppose the VBR speech source is allocated the service rate C in the cellular network (see Figure 3.3). Let us denote by R the peak rate of the VBR source, and by $\bar{r}$ the average rate. Thus, for example, if the on-off VBR source has an average on duration of 400 ms and average off duration of 600 ms, then with $R = 64$ Kbps, we will have $\bar{r} = \frac{400}{400+600} \times R = 25.6$ Kbps. It is clear that it is a waste of bandwidth to make $C > R$, and that it is necessary that $C \geq \bar{r}$. Now suppose we take $C < R$. Notice that when the voice source is emitting data at the rate R, then the link buffer builds up at the rate $(R - C)$ Kbps. Any byte that arrives when the buffer level is, say, B bits will be delayed by $\frac{B}{C}$ ms. A priori, we do not know for how long this *rate mismatch* will last (the average rate $\bar{r} = 25.6$ Kbps could have been obtained with a 4 sec on time and a 6 sec off time too!). Hence, if we want to bound the delay of the voice bytes in the link buffer, in the absence of any other information about the source, our only recourse is to use $C = R$.

This approach of *peak rate allocation* could be one way in which the cellular network manages its resources, and is typically the approach adopted in FDM-TDMA and CDMA cellular systems. In CDMA systems, even though the peak rate is allocated to a call, the on-off nature of VBR speech is exploited because during the voice silence periods a call does not cause interference (see

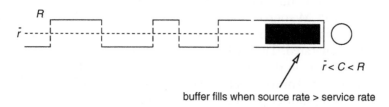

buffer fills when source rate > service rate

Figure 3.3 An on-off VBR source, of peak rate R Kbps and average rate $\bar{r}$ Kbps, being carried by a link of capacity C Kbps, $\bar{r} < C < R$.

Section 2.4.1). On the other hand, if some bounds on source behavior are known, or if there is a reliable statistical characterization of the output of the speech coder, then the network could assign a rate $C < R$, while still meeting the QoS objectives (techniques for carrying out such a design can be found in [89, Chapters 4 and 5]). Such a design approach requires a characterization of the source output, but can lead to a more efficient utilization of the resources of the cellular network.

3.3.3   Speech Playout

Consider VBR coded packetized speech telephony between the devices B and D, or B and C. At the devices C and D there is the problem of playing out the individual packets in a way that the original voice patterns are reproduced, in spite of the random network delays introduced by the packet network. Suppose that the cellular network handles only CBR speech; then the problem of converting the speech packets, arriving asynchronously over the Internet, to CBR speech over the cellular network will be a task of the gateway, GW, between the Internet and the cellular network.

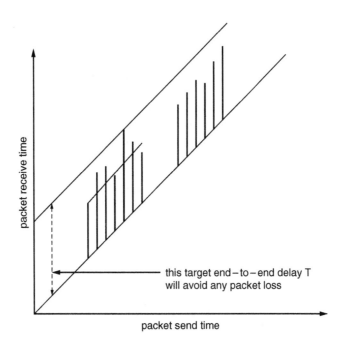

Figure 3.4   Jitter in the delay of voice packets, and the need for playout delay at the receiver. The height of each bar is the delay of the corresponding packet in the network. On the y-axis, the bottom of each bar is the packet's send time, and the top of the bar is the packet's receive time.

To understand the problem of playing out asynchronously arriving speech packets, let us look at Figure 3.4. Consider a packet voice call between the end-points B and D. If each packet that left B arrived at D instantaneously, then the packet send times and packet receive times would lie along the $y = x$ line in the figure. However, the voice packets, in being transported across the wide area network, will encounter transmission, queuing, and propagation delays. In addition, the queuing delays will be random, leading to delay variation, also called *delay jitter*. These random network delays are depicted in Figure 3.4 by the vertical bars standing on the slanting line. Each such bar corresponds to one packet, and is positioned on the x-axis at the send time of the packet. The bursts of packets from talk-spurts can be identified as consecutive periodically occurring vertical bars. The gaps between these bursts are the silence periods. The height of each bar represents the network delay experienced by the packet. Suppose the receiver adopted the policy of playing out each packet as soon as it was received. We look at the first talk-spurt and notice that if the first packet was played out beginning from the time that it was received, its bits would be played out along the slanting line (parallel to the $y = x$ line) starting from the tip of the first vertical bar. Where this slant line meets the second bar is where the playout of the first packet would complete, and the speech decoder would be ready for the next packet. This packet, however, would be too late. It follows that if an *immediate playout* policy is adopted at the receiver, then in the first talk-spurt four out of the seven packets would arrive too late for playout. Although the coder could attempt to *interpolate* these lost packets, such a high frequency of interpolation would lead to very poor speech quality.

An obvious alternative is to adopt the policy of *deferred playout*. A *playout delay* is applied to each packet to allow trailing packets to "catch up." From Figure 3.4 it is clear that if all packets are played out starting from the uppermost slant line, then (for the fragment of the packet process shown) no packet would be late. Obviously, this naive approach has two practical problems. We do not know in advance the maximum delay that any packet in the connection will encounter, and in any case this worst-case delay could be very large. Notice that playout delays add to the MtoE delay. Suppose, however, that we are able to determine a value T such that the packet delay rarely exceeds T. Then, as shown in Figure 3.5,

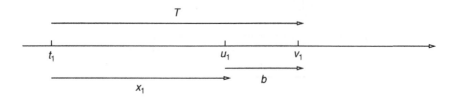

Figure 3.5 Arriving voice packet delays are stretched to a target delay T. A voice packet that left the source at time t_1 arrives at the receiver at time u_1, thus incurring a network delay of X_1. It is buffered for a time b so that its end-to-end delay is T. If $X_1 > T$ then the packet is discarded.

the receiver can stretch out the delay of each arriving packet to T; thus packets that are delayed more than T are lost and may be interpolated.

We are still left with the problem of determining a value for T. There are two alternatives. The packet network may have the ability to provide a delay guarantee at call setup (e.g., $\Pr(X > T) < \epsilon$ where X is the delay of packets in the network; see Figure 3.5). In such a case, the endpoints specify the traffic characteristics of the source they want to be carried, and the values of T and ϵ. The network evaluates whether the call can be accepted, and, if the call is accepted, the network sets up the appropriate mechanisms along the path of the call so that the delay objective is met. Now T is known to the receiver at call setup time, and the procedure shown in Figure 3.5 can be performed. If the network cannot provide delay guarantees, then the receiver would need to estimate T as the call progresses. Time stamps carried by the voice packets in their headers would be used to obtain a statistical estimate of T. This estimate can then be used to set the playout delays of arriving packets. Since there is no guarantee, the value of T could be larger than desired and could vary over time as congestion in the network varies.

3.3.4 QoS Objectives

We gather that the MtoE delay for a packet voice call is the sum of several terms as shown in the following equation:

$$\text{MtoE Delay} = \text{coding delay} + \text{packetization delay}$$

$$+ \text{network propagation delay}$$

$$+ \text{network transmission and queuing delay}$$

$$+ \text{receiver playout delay} + \text{decoding delay}$$

In this expression, the network propagation delay is just the signal propagation delay over the various media interconnecting the routers and switches in the path of the call. A rule of thumb is to compute this fixed delay as 5 ms per 1000 Km of cabled transmission. Thus, for example, between points in the continental United States and India separated by a distance of about 20,000 Km, the one-way WAN propagation delay would be about 100 ms. For a geostationary satellite link, the one-way propagation delay is computed as the time taken for radio waves to travel from the transmitter up to the satellite and then down to the receiving ground station, or about 250 ms.

In addition to the MtoE delay, some voice packets can be lost, either owing to buffer overflows in routers, or because they arrive after their scheduled playout time at the receiver. Thus, an example of the QoS expected by a voice call could be:

$$\Pr(\text{MtoE Delay} > 200\ \text{ms}) < 0.02$$

and

$$\Pr(\text{Packet Loss}) < 0.05$$

We recall that packet loss includes loss due to late arriving packets, as well as loss due to buffer overflows and errors in the network. All such missing packets will need to be interpolated by the voice decoder, which leads to degradation in voice quality.

Notice that the MtoE delay has some fixed parts (coding and decoding delay, packetization delay, and propagation delay), and some variable parts (transmission and queueing delay, and the playout delay). It is these delays that are governed by the characteristics of the traffic emitted by the source, and the way the traffic is handled in the network (i.e., the other traffic with which it is multiplexed, and the resource allocation decisions made by the network). Letting X denote the variable (random) delay, and then subtracting the fixed delays from the MtoE delay target, the network performance requirement can be reduced to

$$\Pr(X > T) < \epsilon$$

where, for example, $\epsilon = 0.02$. Now consider a call between the devices B and D. The computer B is attached to the Internet by a high-speed enterprise or campus LAN, whereas D is attached to the Internet by a contention-based WLAN. One approach to the analysis of such a situation is to break down the end-to-end QoS objective into subnetwork-wise objectives. Thus, one could break up the end-to-end delay bound T as $T = T_1 + T_2$, and the probability of violating the delay bound can be split up as $\epsilon = \epsilon_1 + \epsilon_2$. We can call T_1 and T_2 the *delay budgets* in the respective subnetworks. Let the stationary random delay over the Internet segment of the call be denoted by X_1 and that over the WLAN be denoted by X_2. Suppose we ensure that, for each $i = 1, 2$,

$$\Pr(X_i > T_i) < \epsilon_i$$

It will then follow that

$$\Pr(X > T) = \Pr(X_1 + X_2 > T_1 + T_2)$$

$$\leq \Pr(\{X_1 > T_1\} \cup \{X_2 > T_2\})$$

$$< \epsilon_1 + \epsilon_2 = \epsilon$$

where the first inequality follows from the simple observation that if $X_1 + X_2 > T_1 + T_2$, then it cannot be that $X_1 \leq T_1$ and $X_2 \leq T_2$; the last inequality is just the union bound. The resource allocation in the WLAN can then be performed so as to ensure that the voice packet delay exceeds T_2 with a probability less than ϵ_2. The same approach can be used if end-device D is attached to the Internet via a WMAN (see Figure 3.1).

3.3.5 Network Service Models

In the previous section we showed how to derive, from the end-to-end delay QoS problem for a connection, a QoS objective for the access network. If the access network accepts calls based on peak rate allocation, then the only issue is to design the network for a desired call blocking probability. For this purpose, the Erlang blocking model (see Appendix D) can be used. If the access network assigns a fixed rate less than the peak rate of a VBR call, then the model depicted in Figure 3.3 can be used. The analysis of such models has been discussed at length in [89, Chapter 5].

In some access networks, however, the service rate applied to a connection may not be constant. For example, in OFDMA systems, the number of bytes to be served from a queue can vary from frame to frame depending on the fading in the various carriers, the competing traffic, and the power constraint. Thus, in this case we have a dynamically controlled server; see Chapter 6. Detailed analysis of such systems to obtain buffer occupancy distributions or delay distributions is difficult. Wireless LANs, based on the IEEE 802.11 standard, are contention-based systems. Hence the service applied to a queue is time varying because the number of contending nodes varies over time, as some queues empty out while others receive new traffic. Some progress has been made on developing analytical models for the performance analysis of wireless LANs; some of these approaches will be discussed in Chapter 7.

3.4 Elastic Transfers: Feedback Control

Elastic traffic is generated by applications whose basic objective is to move chunks of data between the disks of two computers connected to the network. Elastic flows can be speeded up or slowed down depending on the number of flows contending for the capacity of the network. Figure 3.6 shows that, at the most basic level, an elastic session simply involves the transfer of some files from one host attached to the packet network to another host. For example, the two hosts could be e-mail relays; each file transfer would then correspond to an e-mail being forwarded toward its destination mail server. Alternatively, the source host could be an FTP archive, and at the destination host, a user is downloading several files during an FTP session. A similar example would be that of the source being a web server, and the destination being a client with a web browser, using the HTTP (Hyper-Text Transfer Protocol) protocol to browse the files at the server. In an internet, for example, when a user requests a web *page* (using an HTTP GET request), first, a *base* file is downloaded, which in turn may trigger the transfer of several *embedded objects*, such as images. When there are embedded objects, the exact mechanism for downloading the objects depends on the version of HTTP in use. In HTTP 1.0, for the transfer of the base file, and for each embedded object file, a separate TCP connection would be set up between the client and the server.

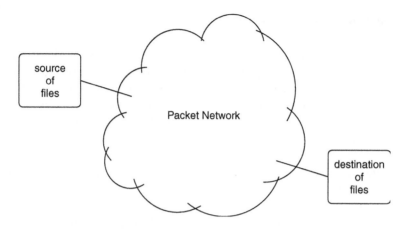

Figure 3.6 An elastic session simply involves the transfer of one or more files from one host to another.

In HTTP 1.1, in order to reduce connection set up overheads, a TCP connection once set up would be reused for several file transfers between the same client and server.

In all these cases the basic problem is to transfer each file in its entirety from the source machine to the destination machine. This is the primary objective. There is no intrinsic rate at which the files *must* be transferred. In fact the transfer rate can vary as a file is transferred. Although the user downloading the files would want to receive the files quickly, this requirement is not really a part of the service definition of an elastic session.

Further, we note that there is no intrinsic packet size that the files need to be segmented into during their transfer. The transfer protocol can just view each file as a byte stream, and transfer varying amounts of it in each packet.

We will deal primarily with point-to-point elastic sessions, and this is what we will mean when we use the term "elastic session." Thus an elastic session involves a connection between two endpoints. The network determines a route between the endpoints in each direction; if the session lasts long enough, there is a possibility that the route, in either direction, changes during the session. During a session, data transfers may take place in either direction, with possible gaps between the successive transfers. For example, if a user at a computer logs on to an FTP server, then FTP's *get* or *put* commands can be used to download or upload files. The user may need to do some other activities in between the file transfers (e.g., read what is downloaded and make notes); in user models, these gaps often are referred to as *think times*. Similarly, a user browsing a web server would download a web page, and spend some time looking at it, before downloading another web page from the same site. If the user shifts to browsing another web server, then we view this as another session starting, typically over a different pair of network routes.

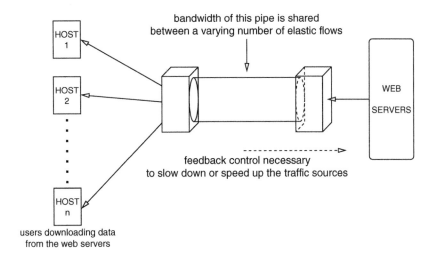

Figure 3.7 Several users dynamically share a link to download files from web servers.

3.4.1 Dynamic Control of Bandwidth Sharing

Figure 3.7 shows a very simple "network" comprising a single link over which several users, on their respective hosts, are downloading files from some web servers. Let us take the link capacity to be C bps, and assume that the local networks attaching the users and the web servers to this link are infinitely fast. We use this simple scenario to illustrate and discuss some basic issues that arise when several elastic sessions share the network bandwidth. In fact, the situation depicted in Figure 3.7 is similar to what happens when several mobile users download files (text, music, video, etc.) from a server attached to a cellular operator's own high speed local area network. The important difference is that in cellular systems the system bandwidth is not managed as one "fat pipe."

Suppose, to begin with, a single user initiates a download from a web server over the link. It is reasonable to expect that an *ideal* data transfer protocol will (and should) provide this file transfer with a throughput of C bps. This much bandwidth is available, and if all of it is provided to the transfer, the session will be out of the system as early as possible. Now suppose another user starts a session, while the first file transfer is still progressing. When the corresponding web server starts transferring data toward the user, the total input rate into the link (from the web server's direction) will exceed C bps. If the first file transfer is proceeding at C bps, then the link's service rate will be exceeded no matter how slowly the second server sends its data. This will lead to link *congestion*. The network device that interconnects the server's LAN to the backbone link will have buffers "behind" this link. These buffers can absorb excess data that accumulate during this overload, provided the situation does not sustain for long. In addition, this situation is clearly *unfair*, with one user getting the full link rate, and the other user getting practically no throughput; this situation should not be allowed to persist.

Both of these issues (congestion and unfairness) require that there be some kind of *feedback* (explicit or implicit) to the data sources so that the rate of the first transfer is reduced, and that of the second transfer is increased, so that ultimately each transfer obtains a rate of $\frac{C}{2}$ bps. Now suppose the first file transfer completes; then, if the second transfer continues to proceed at $\frac{C}{2}$ bps, the link's bandwidth is wasted, and the second session is unnecessarily prolonged. Hence, when the first session departs, the source of the second session should increase its transfer rate so that a throughput of C bps is obtained.

In summary, from this discussion we conclude that an explicit or implicit feedback control mechanism needs to be in place so that as the number of sessions varies, the transfer rate provided to each session varies accordingly. By an *explicit feedback* we mean that control packets flow between the traffic sources, sinks, and the network, and these packets carry information (e.g., an explicit rate or a rate reduction signal) that is used by the sources to adapt their sending rates. On the other hand, *implicit feedback* can be provided by packet loss or increase in network delay; that is, a source can reduce its rate on sensing that one of the packets it sent may not have reached the destination. For example, in much of the current Internet, TCP (Transmission Control Protocol) uses an implicit feedback-based congestion control mechanism. Explicit rate control was proposed for the ABR (available bit rate) service in ATM (asynchronous transfer mode) networks. The idea was to associate with each ABR session a flow of control cells (called Resource Management (RM) cells) generated by the source. As the RM cells of a session flow through the network, the ATM switches in their path could set an explicit rate value in these cells. The sink would return the RM cells back to the source, and the source could use the explicit rate in the returned RM cells to adjust its cell emission rate. On the other hand, TCP uses a window-based transmission protocol. For a fixed round trip delay, the TCP throughput is proportional to the average window size. Thus, the window can be adapted to vary the TCP transfer rate. Window adaptation works by a TCP source detecting lost packets, taking these as indications of rate mismatch and network congestion, and voluntarily reducing its transmission rate by reducing the transmission window.

3.4.2 Control Mechanisms: MAC and TCP

As mentioned in the previous section, and as is clear from our discussions in Chapter 2, in wideband cellular wireless networks, the entire system bandwidth is not used as one fat pipe. Instead, radio resource allocation is done on an MS by MS basis, depending on the channel conditions to the mobiles. Thus, even at the medium access layer it is possible to implement control strategies that achieve some sort of a rate allocation objective over the MSs. For example, the objective could be *equal rate allocation* (this is an example of a more general fairness objective called *max-min fairness*); such an approach might be very inefficient as the MS with the weakest link will determine the rate that all MSs obtain. Another objective could be to allocate rates so as to maximize the total rate over the MSs; such an approach might be very unfair as MSs with poor connectivity might obtain no throughput. We will examine MAC level rate allocation for elastic traffic in CDMA cellular systems in Chapter 5.

In CSMA/CA based wireless LANs, the medium access control protocol results in some default bandwidth sharing. If only downlink file transfers are considered then it is found that the IEEE 802.11b standard results in equal rate sharing, irrespective of the physical rate at which MSs are connected. We will provide an analytical model for understanding this in Chapter 7.

Bandwidth sharing in the wide area Internet is controlled by the Transmission Control Protocol (TCP) that resides in all end-systems attached to the Internet, including Internet-enabled cellular phones. In OSI terminology, TCP is a Transport Layer protocol. Thus TCP sits between the applications and IP, the Internet's packet routing and forwarding protocol. TCP is connection oriented, which means that a connection has to be established between the endpoints before data transfer can start, and this connection is taken down when the data transfer completes. TCP enhances the unreliable, nonsequential packet transport service provided by IP to a reliable and sequential packet transport service. It uses a window-based packet loss recovery mechanism to achieve this function. In addition, the window-based mechanism is employed for two other major functions that TCP provides: (1) sender-receiver flow control, which prevents a fast source of packets (at the application level) from overwhelming a slow sink, and (2) adaptive bandwidth sharing in the network. The TCP transmitter maintains a *congestion window* that increases if packets are acknowledged in sequence. On the other hand if the desired acknowledgment (ACK) fails to show up then the transmitter takes this as an indication of congestion, and reduces the transmission window. The transmission window can also be controlled by the receiver, by a *window advertisement* in the ACK packets. By the latter means, the receiver can exercise flow control over the transmitter. For a connection, the number of packets in the network is roughly related to the TCP window, and the average window divided by the mean round trip packet delay is an estimate of the TCP throughput.

The adaptive window-based congestion control mechanism of TCP has evolved over several versions. In the earliest version, any packet loss resulted in a transmitter time-out, and the reduction of the congestion window to one. A TCP receiver continues to accept packets even if previous packets are missing. For all such out-of-order packets, the transmitter returns an ACK packet "asking" for the first missing packet. These ACKs are called *duplicate ACKs*. Thus, duplicate ACKs are an indication of out-of-order packets at the receiver, and multiple duplicate ACKs are indicative of packet loss. In a later version, called *TCP Tahoe*, loss recovery was initiated at the transmitter by the receipt of three duplicate ACKs; this was called *fast retransmit*. However, the transmitter dropped the congestion window to one. Thus, although loss recovery started earlier than the older version (say, *OldTahoe*), the window had again to be built up from one. TCP Tahoe was followed by *TCP Reno* in which, on the receipt of three duplicate ACKs, the congestion window was cut by half, and loss recovery was initiated; this was called *fast retransmit and fast recovery*. A more aggressive loss recovery than TCP Reno was implemented in *TCP NewReno*. This version was followed by another improvement in *TCP SACK*, in which the TCP transmitter uses the ACK packets to send the pattern of missing TCP packets back to the TCP transmitter.

From the foregoing, we see that the rate allocation achieved by elastic transfers to or from devices that attach to the Internet via wireless access networks will be governed by the interaction between the rate allocation strategies in the wireless MAC, the behavior of the wireless channel, and TCP's window-based end-to-end control mechanisms.

3.4.3 TCP Performance over Wireless Links

In [89, Chapter 7] we have discussed the TCP protocol at length and have studied models for evaluating the performance of TCP-controlled file transfers in several situations. We studied a model that can be used to obtain the performance of TCP controlled file transfers with random packet loss. We saw that the performance of TCP can be significantly affected by packet loss. In these discussions, the concern was with congestion-related loss; that is, either a packet was lost owing to buffer overflow, or a packet was deliberately dropped at a router queue owing to imminent congestion. We were not concerned with the possibility of packet loss in the physical bit carriers. In a sense, we were assuming a wired physical infrastructure. Wired links can be properly established so that they have small BERs. On the other hand, mobile wireless links can have high packet loss rates, and are subject to random variations in their quality. Also, in CSMA/CA based wireless LANs, it is unrealistic to model the service provided to a flow as being at a constant bit rate. It is therefore of interest to study the performance of TCP transfers over wireless access networks, particularly in light of the growing importance of mobile wireless access to the Internet.

It is well known that the *bandwidth delay product (BDP)* (normalized to the packet length) along a path in a network is defined as

$$\frac{2C\delta}{L}$$

where C is the bottleneck link rate along the path of the TCP connection, 2δ is the round-trip propagation delay (RTPD), and L is the packet length. If the TCP window grows to the BDP and stays at that value, then the bottleneck link can be kept fully occupied. This yields the highest possible TCP throughput on that path. With this in mind, let us consider elastic transfers from the server E to an MS associated with the cellular network in Figure 3.1. Let us suppose that the cellular system assigns a fixed service rate to each transfer, where the rate depends on the condition of the channel to the MS and on the other MSs that are being served by the system. Typical service rates would be in 100s of Kbps, and hence the cellular network would be the bottleneck in the path of the TCP connection. Even if the server E is 20,000 Km away (halfway around the earth), thus yielding an RTPD of about 200 ms, we have a BDP of about four packets for a bottleneck rate of 250 Kbps and $L = 1500$ bytes. The TCP maximum window size implemented in various operating systems is 20 packets or more. Hence, assuming that the wide area Internet has negligible packet loss, the TCP window is well above the minimum to keep the bottleneck link busy.[1] This observation permits us to make the simplification that we may ignore the wide area packet network, and study only the interaction between TCP and the behavior of the wireless link. It is as if the server E was attached to the local area network of the cellular operator.

Independent Packet Losses

With this discussion in mind, Figure 3.8 shows a simple scenario in which a mobile host is doing a TCP controlled file transfer from a file server on a wired LAN. The LAN wireless router network would be located at the base station. The propagation delay between the base station and the mobile host is negligible. The BER on the wireless link is such that packets are lost with probability p. The packets are lost independently; correlated losses owing to channel fading are not modeled here. Only ACKs are sent from the mobile host to the LAN, and since these are small (40 bytes), their loss probability is ignored; recall that TCP uses cumulative ACKs, which further limits the effect of ACK loss. The link layer

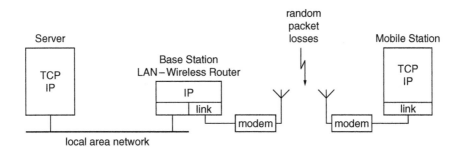

Figure 3.8 A mobile station transferring data over a wireless link from a server on the LAN attached to the base station.

---

[1]A simple way to quantify the effect of random losses in the wide area Internet is to use the *square root* formula $\sqrt{\frac{1.5}{p}}$ where p is the packet loss probability. This formula gives an approximation to the mean window size of a TCP connection over a wide area network, if the loss probability is p and the connection stays in congestion avoidance. Typical values of p in a well-engineered ISP network would be 0.001 or 0.005. The resulting average window is well above the 4 required to keep the bottleneck link busy.

protocol is unable to recover all the wireless packet losses; hence any residual packet losses have to be recovered by TCP. As the TCP transmitter on the file server grows its window, the wireless link buffer builds up. The buffer can hold as many packets as needed; that is, there is no buffer loss. Eventually, a loss occurs in the wireless link, and one of the loss recovery mechanisms is invoked.

The throughput of a large file transfer can be analyzed via a stochastic model of the TCP protocol with random packet losses. A sample of results obtained from this analysis is shown in Figure 3.9. The parameters and the results are normalized. We plot the file transfer throughput versus the packet loss probability. The throughput is normalized to the bit rate of the wireless link. One set of parameters that would correspond to the results is LAN speed; 10 Mbps wireless link bit rate: 2 Mbps; TCP packet length: 1500 bytes (hence the packet transmission time is 6 ms); time-out granularity: 420 ms; minimum time-out: 600 ms; and $W_{max} = 24$ packets, where W_{max} is the maximum TCP window. The performance of four versions of TCP is compared: OldTahoe (which is the name we give to the version of TCP that predates Tahoe and always requires time-outs to recover losses), Tahoe, Reno, and NewReno. We observe that even with a packet loss probability of 0.001, the throughput with OldTahoe is less than the full link rate,

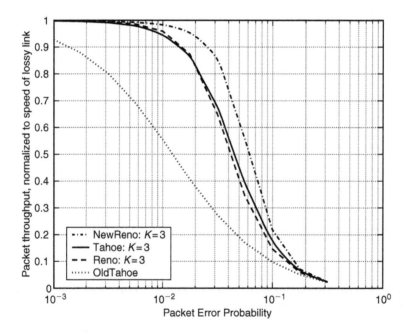

Figure 3.9 File transfer throughput (normalized to the link's bit rate) vs. packet loss probability for various versions of TCP; OldTahoe refers to a version that recovers losses only by timeout. K is the duplicate ACK threshold for fast-retransmit in the TCP loss recovery protocol. Adapted from Kumar [85].

and drops to just over 50 percent of the link rate for a packet loss probability
of 0.01. The other three TCP versions implement fast-retransmit and they yield
100 percent throughput at $p = 0.001$, and better than 95 percent throughput up
to $p = 0.01$. Beyond 1 percent packet loss, the performance of these versions too
begins to drop, and is not much better than OldTahoe for a 10 percent packet
loss rate. Reno is slightly better than Tahoe up to $p = 0.02$, but becomes worse
for large loss rates since multiple losses cause it to waste more time than Tahoe.
The more aggressive fast-recovery of NewReno results in this version yielding
almost 90 percent throughput up to $p = 0.03$. We can make a broad observation
that random packet loss probabilities larger than 1 percent can significantly affect
the performance of TCP, with these parameters. Note that the coarse time-out
and minimum time-out values are large in this example. Smaller values for these
parameters will yield better performance, as losses will then result in less wastage
of link capacity.

   Thus, we see that there is a maximum packet loss probability below which
the TCP throughput is just the bit rate of the wireless link. If the packet loss
probability is ensured to be less than this maximum then the effect of TCP can be
ignored, and we can just take the bit rate provided by the MAC mechanisms as
the transfer rate obtained by the elastic application.

   If a packet loss probability of p is desired, and the packet length is L bits,
then the BER on the wireless link ϵ should satisfy the requirement $p = 1 - (1 - \epsilon)^L$.
An upper bound on p thus yields an upper bound on the BER. Hence we see that
the performance of the application we wish to carry on the wireless link puts a
requirement on the performance of the link. We saw in Chapter 2 that the BER
on a wireless link is a function of the SNR. Hence, for a given modulation and
coding scheme, the desired BER places a requirement on the minimum SNR at
which the link can operate. We also note that a desired packet loss probability can
be obtained by using an ARQ protocol over a physical link with a higher BER than
calculated from the formula above. Since the propagation delay on cellular links
is negligible (i.e., the number of bits "in flight" is much smaller than the packet
length), a *stop-and-wait* ARQ suffices. The overall effect of using an ARQ protocol
is that we have a lower bit rate link (due to ARQ overheads and retransmissions)
with the desired packet loss rate.

Correlated Packet Losses

We now turn to the performance of TCP controlled file transfers over a fading
channel. In Section 2.1.4 we discussed models for channel fading. We pointed
out that the fading is correlated in time. Thus for a given average BER there
would be periods when the BER is greater than the average, and periods during
which the BER is less than the average. A similar statement can be made for the
packet error rate if fixed length packets are being used, as is typically the case
with large file transfers over TCP. A simple approach is to model the channel as
being in one of two states: a Good state (during which a packet transmission is

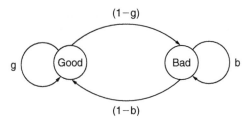

Figure 3.10 Transition structure of the two-state Markov model for a fading channel.

successful), and a Bad state (during which a packet transmission is unsuccessful). A further simplification is to model the state process as a two-state Markov process, on the state space {*Good, Bad*} (see Figure 3.10). The durations in each state are taken to be multiples of the packet transmission time. The transition probabilities of the Markov chain are obtained by specifying the amount of fading that leads to a bad transmission (at other times a good transmission is assumed). The marginal distribution of the fading process and results about correlations in the fading process can be used to obtain the transition probabilities. For a packet length L, channel bit rate C, and Doppler frequency f_d, the parameter $f_d \frac{L}{C}$ is a measure of the fade durations relative to the packet transmission time; thus $f_d \frac{L}{C} = 0.01$ means that channel coherence time is roughly 100 packet transmission times.

Using this Markov model for the channel state, the analysis of the throughput of a long file transfer under TCP can be performed by developing a certain stochastic model. In performing this analysis, in addition to the state of the TCP window adaptation process, the state of the channel will also need to be maintained. Figure 3.11 shows some typical numerical results with Rayleigh fading. The normalized throughputs with TCP Tahoe and Reno are plotted versus the average packet error probability, with and without fading. For the results with fading, the parameter $f_d \frac{L}{C} = 0.01$. The other parameters are the same as in Figure 3.9, except that the local area network is taken to be infinitely fast. Notice that the performance without fading is similar to that depicted in Figure 3.9. With fading, the performance is significantly different. For the same probability of error, we find that the performance of TCP Tahoe increases substantially, whereas that of TCP Reno drops for $p < 3 \times 10^{-2}$, and improves for large packet loss probabilities. This can be understood as follows. With independent losses, the repeated reductions in the window lead to a small effective window; hence when a loss occurs there are not enough packets in circulation to generate the number of duplicate ACKs required for a fast retransmit. Thus with uncorrelated losses, time-outs are more frequent. When packet errors are clustered (as in the case of fading), the durations between packet loss events are larger. Hence with correlated packet losses, the TCP transmitter is able to grow its window to larger values than

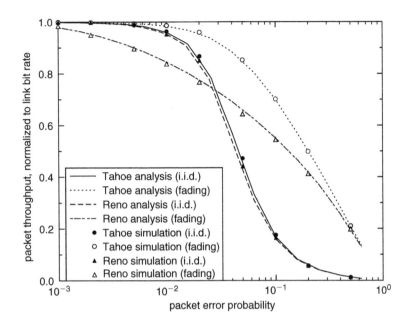

Figure 3.11 File transfer throughput (normalized to the link's bit rate) vs. packet loss probability for TCP Tahoe and Reno; with independent losses (denoted as i.i.d.), and with Rayleigh fading with $f_d \frac{L}{C} = 0.01$. Adapted from Zorzi et al. [144].

in the independent packet loss case (for the same average packet error probability). When a loss does occur, it is more likely that there are enough successful packets sent subsequently in the window to trigger a fast retransmit. Even if a time-out does occur, it is long enough to last out the fade, so that when transmission resumes, the channel is likely to be in the Good state.

For small values of p, the performance of Reno is worse since Reno requires additional duplicate ACKs for recovering each lost packet. With correlated losses, multiple losses are more likely and this results in Reno wasting more time than Tahoe. Reno attempts to perform a fast-retransmit for each lost packet, spends time in this process waiting for duplicate ACKs, and then times out anyway. For large values of p, the two protocols have similar behavior since with the high loss rate the window grows to small values, the number of duplicate ACKs are insufficient to trigger a fast-retransmit, and hence it is very likely that both protocols recover with a time-out.

Although this discussion illustrates the effect of correlated errors on TCP controlled file transfer performance, it is important to make a comparison by fixing the average SNR. The same two-state Markov model can be used. The SNR that corresponds to a Bad state is first fixed. Then for each SNR and Doppler frequency, the two-state Markov model can be parametrized. Sample results

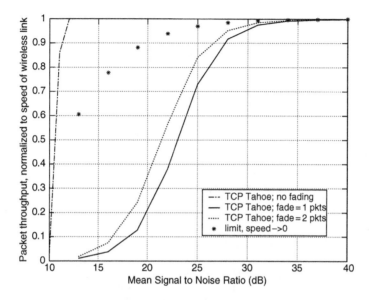

Figure 3.12   File transfer throughput (normalized to the link's bit rate), with TCP-Tahoe, vs. SNR in dB, with no fading (AWGN only) and with Rayleigh fading. The legend fade = _n_ pkts means that the mean Bad state duration is _n_ packets, where a Bad state occurs if the SNR < 10 dB. Adapted from Kumar and Holtzman [87].

for TCP Tahoe are shown in Figure 3.12. In order to compare with the results presented earlier, no channel coding or link level retransmissions are taken into account. The normalized throughput is plotted against the average SNR in dB. We observe that without fading, an SNR of about 12 dB suffices to obtain a TCP throughput of over 90 percent of the link rate. This is because the packet error probability itself is very small without fading. With fading, however, much larger Rayleigh faded SNRs are required; between 25 to 30 dB for a throughput of 90 percent of link rate. We notice that slower fading and hence more correlated errors improves the TCP throughput, but the throughput even with this improvement is much worse than that without fading. We also show the case of speed → 0. This corresponds to the fade level being constant during the entire TCP transfer; either the channel is good throughout the transfer, or is bad throughout. This is a bound on the achievable throughput with fading. As the faded SNR decreases, the probability of the Good state reduces and hence the bound rapidly decreases for decreasing average faded SNR.

3.5 Notes on the Literature

Extensive analytical treatments of QoS issues and models is provided in [89], [133], and [119]. References [81] and [54] provide a discussion of issues in transporting voice over packet networks. The material on analysis of TCP controlled file transfer throughput over lossy wireless links has been taken from the papers by Kumar ([85], which assumes i.i.d. packet loss) and by Zorzi et al. ([144], which accounts for correlated packet losses). An approach for two-state Markov modeling of a fading channel is provided by Zorzi et al. in [146]. Additional references on TCP throughput analysis with correlated packet losses in the wireless setting are Kumar and Holtzman [87], Zorzi and Rao [145], and Anjum and Tassiulas [3].

CHAPTER 4

Cellular FDM-TDMA

T he FDM-TDM technique for allocating spectrum and time resources to calls is the most classical one, and systems based on this technique carry a substantial majority of the cellular telephony traffic around the world. Many of the basic techniques of cellular telephony emerged from the design of such systems; for example, techniques such as frequency reuse management, cell sectorization, power control, and handover management.

Overview

One of the main ideas developed in this chapter is that of *spatial reuse* by partitioning the available *FDM carriers* into *reuse groups*, and then allocating these reuse groups to *cells* in such a way that *cochannel interference* is within acceptable limits. It is shown that the cochannel interference constraint places a constraint on the $\frac{D}{R}$ ratio, the ratio between the shortest distance between *cochannel cells* and the *cell coverage radius*. This analysis is based on *signal-to-interference ratio (SIR)* modeling, where we use a power attenuation model that includes path loss and shadowing. Assuming that the cells form a hexagonal tessellation of the plane, the $\frac{D}{R}$ ratio is related to the number of cochannel reuse groups into which the cells must be partitioned. It is shown how the partitioning of the channels and other system parameters affect the *spectrum efficiency*. Once *channel allocation constraints* are understood, various channel allocation strategies are considered and a *call blocking* analysis is developed. Finally, we consider intercell *handovers*. We show how signal strength measurements from neighboring BSs are used to determine that a call needs to be handed over between cells. An approximate *handover blocking* analysis is also shown. The chapter ends with an overview of call handling in GSM, the most widely deployed FDM-TDMA cellular system.

4.1 Principles of FDM-TDMA Cellular Systems

Suppose that a system bandwidth of W_{system} is to be used for providing FDM-TDMA based telephony services in a certain coverage area, say a town in some country. The operator would have to pay a substantial fee to the authority managing the spectrum in that country, and hence it is in the operator's interest to maximize the revenue from operating the service while keeping costs down. We begin by providing an understanding of the issues involved in the efficient design of an FDM-TDMA cellular telephony system. We will refer to a mobile handset as an *MS (mobile station)* and to the fixed stations that are connected to the wire-line network as *base stations (BSs)*.

There are several commercial implementations of FDM-TDMA technology for mobile telecommunication, the one with the most widespread deployment being the GSM system (see Section 4.7). In any of these implementations, the system bandwidth, W_{system}, is partitioned into several nonoverlapping FDM channels, each of which is then digitally modulated, and then time slotted to yield FDM-TDM channels. A *guard band* is left vacant at either end of an operator's spectrum allocation to prevent power from one operator's system from interfering with another system. For example, in the GSM system the FDM channel spacing is 200 KHz. After digital modulation, each such channel is time slotted to provide eight TDM channels, each of which can carry one direction of a digitized voice call.

Each voice call has two directions, and hence for each call we need two links to be established, one from the MS to the BS, and one from the BS to the MS. The way these two links are established is called the *duplexing* technique. In FDM-TDMA systems the common duplexing mechanism employed is frequency division duplexing (FDD); that is, two separate FDM carriers are used to carry the two directions of a call. Thus, the operator actually gets two nonoverlapping segments of the radio spectrum, each of bandwidth W_{system} (see Figure 4.1). One of these is the *uplink band* and the other is the *downlink band*. Each band is partitioned into an equal number of nonoverlapping FDM channels. The FDM channels in the uplink and downlink bands are then paired, as shown in Figure 4.1, for two FDM channels j and k. Thus, when we say that FDM channel j is assigned to an MS, for the purpose of making a telephone call, then actually two TDM slots, one in each of the two FDM channels with center frequencies f_j^u and f_j^d, are assigned to the call, for the entire duration of the call.

For example, if the operator leases a W_{system} of 5 MHz, then (allowing for a total guard band equal to the bandwidth of one FDM channel), the system can be used to carry $24 \times 8 = 192$ simultaneous calls. Let us denote the FDM channel

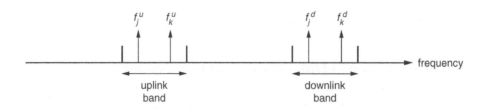

Figure 4.1 Frequency division duplexing in FDM systems: The FDM channels f_j^u and f_j^d are paired, as are the channels f_k^u and f_k^d.

bandwidth by W, the number of FDM channels in the system bandwidth by C (24 in the preceding example), and the number of traffic carrying TDM slots per FDM channel by s (8 in the preceding example). Let $N = C \times s$ be the number of calls that can be carried simultaneously. We will further denote the set of FDM carriers by $\{f_1, f_2, \ldots, f_C\}$. In view of our earlier discussion on duplexing, each element of the set of carriers $\{f_1, f_2, \ldots, f_C\}$ actually denotes an uplink-downlink pair.

The system can now be set up as shown in the left panel of Figure 4.2. We observe that a large power will need to be used in order to serve the MSs at the periphery of the coverage area, in order to ensure that the power received at either end of an MS-BS link is such that the SNR exceeds the minimum required for the desired bit error rate for the voice coder being used. Such MSs will quickly drain their batteries, and will also cause interference to systems in neighboring coverage areas. Also, the maximum number of users that can be simultaneously served in this simple system is N.

Let $B(\rho, n)$ denote the blocking in an Erlang blocking system with a load of ρ Erlangs and n servers (see Appendix D, Section D.5.1). If the arrival rate of new calls into the system is λ per second, and the mean holding time of a call is h seconds, then for this system $\rho = \lambda h$, and the blocking probability becomes $B(\rho, N)$. Table 4.1 shows a sample *Erlang table* from which the number of servers that would be required to obtain a specified blocking probability for a given traffic intensity can be obtained.

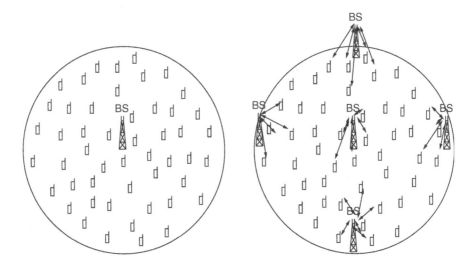

Figure 4.2 Spatial reuse: In the left panel, all the MSs associate with one BS, and the entire band is used to serve all the calls in the desired coverage area. In the right panel the band is reused at multiple BSs.

$n \downarrow$	\	\	\	Loss Probability	\	\	\	\	\	$n \downarrow$
	0.0001	0.001	0.002	0.005	0.01	0.02	0.05	0.1	0.2	
15	4.7812	6.0772	6.5822	7.3755	8.1080	9.0096	10.633	12.484	15.608	15
16	5.3390	6.7215	7.2582	8.0995	8.8750	9.8284	11.544	13.500	16.807	16
17	5.9110	7.3781	7.9457	8.8340	9.6516	10.656	12.461	14.522	18.010	17
18	6.4959	8.0459	8.6437	9.5780	10.437	11.491	13.385	15.548	19.216	18
19	7.0927	8.7239	9.3515	10.331	11.230	12.333	14.315	16.579	20.424	19
20	7.7005	9.4115	10.068	11.092	12.031	13.182	15.249	17.613	21.635	20
21	8.3186	10.108	10.793	11.860	12.838	14.036	16.189	18.651	22.848	21
22	8.9462	10.812	11.525	12.635	13.651	14.896	17.132	19.692	24.064	22
23	9.5826	11.524	12.265	13.416	14.470	15.761	18.080	20.737	25.281	23
24	10.227	12.243	13.011	14.204	15.295	16.631	19.031	21.784	26.499	24
25	10.880	12.969	13.763	14.997	16.125	17.505	19.985	22.833	27.720	25

Table 4.1 Part of the Erlang table showing the traffic intensity that can be offered to a link of capacity n (rows) circuits for specified blocking probabilities (columns).

Telephony systems typically are designed for blocking probabilities such as 0.01 or 0.02. If a single phone is expected to provide a load of 0.1 Erlangs (e.g., two calls per hour, with an average holding time of 3 min), then for a coverage area with 2500 MSs the Erlang load is 250, and we see that this proposed system (with $N = 192$) will yield an unacceptable probability of call blocking $B(250, 192)$.[1]

The right panel of Figure 4.2 shows the idea of *spatial reuse*. The MS-BS communication is done using smaller powers. The same channel can be used in several places in the system, provided that the *cochannel interference* is such that the *signal-to-interference plus noise ratio (SINR)* of any MS-BS link is maintained above a threshold. Suppose that, by exploiting spatial reuse, each channel could be reused, say, five times in the same coverage area; then we would have effectively multiplied the number of calls that can be simultaneously handled by a factor of 5.

On further thought, however, a problem becomes evident with the simple arrangement in Figure 4.2. If an MS is not close to any BS, then in order to serve it from some BS, on any channel, a large transmission power will need to be used. This will cause cochannel interference if the same channel is reused elsewhere in the coverage area, thus rendering spatial reuse less effective.

In order to address this problem, the cellular FDM-TDMA approach is to tessellate the coverage area into *cells*, each of which has a BS. The set of FDM carriers is partitioned into subsets called *reuse groups*. These channel groups are then assigned to the cells in such a way that cells with the same group of channels (called *cochannel cells*) are not close together. How close cochannel cells can be depends on the SINR required for reliable communication. Each cell then acts as an Erlang blocking system for the calls that require a channel in it.

We observe that, if the SINR required is large, then the cochannel cells will need to be kept far apart. This will require more channel reuse groups, and hence fewer channels per reuse group. This brings us to another issue. If the reuse groups have only a small number of carriers in them, then *trunking efficiency* is lost. By this we mean the following. For a fixed probability of blocking, ϵ, let $\rho_\epsilon(n)$ denote the Erlangs that can be carried when the number of servers is n,

$$B(\rho_\epsilon(n), n) = \epsilon \qquad (4.1)$$

To understand this, notice that each column of Table 4.1 corresponds to a value of ϵ, and each element in that column gives the value of $\rho_\epsilon(n)$ for the corresponding n in the first column. Now define $g_\epsilon(n) = \frac{\rho_\epsilon(n)}{n}$, that is, $g_\epsilon(n)$ is the Erlangs per server that can be offered, when the number of servers is n and the target blocking

---

[1]Note that the maximum load that 192 servers can carry is just 192, so a load of 250 Erlangs will give a blocking probability close to 1. As a rule of thumb, for $B(\rho, n)$ to be as small as 0.01 or 0.02, it is necessary that $\rho < n$. This follows because, with a blocking probability of ϵ, the rate of calls that are carried is $(1-\epsilon)\lambda$, and time average number of busy servers is $(1 - \epsilon)\lambda h = (1 - \epsilon)\rho$ (by Little's Theorem; see Appendix D). For low blocking, however, the average number of busy servers will be substantially less than n (see Table 4.1).

probability is ϵ. For $n = 1$, note that $B(\rho, n) = \frac{\rho}{1+\rho}$, which yields $g_\epsilon(1) = \frac{\epsilon}{1-\epsilon}$. Thus, a very small load (per server) can be handled if there is just one server. However, $g_\epsilon(n)$ increases monotonically to 1 as n increases. Thus, we see that it is beneficial to not partition the set of carriers into small groups, as this reduces trunking efficiency.

We conclude that a larger target SINR results in smaller reuse groups, which results in lower trunking efficiency. It follows that there is a trade-off between keeping the SINR above a required threshold and keeping the trunking efficiency high. The number of reuse groups we use in a system will be denoted by N_{reuse}.

We observe from this discussion that the SINR requirements, the spatial reuse, and the system efficiency are intimately linked, and some analysis is required to evaluate the trade-offs.

4.2   SIR Analysis: Keeping Cochannel Cells Apart

In Figure 4.3, we depict uplink and downlink cochannel interference in a configuration in which a channel is reused at the five BSs shown. The circular boundaries indicate the *coverage* of each BS; these are assumed to be of radius R. The distance between the centers of each of the outer BSs and the one in the middle is D. It is intuitively clear that a large $\frac{D}{R}$ ratio will be required if the cochannel interference has to be kept very small. In this section we will study how to carry out the cochannel interference analysis with a target SINR, in order to determine the required $\frac{D}{R}$ ratio.

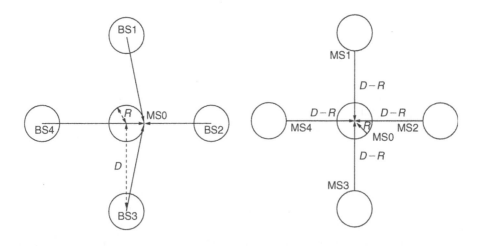

Figure 4.3 Depiction of downlink (left panel) and uplink (right panel) cochannel interference. In each case the MS0 is taken to be in the most unfavorable position, such that the desired signal will suffer the maximum attenuation and the interference will suffer the least attenuation. D is the shortest distance between BSs of cochannel cells, and R is the coverage radius of each BS.

For the purpose of studying the cochannel interference, the MS whose signal performance is being analyzed is considered to be in the most unfavorable position (at the periphery of its BS's coverage area), and the interferers are also assumed to be in the most unfavorable position, as close as possible to the receiver of the desired transmission. For example, in the right panel of the figure, the uplink is being considered, and therefore the interferers are cochannel MSs in the other cells. Notice that these are being assumed to be at the peripheries of their own cells, and placed so that they are as close as possible to BS0. Such worst case configurations are used to determine how far away cochannel cells need to be kept.

At this point we recall the material in Section 2.1.4. Let H denote the channel power gain (actually, an attenuation) between the transmitter of the desired signal and its receiver, and let H_i denote the power gain from the i-th cochannel interferer to the receiver of the desired signal. Let there be N_I interferers. We will view all transmitter powers as being the Rayleigh faded mean values at the reference distance d_0 (see Section 2.1.4). With this convention, let P be the power used by the transmitter of the desired signal, and $P_i, 1 \le i \le N_I$ be the powers of the interfering transmitters. It then follows that the SINR at the receiver is given by

$$\Psi = \frac{PH}{N_0 W + \sum_{i=1}^{N_I} P_i H_i}$$

We note that these FDM-TDMA systems use narrowband modulation, and hence the SINR requirements are in the range of 10 dB to 20 dB. Also the noise power, $N_0 W$, is very small; approximately -120 dBmW (i.e., 10^{-12} mW). It is, therefore, assumed that the noise power is much less than the received signal power, and we neglect this term in the denominator. Let d be the distance between the transmitter and its receiver, and d_i the distance between the i-th interfering transmitter and the receiver. We can then write (see Section 2.1.4)

$$H = \left(\frac{d}{d_0}\right)^{-\eta} 10^{-\frac{(\xi + \xi_0)}{10}}$$

$$H_i = \left(\frac{d_i}{d_0}\right)^{-\eta} 10^{-\frac{(\xi_i + \xi_0)}{10}}$$

where ξ, ξ_0, and $\xi_i, 1 \le i \le N_I$, normally are distributed and correspond, respectively, to the shadowing at the transmitter of the desired signal, at the receiver, and at the N_I interferers. Here the $\xi, \xi_0, \xi_i, 1 \le i \le N_I$, are i.i.d. normally distributed, 0 mean, and with variances $\frac{\sigma^2}{2}$; thus, the lognormal shadowing standard deviation on any path is σ dB. This form of the lognormal shadowing is used since shadow fading comprises a part due to the shadowing near the receiver (which is common to all paths to the receiver), and a part near the transmitters (which is assumed to be independent for widely separated transmitters). Hence, we can write the SINR (or, simply, the SIR) Ψ as

$$\Psi = \frac{P\left(\frac{d}{d_0}\right)^{-\eta} 10^{-\frac{(\xi+\xi_0)}{10}}}{\sum_{i=1}^{N_I} P_i \left(\frac{d_i}{d_0}\right)^{-\eta} 10^{-\frac{(\xi_i+\xi_0)}{10}}}$$

Notice that the terms ξ_0 all cancel. We can then rewrite the SIR expression in the following form:

$$\Psi = \frac{10^{-\frac{1}{10}\left(-10\log_{10}P + 10\eta\log_{10}\frac{d}{d_0} + \xi\right)}}{\sum_{i=1}^{N_I} 10^{-\frac{1}{10}\left(-10\log_{10}P_i + 10\eta\log_{10}\frac{d_i}{d_0} + \xi_i\right)}} \tag{4.2}$$

Notice that in the numerator we have a log-normally distributed random variable of the form $10^{-\frac{1}{10}Q}$, where Q has units of dB, and is normally distributed with

$$E(Q) = m := \left(-10\log_{10}P + 10\eta\log_{10}\frac{d}{d_0}\right)\,\mathrm{dB}$$

$$VAR(Q) = v := \frac{\sigma^2}{2}$$

and in the denominator we have a sum of N_I log-normally distributed random variables of the form $10^{-\frac{1}{10}Q_i}$, where Q_i also has units of dB, and is normally distributed with

$$E(Q_i) = \left(-10\log_{10}P_i + 10\eta\log_{10}\frac{d_i}{d_0}\right)\,\mathrm{dB}$$

$$VAR(Q_i) = \frac{\sigma^2}{2}\;(= v)$$

Thus, we have

$$\Psi = \frac{10^{-\frac{1}{10}Q}}{\sum_{i=1}^{N_I} 10^{-\frac{1}{10}Q_i}}$$

where $Q, Q_i, 1 \leq i \leq N_I$, are independent normally distributed random variables that essentially model the shadowing. Since shadow fading is correlated over distances of several 10s of meters, we assume that the shadowing is "sampled" once during a call, and independent samples of the shadow fading random variables are taken from call to call. We also assume that a call, during its holding time, samples the entire distribution of the Rayleigh fading; it does not get "stuck" in a deep fade. This corresponds to our use of mean transmit powers averaged over Rayleigh fading.

We are now interested in ensuring that the SIR exceeds a threshold γ with a high probability, say, $1 - \epsilon$. Note that, given a target BER, γ will be obtained from an analysis of the underlying modulation scheme under Rayleigh distributed flat fading and additive white Gaussian noise; see Section 2.1.4. Then ϵ would be the *outage* probability. What does this probability mean? Consider instances of calls from or to MSs at the boundaries of the coverage areas of the cells in which they are handled. Then the fraction of such calls that will experience a BER higher than the target will be less than ϵ. This is because for such calls, we have assumed that the cochannel interferers are placed at the most unfavorable locations; refer back to Figure 4.3.

One approach to carrying the analysis forward is to approximate the distribution of $\sum_{i=1}^{N_I} 10^{-\frac{1}{10}Q_i}$ by a log-normal distribution. Such an approximation is known to work well. Also, the resulting SIR threshold analysis becomes simple, since the ratio of two independent log-normal random variables is obviously log-normal. So, let us write

$$\sum_{i=1}^{N_I} 10^{-\frac{1}{10}Q_i} \approx 10^{-\frac{1}{10}Q_I}$$

The approximation is performed by matching the mean and second moment of the random variables on the two sides. This is called the Fenton-Wilkinson method, the details of which can be found in standard texts on wireless digital communication (see, for example, [123]). Let us suppose that this procedure yields Q_I as normally distributed with mean m_I and variance v_I. Then, we have

$$\Psi = 10^{-\frac{1}{10}(Q-Q_I)}$$

or, equivalently,

$$(\Psi)_{dB} = Q_I - Q$$

where $Q_I - Q$ is in dB, and is normally distributed with mean $m_I - m$ and variance $v + v_I$. Thus, $m_I - m$ is the mean SIR in dB, and the SIR variance is $v + v_I$. We need that $Q_I - Q > (\gamma)_{dB}$ with a large probability, where, as usual, $(\gamma)_{dB} = 10 \log_{10} \gamma$.

From Figure 4.3, we notice that in the downlink worst case situation (left panel of the figure), the BSs are all taken as transmitting to MSs at the peripheries of their coverage areas. Similarly, in the uplink worst case situation (right panel) the BSs are all receiving from MSs at the peripheries. We thus assume that the transmission powers $P, P_i, 1 \leq i \leq N_I$, are all equal. It follows from (4.2) that the transmitter powers cancel in the SIR expression. The mean value, m, corresponds to the path loss from the transmitter of the desired signal to its receiver. Further, m_I depends on the path losses from the interferers to the receiver, and is the effective path loss of the interfering transmitters, in dB. We can thus require that $m_I - m > 0$, that is, the interferers have a larger effective power attenuation to the receiver than does the desired transmitter. Figure 4.4 depicts a typical situation, when the target probability of exceeding γ is being met. The normal density of $(\Psi)_{dB}$ has been plotted in this figure. It follows that for an outage probability ϵ (i.e., to ensure that $\Pr((Q_I - Q) < \gamma) < \epsilon,)$ there is a τ_ϵ, such that we need to ensure that

$$m_I - m > \gamma + \tau_\epsilon \sqrt{v + v_I}$$

Such a τ_ϵ will be obtained from a table of the tail of the normal distribution. For example, $\tau_{0.01072} = 2.3$, as can be seen from Table 4.2. This inequality provides the insight that shadowing variance of the signal and of the interference add up, and a larger value of this total variance requires the cochannel reuse to be designed so that there is a larger difference between the mean interference attenuation and the signal attenuation, $m_I - m$.

As an application of this analysis, consider the uplink configuration shown in the right panel of Figure 4.3. Since, in this case, the interferers are as close as possible to the receiver of the desired signal, this situation is worse than the downlink situation shown in the same figure. It can be shown that the Fenton-Wilkinson analysis yields

$$m = 10\eta \log_{10} R$$

$$m_I = 10\eta \log_{10}(D - R) - \frac{1}{2a} \ln\left(\frac{e^{a^2v}N_I^3}{e^{a^2v} + (N_I - 1)}\right)$$

$$v_I = \frac{1}{a^2} \ln\left(\frac{e^{a^2v} - 1}{N_I} + 1\right)$$

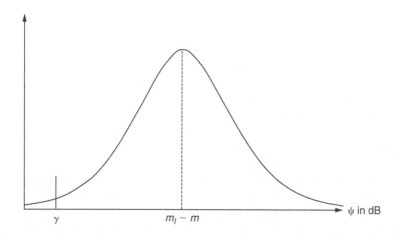

Figure 4.4 A sketch of the normal probability density of the SIR, Ψ, in dB. γ is the target SIR, and $m_I - m$ is the mean SIR.

z	$Q(z)$	z	$Q(z)$
0.0	0.50000	2.0	0.02275
0.1	0.46017	2.1	0.01786
0.2	0.42074	2.2	0.01390
0.3	0.38209	2.3	0.01072
0.4	0.34458	2.4	0.00820
0.5	0.30854	2.5	0.00621
0.6	0.27425	2.6	0.00466
0.7	0.24196	2.7	0.00347
0.8	0.21186	2.8	0.00256
0.9	0.18406	2.9	0.00187
1.0	0.15866	3.0	0.00135
1.1	0.13567	3.1	0.00097
1.2	0.11507	3.2	0.00069
1.3	0.09680	3.3	0.00048
1.4	0.08076	3.4	0.00034
1.5	0.06681	3.5	0.00023
1.6	0.05480	3.6	0.00016
1.7	0.04457	3.7	0.00011
1.8	0.03593	3.8	0.00007
1.9	0.02872	3.9	0.00005

Table 4.2 The probability under the right "tail" of the normal (Gaussian) distribution: $Q(z) = \frac{1}{\sqrt{2\pi}} \int_z^{\infty} e^{\frac{-x^2}{2}} dx$.

where $a := \frac{\ln 10}{10} \approx 0.23026$. We thus get the requirement

$$10\eta \log_{10}\left(\frac{D}{R} - 1\right) > \gamma + \tau_\epsilon \sqrt{v + \frac{1}{a^2}\ln\left(\frac{e^{a^2 v} - 1}{N_I} + 1\right)}$$

$$+ \frac{1}{2a}\ln\left(\frac{e^{a^2 v} N_I^3}{e^{a^2 v} + (N_I - 1)}\right) \tag{4.3}$$

We notice that this inequality places a constraint on the ratio between D, the distance between cochannel cells, and R, the cell radius. Let us look at two numerical examples, both with $\eta = 4$. Consider first $v = 0$, that is, there is no log-normal shadowing, just path loss. The $\frac{D}{R}$ constraint reduces to

$$40 \log_{10}\left(\frac{D}{R} - 1\right) > \gamma + 10 \log_{10} N_I \tag{4.4}$$

Exercise 4.1

Obtain the expression in (4.4) directly from the SIR expression in (4.2).

For $N_I = 6$, we find

$$40\log_{10}\left(\frac{D}{R} - 1\right) > \gamma + 7.78$$

On the other hand, if the shadow fading standard deviation is $8\,\mathrm{dB}$, then $v = \frac{\sigma^2}{2} = 32$. For $\eta = 4$, $N_I = 6$, and outage probability $\epsilon = 0.01$, (4.3) yields

$$40\log_{10}\left(\frac{D}{R} - 1\right) > \gamma + 25.25$$

We conclude that shadow fading has a significant effect on the $\frac{D}{R}$ ratio.

Discussion

a. We observe, from the preceding analysis, that as long as the transmitter powers are all assumed to be equal, the required $\frac{D}{R}$ ratio does not depend on the actual values of the transmit powers.

b. We notice also that only the ratio $\frac{D}{R}$ is determined, but not the absolute values of D and R. This provides the important insight that the cell sizes can be shrunk while retaining the $\frac{D}{R}$ ratio. This increases the system call handling capacity, since the channel groups in each cell are used to serve a smaller cell area. This approach to increasing the system capacity has its limitation, however. As the cell size decreases the MSs tend to more frequently require *intercell handovers*. Since the blocking of handover requests leads to the *dropping* of ongoing calls, an increase in handover rates needs more channels to be reserved for handover handling (see Section 4.6), thus leading to a possible reduction in call handling capacity. In addition, the higher frequency of handovers results in more signaling load, thus possibly overloading the call handling processors in the system.

4.3   Channel Reuse Analysis: Hexagonal Cell Layout

It is evidently not practical to use the cell configuration shown in Figure 4.3, as this leaves large portions of the service area uncovered. Hence, as explained in Section 4.1, the service area is tessellated with cells. The set of FDM carriers is partitioned into disjoint sets, which are assigned to subsets of the cells, in such a way that cochannel cells respect the $\frac{D}{R}$ ratio. In order to analyze such a system, it is convenient to take the cells to be hexagons of equal size. This permits an easy visualization of the tessellation in the two-dimensional plane. It is then useful to recall the simple geometrical concepts shown in Figure 4.5.

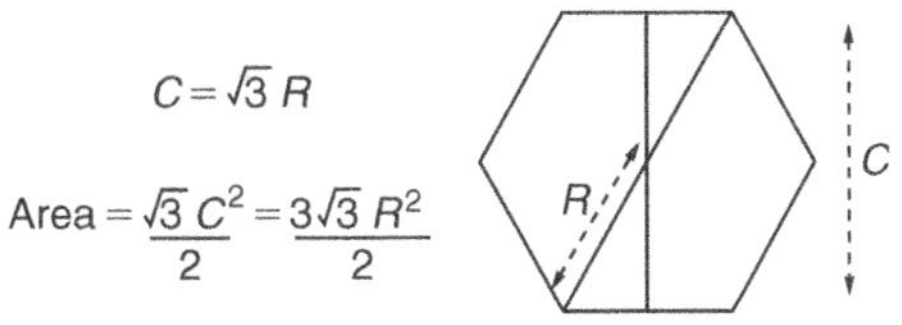

$$C = \sqrt{3}\,R$$

$$\text{Area} = \frac{\sqrt{3}}{2}\,C^2 = \frac{3\sqrt{3}}{2}\,R^2$$

Figure 4.5 Hexagon geometry: relations between the cell width, *C*, the cell radius, *R*, and the area of the hexagon.

4.3.1 Cochannel Cell Groups

In Figure 4.6 we show a tessellation of the plane with hexagonal cells. The FDM channels are partitioned into reuse groups. One of these groups is assigned to Cell 0, shown at the center of the cell layout. This will be our reference cell in the following discussion. We next wish to determine which other cells in the layout should use the same group of carriers. For this purpose it is convenient to work with a coordinate system with axes inclined at 60° to each other, as shown by the axes *u* and *v* in the figure. For simplicity in our description, we draw a third "axis," *w*. The axes pass through the center of the reference cell. There is an angular separation of 60° between *u* and *v*, and the same between *v* and *w*. Notice that moving a cell width, *C*, along any of the axes takes us to the center of a neighboring cell. Thus, let *C* be unit length along the axes. Now, starting from the origin of this system (the center of Cell 0), move *i* units along the *u* axis and then *j* units along the *v* axis. Observe that for $i = 3$ and $j = 2$ this brings us to the cell labeled 1. Let the Euclidean distance between the centers of Cell 0 and Cell 1 be $D(i, j)$, the distance between two cells whose relative positions depend on (i, j) in the manner just explained. The following calculation follows from simple geometry.

$$D(i, j) = \sqrt{\left(j\frac{\sqrt{3}}{2}\right)^2 + \left(i + j\frac{1}{2}\right)^2}$$

$$= \sqrt{j^2\frac{3}{4} + i^2 + ij + j^2\frac{1}{4}}$$

$$= \sqrt{i^2 + ij + j^2}$$

In a similar manner, fixing $i = 3$ and $j = 2$, we can identify Cells 2, 3, 4, 5, and 6, as shown in Figure 4.6. These will be the cochannel cells in relation to Cell 0. Observe that if we carry out the same procedure, for $i = 3$ and $j = 2$, for Cell 1 in the figure, then we will obtain Cells 2, 0, and 6, and three other cells, above and to the right of Cell 1; these cells are not shown in the figure. Thus, this process yields

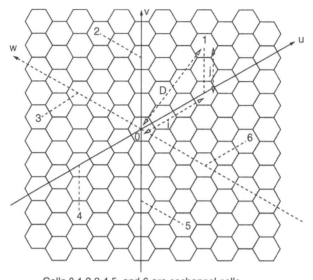

Cells 0,1,2,3,4,5, and 6 are cochannel cells
to locate a cochannel cell w.r.t. to a cell:
move i cells along an axis, then turn clockwise and move j cells

Figure 4.6 Tessellation of the coverage area by hexagonal cells. Cells 0, 1, 2, 3, 4, 5, 6 are cochannel cells for $(i, j) = (3, 2)$.

a subset of the hexagons that tessellate the plane. Applying the procedure starting from any element of this subset yields the same set of cells. Notice, however, that if we start from one of the cells adjacent to Cell 0, and use the same (i, j), then we will get a subset of cells that is disjoint from the previous one. In fact, looking at Figure 4.7, for each cell in the large dashed hexagon with Cell 0 at its center, we will obtain a different subset of hexagons, and all these subsets (19 for $(i, j) = (3, 2)$) are mutually disjoint and together they form a partition of the tessellation. We can call each of these subsets of cells a *cochannel group*.

4.3.2   Calculating N_{reuse}

The number of cochannel groups (which we had denoted earlier by N_{reuse}) thus depends on the choice of (i, j). For example, with $(i, j) = (1, 0)$ there is only one cochannel group. What is the general relation between (i, j) and the number of cochannel groups? This can be worked out as follows. In Figure 4.7 the area of the large dashed hexagon is $\frac{\sqrt{3}}{2} D^2$ (where we recall that the unit of length is the cell width, C). There are as many cochannel groups as the number of cells in this large hexagon. Exactly one cell from any of the cochannel groups lies in this large hexagon. Hence, given a large coverage area A, the number of cells in a cochannel group is $\frac{A}{(\sqrt{3}/2)D^2}$ (see Figure 4.5). The total number of cells is $\frac{A}{\sqrt{3}/2}$. Thus, the

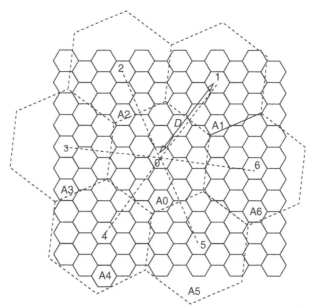

The large dashed hexagons are centred at cochannel cells

Figure 4.7 Tessellation of the plane by hexagonal cells. There is one cell from each cochannel group in each of the large dashed hexagonal areas. Notice that Cells A0, A1, A2, A3, A4, A5, and A6 belong to a different cochannel group than Cells 0, 1, 2, 3, 4, 5, and 6.

number of cochannel groups is D^2. Obviously, the number of cochannel groups has to be the same as the number of groups into which we partition the set of FDM carriers (i.e., N_{reuse} (as defined earlier)). Thus

$$N_{reuse} = D^2 = i^2 + ij + j^2$$

We also observe that for a given (i, j), following this procedure for fixing the cochannel cells, we have also fixed the $\frac{D}{R}$ ratio to the value (see Figure 4.5):

$$\frac{D(i, j)}{R} = \sqrt{3(i^2 + ij + j^2)}$$

$$= \sqrt{3 N_{reuse}}$$

Table 4.3 shows the values of N_{reuse} and $\frac{D(i,j)}{R}$ that are obtained for various values of (i, j).

Recall that the SIR analysis in Section 4.2 yielded a constraint on the $\frac{D}{R}$ ratio. For example, (4.3) provided a constraint on the $\frac{D}{R}$ ratio when a reference

i	j	N_{reuse}	$\frac{D(i,j)}{R}$
1	0	1	1.73
1	1	3	3.00
2	0	4	3.46
2	1	7	4.58
3	0	9	5.20
2	2	12	6.00
3	1	13	6.24
4	0	16	6.93
3	2	19	7.55
4	1	21	7.94
4	2	28	9.17

Table 4.3 N_{reuse}, and $\frac{D(i,j)}{R}$ ratio, for relative locations of cochannel cells, (i, j).

cell is surrounded by N_I cochannel cells whose centers are all at distance D from the center of the reference cell. We say that the analysis considered only the *first tier interferers*, the nearest cochannel cells. In general, in large cellular networks, there will be second and third tier interferers and even more beyond. Of course, the interference from second, third, and higher tiers is substantially lower than that from the first tier, especially when the path loss exponent, η, is large. In any case, the SIR analysis yields a $\frac{D}{R}$ ratio, and then Table 4.3 can be used to determine the value of N_{reuse} that provides this $\frac{D}{R}$ ratio, and the corresponding (i, j) to be used to lay out the cells. For example, if the required $\frac{D}{R}$ ratio is 7, then we must take $N_{\text{reuse}} = 19$, which is achieved with $(i, j) = (3, 2)$.

4.3.3 $\frac{D}{R}$ Ratio: Simple Analysis, Cell Sectorization

It is instructive to compare various cases of first tier cochannel interference while ignoring shadowing, and accounting only for path loss. Such analysis provides quick insight into the comparisons between the various cases. Thus, accounting only for path loss, and taking the transmitter powers, in the worst-case transmitter-receiver configurations, to be equal (see the discussion in Section 4.2), the following is the general expression for the SIR

$$\Psi = \frac{R^{-\eta}}{\sum_{i=1}^{N_I} D_i^{-\eta}}$$

$$= \frac{1}{\sum_{i=1}^{N_I} \left(\frac{D_i}{R}\right)^{-\eta}}$$

where R is the cell radius, N_I is the number of first tier interferers, and $D_i, 1 \leq i \leq N_I$, is the distance of the i-th interferer from the receiver in the reference cell.

Figure 4.8 shows the forward channel (downlink) worst-case situation, where approximations have been made for the various distances between the interferers and the receiver. We see that

$$\Psi = \frac{1}{2\left(\left(\frac{D}{R}-1\right)^{-\eta} + \left(\frac{D}{R}\right)^{-\eta} + \left(\frac{D}{R}+1\right)^{-\eta}\right)}$$

Suppose we take $N_{reuse} = 9$; then Table 4.3 provides $\frac{D}{R} = 5.20$, from which we find that, for $\eta = 4$, $\Psi = 95.09 = 19.78$ dB.

In Figure 4.9 we show the reverse channel (uplink) worst case situation. Here we see that

$$\Psi = \frac{1}{6\left(\frac{D}{R}-1\right)^{-\eta}}$$

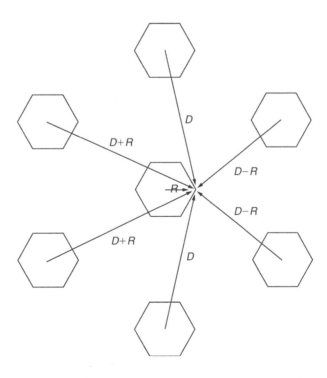

Figure 4.8   Seven cochannel cells, showing the worst-case configuration of first tier downlink interferers. The arrows point at the receiver, and show the direction of the desired signal and the interference. The distances are approximations, and, in general, the cochannel cells may have a different relative orientation from the one shown.

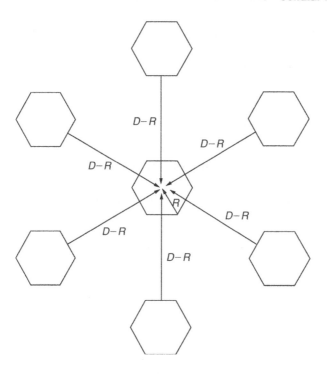

Figure 4.9 Seven cochannel cells, showing the worst case configuration of first tier uplink interferers. The arrows point at the receiver, and show the direction of the desired signal and the interference. The value *D-R* is an approximation, and, in general, the cochannel cells may have a different relative orientation from the one shown.

For $N_{\text{reuse}} = 9$ (i.e., $\frac{D}{R} = 5.20$) we have, for $\eta = 4$, $\Psi = 51.86 = 17.14\,\text{dB}$. Thus we see that the uplink provides a more than 2.5 dB worse performance for the same $\frac{D}{R}$ ratio.

In each of these cases, there are six first tier interferers at a receiver. If directional antennas are used in the BSs then the number of interferers can be reduced. This is achieved by a technique called *sectorization*, which is depicted in Figure 4.10. Each cell is shown divided into three 120° sectors, each with a directional antenna whose angular coverage is designed to coincide with the angular spread of the sector. Thus, an MS in a given sector of a cell is served by the antenna in that sector. Further, the channels are reused only in the corresponding sectors of the cell reuse groups. This is shown in Figure 4.10, where the carrier f is shown being reused in a particular sector of all the cells in a reuse group of seven cells.

To see the advantage of doing sectorization, consider the downlink worst-case situation depicted in the left panel of Figure 4.10. Notice that the MS in Cell 0 sees only two first tier interferers, the two BSs in Cells 2 and 3. The corresponding

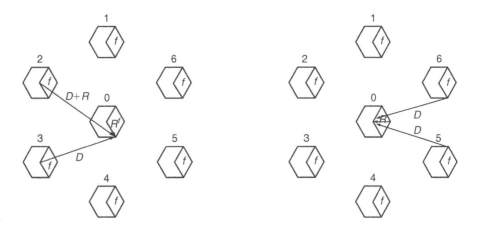

Figure 4.10 Seven cochannel cells, with 120° sectorization, showing the worst case configuration of first tier downlink interferers (left) and uplink interferers (right). The distances shown are approximations.

sectors in Cells 4, 5, and 6 could be using the same channel, but their antenna main lobe is not "visible" to the MS in Cell 0. Making suitable approximations for the distances, the following is the forward channel SIR, ignoring the shadowing.

$$\Psi = \frac{1}{\left(\frac{D}{R}+1\right)^{-\eta} + \left(\frac{D}{R}\right)^{-\eta}}$$

Taking $\eta = 4$, $N_{\text{reuse}} = 9$ (i.e., $\frac{D}{R} = 5.20$) we find that $\Psi = 489.13 = 26.89$ dB, a 7.1 dB improvement over the case without sectorization. The right side of Figure 4.10 shows the worst-case uplink interferers with sectorization. The SIR is given by

$$\Psi = \frac{1}{2\left(\frac{D}{R}\right)^{-\eta}}$$

With $N_{\text{reuse}} = 9$ (i.e., $\frac{D}{R} = 5.20$), taking $\eta = 4$, we find that $\Psi = 365.58 = 25.63$ dB, which is 8.5 dB better than without sectorization. Note, however, that sectorization implies smaller sets of channels in each sector, thus reducing the trunking efficiency.

4.4 Spectrum Efficiency

Let us recall the following system parameters. The RF spectrum allocated to the system is W_{system}, the number of FDM carriers the system bandwidth is partitioned into is C, the number of TDM slots per carrier is s. Assuming equal cell sizes, let

a denote the area of each cell. Further, let K denote the number of sectors in each cell (e.g., $K = 3$ for $120°$ sectors). Recall the definition of N_{reuse}, and the function $g_\epsilon(n)$ (see Section 4.1).

Let us consider the simplest approach of partitioning the C carriers into N_{reuse} subsets. Each subset of carriers is then further partitioned into K sets, each of which is allocated to the same sector in all the cells in a reuse group of cells. Each slot in each carrier in a sector can carry one call. For the present we assume that a call that is initiated in a sector stays in the same sector for its entire duration; that is, there are no handovers in the system. Thus, the $\frac{sC}{N_{\text{reuse}}K}$ slots in a sector, along with the call arrivals to or from MSs in that sector, constitute an Erlang blocking model. It follows that, for a target blocking probability of ϵ, the number of Erlangs that can be offered to a cell is given by

$$g_\epsilon\left(\frac{sC}{N_{\text{reuse}}K}\right) \times \frac{sC}{N_{\text{reuse}}K} \times K$$

where the first term is the number of Erlangs per slot in a sector. Let A denote the coverage area of the system. Then the Erlang capacity of the system, denoted by Λ, is given by

$$\Lambda = \frac{A}{a} \times g_\epsilon\left(\frac{sC}{N_{\text{reuse}}K}\right) \times \frac{sC}{N_{\text{reuse}}}$$

Let us define the *spectrum efficiency* of the system as the Erlang capacity per unit area per Hz of system bandwidth, and denote this by ν. We then have

$$\nu := \frac{\Lambda}{A\,W_{\text{system}}}$$

$$= \frac{1}{a} \times \frac{sC}{W_{\text{system}}} \times g_\epsilon\left(\frac{sC}{W_{\text{system}}}\frac{W_{\text{system}}}{N_{\text{reuse}}K}\right) \frac{1}{N_{\text{reuse}}} \tag{4.5}$$

Notice that $\frac{sC}{W_{\text{system}}}$ is fixed for a given system bandwidth, and depends on the FDM-TDM modulation scheme being used. For example, in the GSM system, the FDM carrier spacing is 200 Khz, and there are eight TDM slots per FDM carrier. Thus, given W_{system}, and allowing for some guard bandwidth on either side, the value of C is determined. In Section 4.3 we saw how N_{reuse} and K can be chosen to achieve the required SIR. Notice that the term $g_\epsilon\left(\frac{sC}{W_{\text{system}}}\frac{W_{\text{system}}}{N_{\text{reuse}}K}\right)\frac{1}{N_{\text{reuse}}}$ decreases with increasing N_{reuse} or K, but we need to set N_{reuse} and K so that the SIR requirements are met while keeping this trunking efficiency term as large as possible. Note that $g_\epsilon\left(\frac{sC}{W_{\text{system}}}\frac{W_{\text{system}}}{N_{\text{reuse}}K}\right)\frac{1}{N_{\text{reuse}}}$ also increases with W_{system}, but having leased a certain amount of the spectrum, the operator will want to work within this leased amount. Finally, the Erlang capacity of the system can be increased by decreasing a; that is, by reducing the cell size. Of course, there are limits to this scaling. As the cell size decreases, there are three issues:

a. As the cell size decreases, we need to consider handovers, and the handover rate increases with decreasing cell size. This will impact the Erlang capacity, as resources need to be reserved for handovers.

b. The signaling load increases due to the increased handover rate. This means that higher capacity call handling systems need to be installed.

c. Reducing cell size requires the installation of more base stations, which can be expensive.

Finally, the design of any given system will have to balance these trade-offs.

4.5   Channel Allocation and Multicell Erlang Models

From the expression for spectrum efficiency in (4.5), we can infer that, apart from reducing the cell size, another way to increase the efficiency is to improve the channel utilization. The earlier analysis assumed a uniform fixed assignment of the FDM carriers to the cells and their sectors. In such an assignment, it is possible that channels are idle in one cell, whereas another cell is overloaded. The trunking efficiency can be improved if the channels are viewed as being in various common pools, from which allocations are made as needed. Of course, such *dynamic channel allocation* must respect the cochannel SIR constraints as the channels are allocated, released, and reallocated to various cells.

4.5.1   Reuse Constraint Graph

A simple model that can be used for designing and analyzing dynamic channel allocation strategies is to specify *pairwise reuse constraints*. Given an array of cells, pairwise reuse constraints specify which pairs of cells cannot use the same FDM carrier at the same time. For example, Figure 4.11 shows a linear array of rectangular cells, such as might be deployed along a highway. The diagrams in the middle and bottom of the figure depict pairwise reuse constraints as *constraint graphs*. In a constraint graph, each cell is represented by a vertex. There is an edge between two vertices if an FDM carrier cannot be simultaneously used in both of the corresponding cells. Thus, the constraint graph in the middle of Figure 4.11 only constrains neighboring cells from reusing the same channel; channels can be simultaneously used in alternate cells. The constraint graph at the bottom, however, permits a channel to be reused only in cells that are separated by at least two other cells.

We note that, in general, representation of reuse constraints by pairwise constraints is conservative. It is possible that among three cells, any two cells can reuse the same channel, but, if the third cell also uses that channel, then, due to the increased interference, the SIR in all the cells may be at an unacceptable level.

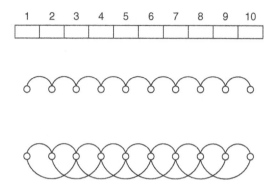

Figure 4.11   A linear array of 10 cells (top), and two sets of pairwise reuse constraints (middle and bottom), shown as constraint graphs.

The modeling of such, more general, constraints requires *hypergraphs*, a generalization of graphs in which edges are subsets of nodes with cardinality greater than two. Such models have been studied in the literature, but we will consider only pairwise constraints in this book.

Formalizing this discussion, let $\mathcal{B} = \{1, 2, \ldots, N\}$ denote the set of cells (or, equivalently, base stations). Let $(\mathcal{B}, \mathcal{C})$ denote the constraint graph with $\mathcal{C}$ being the edge set; that is, for $i \in \mathcal{B}$ and $j \in \mathcal{B}$, $(i, j) \in \mathcal{C}$ if the same carrier cannot be used in Cell i and Cell j simultaneously. We note that the constraint graph is undirected; that is, $(i, j) \in \mathcal{C}$ if and only if $(j, i) \in \mathcal{C}$.

Let $\mathcal{F} = \{f_1, f_2, \ldots, f_M\}$ be the set of FDM carriers that need to be assigned to the cells. Suppose that a certain number of calls need to exist in each of the cells. This will require a certain number of carriers x_j in each of the cells j, $1 \leq j \leq N$, in order to be able to carry those calls. For example, if each carrier has eight TDM slots, then in order to carry 9 to 16 calls in a cell, two carriers are needed. Let x_j denote the number of carriers required in Cell j, $1 \leq j \leq N$. We say that the vector $\mathbf{x} = (x_1, x_2, \ldots, x_N)$ is *feasible* if there exists an allocation of x_k carriers to Cell k such that the reuse constraints are respected. As a simple illustration, if $\mathcal{F} = \{f_1, f_2\}$, and we have the reuse constraints shown in the middle of Figure 4.11, then $\mathbf{x} = (2, 1, \ldots, 1)$ is not feasible. Define

$$\mathcal{X} = \{\mathbf{x} : \mathbf{x} \text{ feasible}\}$$

Recalling some standard concepts from graph theory, we say that a *clique* of $(\mathcal{B}, \mathcal{C})$ is a fully connected subgraph. Thus, a carrier can only be used in exactly one of the cells that form a clique. A *maximal clique* is one that is not contained in any other clique. We will simply refer to maximal cliques also as cliques. Thus, in the bottom diagram of Figure 4.11, the cliques are $\{1, 2, 3\}$, $\{2, 3, 4\}$, and so on.

4.5.2 Feasible Carrier Requirements

Let Q be the number of cliques (i.e., maximal cliques) in $(\mathcal{B}, \mathcal{C})$. Consider the $Q \times N$ matrix $\mathbf{A}$ with

$$a_{ij} = \begin{cases} 1 & \text{if cell } j \text{ is in clique } i \\ 0 & \text{otherwise} \end{cases}$$

We see that a necessary condition for $\mathbf{x} \in \mathcal{X}$ is

$$\mathbf{A} \cdot \mathbf{x} \leq M \mathbf{1}$$

where we recall that M is the number of carriers, and $\mathbf{1}$ is the $Q \times 1$ vector of 1s. Note that this inequality simply says that, for each $i, 1 \leq i \leq Q$, $\sum_{j=1}^{N} a_{ij} x_j \leq M$, where the expression on the left of this inequality is the number of carriers needed in Clique i in order to achieve the carrier allocation given by $\mathbf{x}$. Let us denote

$$\mathcal{X}_{\text{CPA}} = \{\mathbf{x} : \mathbf{A} \cdot \mathbf{x} \leq M \mathbf{1}\}$$

where the suffix CPA expands to *clique packing allocation*. It may appear that $\mathcal{X}_{\text{CPA}}$ is a convenient characterization of $\mathcal{X}$. Since every carrier allocation must satisfy the clique constraints, we see that $\mathcal{X} \subset \mathcal{X}_{\text{CPA}}$. In general, however, $\mathcal{X}$ is a strict subset $\mathcal{X}_{\text{CPA}}$; that is, in general, it can be that $\mathbf{x} \in \mathcal{X}_{\text{CPA}}$, but $\mathbf{x} \notin \mathcal{X}$. An example is shown in Figure 4.12. We can also observe that, if the constraint graph shown in Figure 4.12 is a subgraph of a constraint graph, then $\mathcal{X} \neq \mathcal{X}_{\text{CPA}}$.

Exercise 4.2

Consider a linear array of cells $(1, 2, \ldots, N)$ (as shown in Figure 4.11) with a constraint graph that has the property that if cells i and j, $i \leq j$, are in a clique, then all k such that $i < k < j$ are also in the same clique. Argue that for this situation $\mathcal{X} = \mathcal{X}_{\text{CPA}}$. Show that if $\mathbf{x} \in \mathcal{X}_{\text{CPA}}$ then a feasible carrier assignment is obtained via a *greedy algorithm* that starts by assigning the required carriers to the clique to which the left-most cell belongs, and then moves across the cells from left to right, reassigning carriers as need arises.

4.5.3 Carrier Allocation Strategies

Based on the preceding discussion, we can identify various carrier allocation strategies. We recall that, for a system with N cells, $\mathbf{x} = (x_1, x_2, \ldots, x_N)$ denotes a vector of carrier requirements. Given a set of reuse constraints, a given $\mathbf{x}$ may or may not be feasible. We have defined $\mathcal{X}$ as the set of all feasible carrier requirement vectors: $\mathcal{X} = \{\mathbf{x} : \mathbf{x} \text{ is feasible}\}$.

a. *Fixed Carrier Allocation (FCA).* The carriers are allocated statically to the cells in such a way that the reuse constraints are satisfied. For example, if $\mathcal{F} = \{f_1, f_2\}$, and we have the reuse constraints shown in the middle of Figure 4.11, then $(f_1, f_2, f_1, f_2, \ldots)$ is a valid allocation. With this allocation,

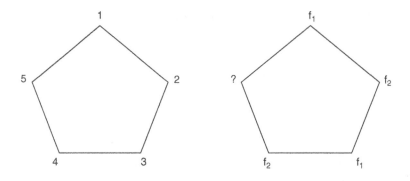

Figure 4.12   A pentagon reuse constraint graph for five nodes is shown on the left. With
M= 2, the vector x=(1,1,1,1,1) satisfies the clique constraints, but there is no feasible
allocation of carriers to cells, as seen in the diagram on the right.

$x = (1, 1, 1, \ldots)$ is feasible, and $x = (2, 1, 1, \ldots)$ is not. For a given fixed allocation of carriers, let $\mathcal{X}_{\text{FCA}}$ denote the set of feasible carrier requirements x. Clearly, $\mathcal{X}_{\text{FCA}} \subset \mathcal{X}$.

b. *Maximum Packing Allocation (MPA).* By definition, for every $x \in \mathcal{X}$ there is a carrier assignment that achieves x. When a call arrives to a cell and is accepted, then this will result in a carrier requirement vector y. Under MPA, if $y \in \mathcal{X}$, then the call is accepted, even if this requires a rearrangement of the carriers. This is not a practical approach as the rearrangement requires a lot of signaling, and the forced handovers of calls as carriers are being swapped. Writing the set of feasible carrier requirements under MPA by $\mathcal{X}_{\text{MPA}}$, we have $\mathcal{X}_{\text{MPA}} = \mathcal{X}$.

c. *Clique Packing Assignment (CPA).* Since the characterization of $\mathcal{X}_{\text{CPA}}$ is simple, for theoretical purposes we may assume that each $x \in \mathcal{X}_{\text{CPA}}$ is acceptable.

In general, we have

$$\mathcal{X}_{\text{FCA}} \subset \mathcal{X} = \mathcal{X}_{\text{MPA}} \subset \mathcal{X}_{\text{CPA}}$$

where, as we have seen, the last containment can be strict. Another channel allocation strategy, which can be viewed as a hybrid of FCA and MPA, is that of *channel borrowing*. Some channels are statically assigned to cells, whereas others are permitted to be borrowed between cells, in order to accommodate local load variations.

4.5.4   Call Blocking Analysis

If a carrier allocation respects the SIR constraints, or if it satisfies certain reuse constraints that, in turn, assure the SIR constraints, then, with a high probability, the accepted calls will experience an acceptable voice quality. This was the purpose of the analysis that we discussed in Section 4.2. Once a particular carrier allocation strategy (denoted CA, generically) is chosen, then the carrier requirement vector $\mathbf{x}$ will remain in $\mathcal{X}_{CA}$. Calls will need to be blocked for this to happen; if acceptance of a new call results in a carrier requirement vector $\mathbf{x} \notin \mathcal{X}_{CA}$, then the arriving call is blocked. In addition to a good voice quality during a call, users also are concerned about the probability of their requests being blocked, or accepted requests being *dropped* because of *handover blocking*. In this section we show how blocking probabilities can be obtained for carrier allocation strategies.

Consider any carrier assignment strategy, and let $\mathcal{X}_{CA}$ denote the set of feasible carrier requirements, $\mathbf{x}$, as discussed earlier. We will assume, for simplicity, that each carrier can carry just one call (rather than, for example, eight in the GSM system). In this section, we also assume that calls stay in the cells into which they arrive, that is, that there are no handovers between cells. Let the arrival rate of calls into Cell j be $\lambda_j, 1 \leq j \leq N$. The arrival processes are assumed to be Poisson processes that are independent from cell to cell. We assume that the time duration for which a call holds a carrier has mean $\frac{1}{\mu}$, and that the holding times from call to call are i.i.d. We also assume that the carrier holding times are exponentially distributed; this assumption can be relaxed, but we will not dwell on that aspect in this discussion (see, however, Appendix D, Section D.5.1).

In this setting, let $X_j(t), 1 \leq j \leq N$, denote the number of carriers utilized in Cell j (equivalently, the number of calls in Cell j) at time t. Then consider the vector random process $\mathbf{X}(t) = (X_1(t), X_2(t), \ldots, X_N(t))$. If the chosen carrier assignment strategy is used then, for all t, $\mathbf{X}(t) \in \mathcal{X}_{CA}$. With the assumptions we have made on the arrival processes and carrier holding time distributions, it can easily be seen that the process $\mathbf{X}(t)$ is a continuous time Markov chain (CTMC; see Appendix D, Section D.2). For finite and positive arrival rates and mean holding times, this CTMC is positive recurrent, since it has a finite number of states. In order to obtain the blocking probabilities we need the stationary distribution $\pi(\mathbf{x}), \mathbf{x} \in \mathcal{X}_{CA}$. Then the blocking probability of calls arriving into Cell j, denoted by $P_{b,j}$ is given by

$$P_{b,j} = \sum_{\{\mathbf{x} \in \mathcal{X}_{CA}: \; \mathbf{x}+\mathbf{e}_j \notin \mathcal{X}_{CA}\}} \pi(\mathbf{x}) \tag{4.6}$$

where $\mathbf{e}_j$ is the unit vector with a 1 in the j-th position. Note that $P_{b,j}$ is the fraction of time during which an arrival into Cell j will be blocked. The fact that this is the same as the fraction of calls arriving into Cell j that are blocked (the quantity on the right-hand side of (4.6)) is a consequence of the Poisson Arrivals See Time Averages theorem (PASTA) (see Appendix D, Section D.4.2). The average blocking over all the cells is then given by

$$P_b = \sum_{j=1}^{N} \frac{\lambda_j}{\sum_{i=1}^{N} \lambda_i} P_{b,j}$$

which can be understood by observing that the probability that a call arrival is for Cell j is $\frac{\lambda_j}{\sum_{i=1}^{N}\lambda_i}$.

It remains to determine the stationary distribution $\pi(\mathbf{x}), \mathbf{x} \in \mathcal{X}_{CA}$. Notice that only the following state transitions are possible in the CTMC $\mathbf{X}(t)$. For $\mathbf{x} \in \mathcal{X}_{CA}$, we can have $\mathbf{x} \to \mathbf{x} + e_j$ for some $j, 1 \leq j \leq N$ (due to an arrival into Cell j), or $\mathbf{x} \to \mathbf{x} - e_j$ (due to a call completion in Cell j; here we require $x_j > 0$ in $\mathbf{x}$). Let $\rho_j = \frac{\lambda_j}{\mu}, 1 \leq j \leq N$, the Erlang load on Cell j. Define

$$\hat{\pi}(\mathbf{x}) = \Pi_{j=1}^{N} \frac{\rho_j^{x_j}}{x_j!}$$

Now consider the transition $\mathbf{x} \to \mathbf{x} + e_j$, and notice that

$$\hat{\pi}(\mathbf{x}) \times \lambda_j = \hat{\pi}(\mathbf{x} + e_j) \times (x_j + 1)\mu_j$$

Also, for the transition $\mathbf{x} \to \mathbf{x} - e_j$, where $x_j > 0$, we have

$$\hat{\pi}(\mathbf{x}) \times x_j\mu_j = \hat{\pi}(\mathbf{x} - e_j) \times \lambda_j$$

With these observations, and defining

$$G_{CA} = \sum_{\{\mathbf{x}:\mathbf{x}\in\mathcal{X}_{CA}\}} \hat{\pi}(\mathbf{x}) \tag{4.7}$$

it can be shown that (see Exercise 4.3) the stationary distribution is given by

$$\pi(\mathbf{x}) = \frac{1}{G_{CA}} \Pi_{j=1}^{N} \frac{\rho_j^{x_j}}{x_j!} \tag{4.8}$$

Exercise 4.3
Use Theorem D.8 in Appendix D to prove that what is being claimed in (4.8) is correct.

4.5.5 Comparison of FCA and MPA

Consider the three-cell example, and the corresponding pair-wise constraint graph shown in Figure 4.13. There are M carriers, each of which can handle one call. If we partition the set of carriers into two equal parts, and assign one set to Cells 1 and 3, and the other set to Cell 2, then the reuse constraints are met, and we get the $\mathcal{X}_{FCA}$ shown by the dashed box in Figure 4.13. On the other hand, in MPA, any carrier not used in Cell 2 can be used in both Cells 1 and 3; $\mathcal{X}_{MPA}$ is also shown in the figure. Suppose that the arrival rate of calls is the same in all the cells.

Let us first numerically investigate the blocking probabilities for $M=2$. Figure 4.14 shows the set of states in $\mathcal{X}_{MPA}$. The set of states in which calls to

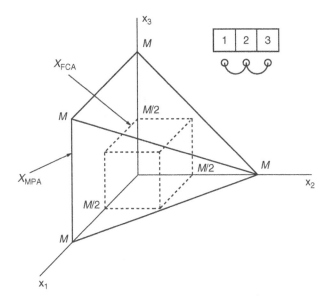

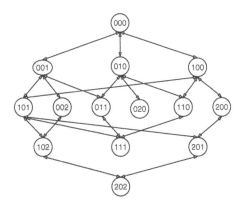

Figure 4.13 On the top right is shown a 3-cell example and the corresponding reuse constraint graph. There are *M* carriers. The sets $\mathcal{X}_{\text{FCA}}$ and $\mathcal{X}_{\text{MPA}}$ are the points with integer coordinates inside the regions shown.

Figure 4.14 The set of states for the three-cell network using maximum packing channel allocation with two channels. The downward transitions occur at rate λ and the upward transitions are at rates that are multiples of μ.

Cell j are blocked are as follows.

For Cell 1: (020), (110), (200), (111), (201), and (202);
For Cell 2: (002), (011), (020), (110), (200), (102), (111), (201), and (202);
For Cell 3: (020), (011), (002), (111), (102), and (202).

Let $\lambda_j = \lambda$ for $j = 1, 2, 3$. The following blocking probabilities are easy to obtain.

$$G_{\text{MPA}} = \frac{1}{4}\rho^4 + 2\rho^3 + \frac{9}{2}\rho^2 + 3\rho + 1$$

$$P_{b,1} = P_{b,3} = \frac{\frac{1}{4}\rho^4 + \frac{3}{2}\rho^3 + 2\rho^2}{G_{\text{MPA}}}$$

$$P_{b,2} = \frac{\frac{1}{4}\rho^4 + 2\rho^3 + \frac{3}{2}\rho^2}{G_{\text{MPA}}}$$

$$P_b = \frac{2}{3}P_{b,1} + \frac{1}{3}P_{b,2}$$

Figure 4.15 shows a plot of blocking probability in each of the cells and the overall blocking probability. For comparison, the blocking probability from a fixed channel allocation is also shown; one channel is allocated to Cell 2, and the

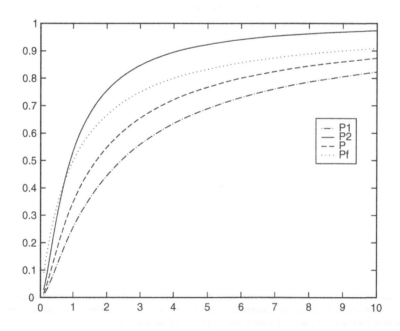

Figure 4.15 Plot of the blocking probability for the 3-cell network shown in Figure 4.14, with $M = 2$, under MPA, in Cell 1 (*P1*), Cell 2 (*P2*), and the overall blocking probability (*P*). P_f is the blocking probability for FCA, with 1 channel allocated to Cell 2, and the other to both Cells 1 and 3.

other to both Cells 1 and 3. Note that in the middle cell, the blocking probability is worse with MPA than with FCA for large ρ. As can be seen from the set of states that block a call to Cell 2, and also the expression for $P_{b,2}$, there are many more states that affect the blocking in the middle cell.

Let us now consider a linear array of $N > 3$ cells, again with the constraint that neighboring cells cannot reuse the same channel. Extending the two-cell analysis via enumeration for $N > 3$ is clearly tedious. We therefore do an asymptotic analysis as $N \to \infty$. Before describing the result, let us see what to expect. With increasing N, the number of middle cells increases as $N \to \infty$, and the blocking probability behavior that we saw for Cell 2 in the numerical example earlier should become typical. We will see that this indeed is what happens.

Let us consider the blocking probability in an interior Cell i. For the same reuse constraints, we see that the set of states $\mathcal{X}_{\text{MPA}}$, is defined by (see Exercise 4.2)

$$\mathcal{X}_{\text{MP}} = \{\mathbf{x} : x_i + x_{i+1} \leq M \quad \text{for } i = 1, \ldots, N-1\}$$

and the blocking states for Cell i are defined by

$$\{\mathbf{x} : x_{i-1} + x_i = M \text{ or } x_i + x_{i+1} = M\}$$

or, equivalently, the set of blocking states for Cell i are

$$\{\mathbf{x} : x_i = M, \text{ or, } (x_{i-1} + x_i = M, 0 \leq x_i < M), \text{ or } (x_i + x_{i+1} = M, 0 \leq x_i < M)\}$$

Thus, using the union bound, the blocking probability at Cell i with MPA is bounded as follows

$$P_{b,i}^{\text{MPA}} \leq \frac{1}{G_{\text{MPA}}} \left(G_1 \frac{\rho^M}{M!} + \sum_{k=0}^{M-1} G_{2,i}(k) \frac{\rho^k}{k!} \frac{\rho^{M-k}}{(M-k)!} + \sum_{k=0}^{M-1} G_{3,i}(k) \frac{\rho^k}{k!} \frac{\rho^{M-k}}{(M-k)!} \right)$$

Here G_1, $G_{2,i}(k)$ and $G_{3,i}(k)$ are given by

$$G_1 := \sum_{\mathbf{x} \in \mathcal{X}_1} \prod_{\substack{n=1 \\ n \neq i}}^{N} \frac{\rho^{x_j}}{x_j!}$$

$$G_{2,i}(k) := \sum_{\mathbf{x} \in \mathcal{X}_{2,i}(k)} \prod_{\substack{n=1 \\ n \neq i-1,i}}^{N} \frac{\rho^{x_j}}{x_j!}$$

$$G_{3,i}(k) := \sum_{\mathbf{x} \in \mathcal{X}_{3,i}(k)} \prod_{\substack{n=1 \\ n \neq i,i+1}}^{N} \frac{\rho^{x_j}}{x_j!}$$

where $\mathcal{X}_1$ is the set of states in which Cell i has M calls, $\mathcal{X}_{2,i}(k)$ is the set of states in which Cell i has k calls and Cell $(i-1)$ has $(M-k)$ calls, and $\mathcal{X}_{3,i}(k)$ is the set of states in which Cell i has k calls and Cell $(i+1)$ has $(M-k)$ calls.

First, let us see what happens for low values of ρ. For low values of ρ, the higher powers of ρ will be insignificant and we can argue that $\frac{G_1}{G_{\text{MPA}}}$, $\frac{G_{2,i}(k)}{G_{\text{MPA}}}$, and $\frac{G_{3,i}(k)}{G_{\text{MPA}}}$ all approach 1 as $\rho \to 0$. Then, as $\rho \to 0$, we can write

$$P_{b,i}^{\text{MPA}} \le \frac{\rho^M}{M!} + 2 \sum_{k=0}^{M-1} \frac{\rho^k}{k!} \frac{\rho^{M-k}}{(M-k)!}$$

$$= \frac{\rho^M}{M!} + 2\frac{\rho^M}{M!} \left(\sum_{k=0}^{M} \frac{M!}{k!\,(M-k)!} - 1 \right)$$

$$= \frac{\rho^M}{M!} + 2\frac{\rho^M}{M!}(2^M - 1) \quad = \quad \frac{2^{M+1} - 1}{M!}\rho^M$$

For fixed channel allocation, each cell would be allocated $\frac{M}{2}$ channels and the blocking probability would be (as before, see Appendix D, Section D.5.1)

$$P_{b,i}^{\text{FCA}} = \frac{\frac{\rho^{M/2}}{(M/2)!}}{\sum_{k=0}^{M/2} \frac{\rho^k}{k!}} \approx \frac{\rho^{M/2}}{(M/2)!}$$

where, since ρ is small, in the denominator we just retain the unit term. We can now see that, for large M, and small ρ, $P_{b,i}^{\text{FCA}}$ decreases as $\rho^{M/2}$, whereas $P_{b,i}^{\text{MPA}}$ decreases faster than ρ^M (for a precise calculation we can use Stirling's approximation for the factorials). Hence, MPA would perform better at low loads.

Let us now see what happens when ρ is large. The stationary probability of there being x active calls in Cell i can be written as

$$\pi_i(x) = \frac{1}{G_{\text{MPA}}} \sum_{\mathbf{x} \in \mathcal{X}_3(x)} \frac{\rho^x}{x!} \prod_{\substack{j=1 \\ j \ne i}}^{M} \frac{\rho^{x_j}}{x_j!} = \frac{1}{G_{\text{MPA}}} \frac{\rho^x}{x!} \phi(i, x)$$

where $\mathcal{X}_3(x)$ is the subset of $\mathcal{X}_{\text{MPA}}$ in which there are x calls active in Cell i and

$$\phi(i, x) := \sum_{\mathbf{x} \in \mathcal{X}_3(x)} \prod_{\substack{j=1 \\ j \ne i}}^{M} \frac{\rho_j^x}{x_j!}$$

We can see that the carried load in Cell i is the average number of active calls in Cell i and is given by $\sum_{x=1}^{M} x\pi_i(x)$. Subtracting the carried load from the offered load (ρ to each cell) and expressing it as a fraction of the offered load, the loss probability, $P_{b,i}^{\text{MPA}}$, is

$$P_{b,i}^{\text{MPA}} = \frac{\rho - \frac{1}{G_{\text{MPA}}} \sum_{x=1}^{M} x \frac{\rho^x}{x!} \phi(i, x)}{\rho} = 1 - \frac{1}{G_{\text{MPA}}} \sum_{x=0}^{M-1} \frac{\rho^x}{x!} \phi(i, x+1)$$

Obtaining $\phi(i,x)$ is involved and we will omit that here. For $M = 2$, and $N \to \infty$, $P_{b,i}^{MPA}$ can be shown to be given by

$$P_{b,i}^{MPA} = \frac{p^2(14 - 10p - 5p^2 + 3p^3)}{2(2 + p^2 - 2p^3)}$$

where p is the solution in $(0,1)$ to the cubic equation $\rho(1 - p)(2 - p^2) = 2p$.

Figure 4.16 shows $P_{b,i}^{MPA}$ and $P_{b,i}^{FCA}$ as a function of the offered load ρ. Notice at about $\rho = 2.6$ the fixed channel assignment outperforms the maximum packing dynamic channel assignment! What is more interesting is that it can be shown that as M increases, the crossover happens at lower values of ρ and the crossover point is asymptotically 0! This indicates that for high capacity cellular networks, the fixed channel allocation scheme will outperform the dynamic channel scheme when the load is time and space homogeneous. This result is definitely counterintuitive; we expect dynamic schemes to be better than static schemes. A heuristic explanation for the effect just described is that the MPA allocation scheme can upset the tight packing of the channels and calls at high loads and spends more time in the many Bad states that are possible with dynamic allocation.

A conclusion that we may draw from this analysis is that it might be better to reject some calls, especially at high loads, to be able to improve the overall system performance. The MPA scheme will accept a call if the channels can be rearranged

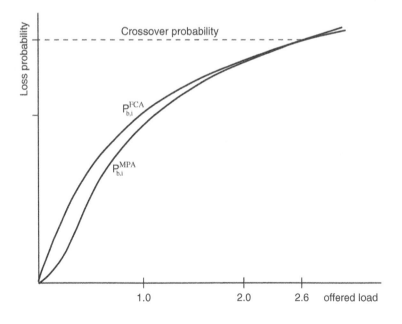

Figure 4.16 $P_{b,i}^{MPA}$ and $P_{b,i}^{FCA}$ as a function of ρ for M = 2, and N $\to \infty$. Adapted from [73].

to accommodate it while the FCA will reject a call if all the channels allocated
to the cell are busy; that is, it will not borrow channels from other cells to fulfill
a request. Thus, it is not automatic that a dynamic channel allocation performs
better than a fixed allocation scheme. However, we cannot conclude that dynamic
channels do not have advantages. Rather that the advantages are realized if the
offered load is nonhomogeneous in space and is time varying, in which case the
dynamic schemes adapt to the changing load.

4.6   Handovers: Techniques, Models, Analysis

In our discussions thus far, we have essentially assumed that mobiles are confined
to the cells in which they initiate their calls. Since typically, neighboring cells do
not reuse a carrier, when a mobile moves to a neighboring cell, it must switch over
to a different carrier in that cell. This is called a handover. Naturally, for cellular
mobile telephony to be a useful service, a handover should be transparent to the
user. This imposes two requirements:

a. An ongoing call should not experience degradation in service when it is at
    the fringes of the cell that is handling it.

b. A handover should rarely fail due to a channel not being available in the
    cell into which a mobile call moves. Such a *handover failure* leads to *call
    dropping*, the constraints on which are more stringent than on call blocking.

Handovers are performed by the MS making signal strength measurements
to neighboring BSs, and conveying this information to the handover management
system, which then decides on the need for a handover and the channel to
be assigned to the new cell. The transfer of such measurements from the MS
to the call handling system became possible in the second generation cellular
systems. Note that the handover strategy basically defines what is meant by a cell's
coverage area.

4.6.1   Analysis of Signal Strength Based Handovers

We consider an MS located on the line joining two BSs, BS 0 and BS 1, as shown
in Figure 4.17. Let $S_i(x)$ = Received signal power from BS $i, i \in \{0, 1\}$, when the
MS is at the distance x from BS 0. Then, recalling the path loss and shadowing
model from Section 2.1.4, we have

$$[S_0(x)]_{dB} = [S_0(d_0)]_{dB} - 10\eta \log \frac{x}{d_0} - \xi_0$$

where η is the path loss exponent, and the shadow fading, $\xi_i, i \in \{0, 1\}$, is normally
distributed with mean 0, and variance σ^2. Also,

$$[S_1(x)]_{dB} = [S_1(d_0)]_{dB} - 10\eta \log \left(\frac{2R - x}{d_0} \right) - \xi_1$$

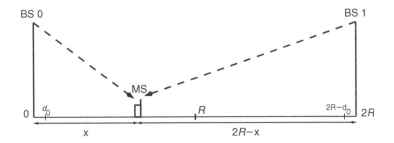

Figure 4.17 Handover: An MS located on the line joining two neighboring BSs that are at the distance *2R*. The MS is located at distance *x* from BS 0. The MS provides signal strength measurements from each of the BSs.

Let us assume that $S_0(d_0) = S_1(d_0)$. Then, for $d_0 \leq x \leq 2R - d_0$,

$$[S_0(x) - S_1(x)]_{\mathrm{dB}} = 10\eta \log\left(\frac{2R - x}{x}\right) + (\xi_1 - \xi_0)$$

where $\xi_1 - \xi_0$ is normally distributed with 0 mean and variance $2\sigma^2$. In Figure 4.18, we show the variation of $[S_0(x) - S_1(x)]_{\mathrm{dB}}$ as the MS moves from BS 0 to BS 1. The solid curve shows the mean $10\eta \log\left(\frac{2R-x}{x}\right)$. The two dashed curves above and below the solid curve represent the variability due to shadowing, and can be viewed as the bounds within which $[S_0(x) - S_1(x)]_{\mathrm{dB}}$ stays, with a high probability. The half-width of the curved strip defined by the two dashed curves is proportional to $\sqrt{2}\sigma$.

Suppose the MS is being served by BS 0. A simple handover approach is to hand over the MS to BS 1 when $[S_0(x) - S_1(x)]_{\mathrm{dB}} < -H$, where H is a design parameter; that is, handover occurs when the signal strength from BS 0 is sufficiently lower than the signal strength from BS 1. For the H shown in Figure 4.18, and with the preceding interpretation of the dashed curves, we can see that that handover will occur with a positive probability when the MS is at a distance greater than R from BS 0 (notice that the lower dashed curve falls below the horizontal line for $-H$, when $x > R$). There are two issues here:

a. If the coverage of either cell extends only up to a distance R, then once the MS is beyond R, the handover should occur with a high probability.

b. With this design, if the MS is moving about in the region around the middle of the line joining the two BSs, then it will be repeatedly handed over between the two BSs, thus increasing the chance of the call being dropped, and also increasing the load on the call management processors.

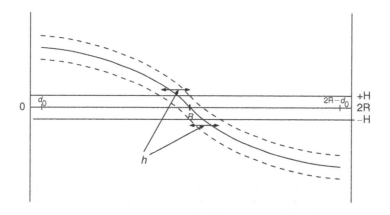

Figure 4.18 Handover: The difference in signal strengths, $S_0(x) - S_1(x)$ (in dB) at an MS that is at position x on the line joining BS 0 and BS 1. For an explanation of h, see the text.

These two issues can be addressed by extending the coverage of each BS beyond R, to an additional distance, say, h. Suppose h is chosen so that

$$10\eta \log \frac{2R - (R+h)}{R+h} + a\sqrt{2}\sigma < -H$$

or

$$10\eta \log \frac{R-h}{R+h} + a\sqrt{2}\sigma < -H$$

where $a\sqrt{2}\sigma$ is the half width of the dashed strip, and a is chosen from the standard normal tables so that the tail probability of the random variable $\xi_1 - \xi_0$ beyond $a\sqrt{2}\sigma$ is small. This choice of h is shown in Figure 4.18, since at $x = R + h$ the upper dashed curve falls below $-H$. Now, when deciding to hand over from BS 0 to BS 1, we check if both of the following tests are true:

$$[S_0(x)]_{dB} < S_{threshold}$$

$$[S_0(x) - S_1(x)]_{dB} < -H$$

for a suitably chosen $S_{threshold}$. Both these tests will succeed beyond $R + h$ with a high probability, and, thus, the handover will take place with a high probability. Further, the reverse handover will take place with a very small probability. Thus, this handover strategy has a *hysteresis* built into it.

Although this design solves the problem of repeated handovers from one cell to the other, the extension of the cell coverage into the neighboring cell impacts the earlier SIR analysis. Let

$$\frac{h}{R} = b$$

so that

$$R + h = (1 + b)R$$

Thus, in the cochannel interference calculations, we now need to use

$$\frac{D}{(1+b)R}$$

It follows that a larger $\frac{D}{R}$ value will need to be used for a given SIR constraint, thus requiring a larger value of N_{reuse}, and lowering the spectrum efficiency. It is thus important to design handover schemes that can reduce the cell expansion factor b.

4.6.2 Handover Blocking, Call Dropping: Channel Reservation

Let us consider a cell in an FDM-TDMA cellular system with new call arrival rate λ_0, and handover call arrival rate (from neighboring cells) λ_h. Define

P_b = new call blocking probability

P_h = handover blocking probability

P_d = call dropping probability

Note that a call may undergo several handovers, and the call gets dropped at the first of its handovers that is blocked. The preceding definitions can be formally expressed as

$$P_h = \lim_{t \to \infty} \frac{\text{number of handovers lost in } [0, t]}{\text{number of handovers in } [0, t]}$$

and

$$P_d = \lim_{t \to \infty} \frac{\text{number of accepted calls dropped in } [0, t]}{\text{number of calls accepted in } [0, t]}$$

Note that P_b and P_d are user perceived performance measures, whereas P_h is a measure internal to the system. We need P_d to be very small (e.g., 0.1%), whereas P_b is typically 1 to 2 percent. Let us assume that the time that a call spends in a cell is exponentially distributed with mean $\frac{1}{\nu}$. The duration of a call is exponentially distributed with mean $\frac{1}{\mu}$. Then, assuming that whether or not a handover is blocked is independent from handover to handover, we can write

$$P_d = \frac{\nu}{\nu + \mu}(P_h + (1 - P_h) \cdot P_d)$$

This can be understood as follows. $\frac{\nu}{\nu+\mu}$ is the probability that a call leaves the cell it is in before it finishes conversation. If it does leave the cell, then the handover attempt is blocked with probability P_b, or if the handover is not blocked (with probability $1 - P_h$), then we have a renewal point (see Appendix D) and the remaining call experiences dropping with probability P_d. This expression yields

$$P_d = \frac{\frac{v}{v+\mu} P_h}{\frac{\mu}{v+\mu} + \frac{v P_h}{v+\mu}}$$

$$= \frac{\frac{v}{\mu} P_h}{1 + \frac{v}{\mu} P_h}$$

The second expression on the right may be approximated by $\frac{v}{\mu} P_h$ when $\frac{v}{\mu} P_h$ is much smaller than 1. Where the approximation works, its interpretation is the mean number of handovers per call multiplied by the handover blocking probability. This calculation yields a target value of P_h, given a target for P_d.

We had defined λ_h as the rate of arrival of handovers into a cell, and we observe that this is not a given. But with the exponential distribution assumptions we have made, we can write the following:

$$\lambda_h = \left((\lambda_0(1 - P_b) + \lambda_h(1 - P_h)) \frac{v}{\mu + v} \right) \cdot \frac{1}{6} \cdot 6 \tag{4.9}$$

This is obtained as follows. $\lambda_0(1 - P_b) + \lambda_h(1 - P_h)$ is the rate of accepted calls into a cell. Each accepted call causes a handover to a neighboring cell with probability $\frac{v}{\mu+v}$. Each cell is surrounded by six cells, and one-sixth of the handovers of each of its neighbors enters the cell. However, P_h and P_b depend on λ_h. The approach is to iterate, starting with $\lambda_h^{(0)}$. This will yield $P_b^{(0)}$ and $P_h^{(0)}$. Given $P_b^{(k-1)}$ and $P_h^{(k-1)}$ we can obtain $\lambda_h^{(k)}$ by using (4.9), and thus the iterations can continue. How $P_b^{(k-1)}$ and $P_h^{(k-1)}$ are obtained from $\lambda_h^{(k)}$ depends on the way channels are assigned to new calls and handover calls in a cell, and is the next topic of discussion.

The remaining question is whether there is a need to discriminate between new calls and handover calls when assigning channels. If they are handled in the same way, then they will get the same blocking probability (i.e., $P_b = P_h$). Since the target value of P_h is much smaller than that of P_b, we will be forced to operate with much too small a value of new call blocking, which will result in a very low Erlang capacity. Hence channel reservation is done for handover calls. The common approach is *dynamic channel reservation*, which means the following. If there are m carriers in a cell, then a number $m_h < m$ is chosen; typically, m_h is just 1 or 2. When a call arrives, if the number of busy carriers is less than $m - m_h$, then every call is accepted. However, if the number of busy carriers is $\geq m - m_h$, then only handover calls are accepted. If we assume that the arrival process of new calls and handover calls into a cell are independent Poisson processes, then the number of busy carriers becomes a positive recurrent CTMC. Returning to the iterative calculation, earlier, the analysis of this CTMC will provide $P_b^{(k)}$ and $P_h^{(k)}$, given λ_0 and $\lambda_b^{(k)}$. At the k-th iteration, let $\pi^{(k)}(i), 0 \le i \le m$, denote the stationary probability distribution of the CTMC. Then

$$P_b^{(k)} = \sum_{i=m-m_h}^{m} \pi^{(k)}(i)$$

$$P_h^{(k)} = \pi^{(k)}(m)$$

where, again, the PASTA theorem is used (see Appendix D, Section D.4.2).

Exercise 4.4

a. Show the transition rate diagram of the CTMC with dynamic channel reservation for handovers, and obtain the stationary distribution $\pi^{(k)}(\cdot)$.

b. Write a computer program to carry out the proposed iteration and obtain the new call arrival rate, λ_0, that can be offered when $m = 16$, and $m_h = 1$, for a target $P_d = 0.01$. Take the mean call duration to be 100 seconds, and the mean time a call stays in a cell to be 50 seconds. Obtain the new call blocking probability, P_b, that is obtained with this value of λ_0.

c. What is the new call arrival rate that can be handled if no special treatment is provided to handovers, but we still require that $P_d = 0.01$?

4.7 The GSM System for Mobile Telephony

After about 15 years of deployment, the FDM-TDMA-based GSM system (Global System for Mobile communications) is the most popular cellular system for mobile telephony and related services. Figure 4.19 shows the components of a GSM cellular network. The wireless links are only between the mobile stations (MSs; shown as cellular phone handsets in Figure 4.19) and the Base Transceiver Stations (BTSs). An MS can be in the vicinity of several BTSs, but at any point in time, an active MS is associated with one BTS, the one with which it is determined that it has the highest probability of reliable communication. Several BTSs are linked to Base Station Controllers (BSCs) by wired links. Together, the BTSs and the associated BSC is called a BSS (Base Station Subsystem). The BTSs provide the fixed ends of the radio links to the MSs; it is the BSC that has the intelligence to participate in the signaling involved in connection handovers. In turn, the BSCs are connected to the Mobile Switching Center (MSC), which connects to the fixed network infrastructure.

Worldwide, several bands have been used for the operation of GSM networks. The 900 MHz or 1800 MHz bands are the ones commonly used in most countries. In the 900 MHz band the uplink carriers are in the 890–915 MHz frequency band, and the downlink carriers are in the 935–960 MHz frequency

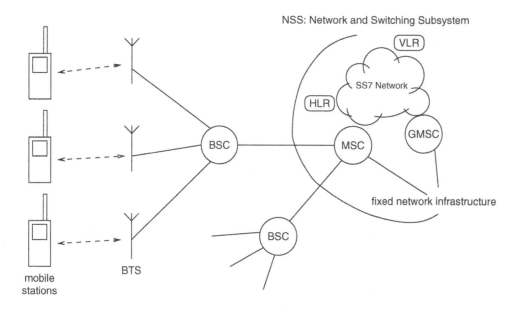

Figure 4.19 The components of a GSM cellular network.

band. As explained earlier in this chapter, if an operator obtains W MHz of spectrum, actually W MHz is provided from the uplink band and another W MHz is provided from the downlink band. This is for the purpose of frequency division duplexing of bidirectional calls. The bandwidth of a GSM operator, in each direction, is then divided into FDM carriers with a spacing of 200 kHz. These FDM carriers are digitally modulated to create a hierarchical TDM carrier. The basic frame time in this TDM carrier is 4.615 ms, which contains eight slots, each of which can be assigned to a different voice call. Each TDM slot can carry 114 bits of payload. Notice that the coded payload bit rate on each carrier is about 200 Kbps. For one standard GSM voice codec, after channel coding, blocks of 456 bits are emitted, which are accommodated in four TDM slots.

Since the resources (i.e., the spectrum) of a cellular wireless network are limited, an MS cannot have permanent access to the network, but has to make a request for a connection. Thus, since an MS is not always connected to the network, there are two problems that need to be addressed:

a. Between the time that an MS last accessed the network and the time that it next needs to access, the MS may have moved; hence, it is first necessary to locate the MS and associate it with one of the cells of the network.

b. Since the MS initially does not have any access bandwidth assigned to it, some mechanism is needed for it to initiate a call or to respond to an incoming call.

Location management and *call set up* are the major activities that need to be overlaid on the basic cellular wireless infrastructure in order to address the first problem. In Figure 4.19 we show the additional components that are needed. Together these are called the Network and Switching Subsystem (NSS), and comprise the MSC, the GMSC (Gateway MSC), the HLR (Home Location Register), the VLR (Visitor Location Register), and the signaling network (standardized as Signaling System 7 (SS7), by the ITU). The SS7 signaling network already exists where there is a modern circuit switched phone network. As their names suggest, the HLR carries the registration of an MS at its home location, and a VLR in an area enters the picture when the MS is *roaming* in that area. Each operator has a GMSC at which all calls to MSs that are handled by the operator must first arrive. The GMSC, HLR, and VLR exchange signaling messages over the SS7 network, and together help in setting up a call to a roaming user.

Location management is done as follows. An MS will be registered with an operator in its home area. A roaming mobile that is turned on briefly associates itself with a nearby BTS and provides the network the information that it is now in the area. If this happens to be an area other than where the MS normally is registered, then the MS's identity is used to determine its home location, and the HLR at this location is informed of the whereabouts of the MS. The VLR at the location that the MS is visiting then receives confirmation from the MS's HLR that this MS is a valid user. Suppose now that someone somewhere in the world calls this MS. The MS's number is used to determine the GMSC of its home operator. A signaling message is sent over the SS7 network to this GMSC, which determines the HLR where the MS is registered, and sends a message to this HLR. The HLR then, knowing that the MS is roaming, queries the VLR in the area where the MS is roaming. The VLR knows which local MSC the MS is in the control of, and provides this information to the HLR. The HLR forwards this information to the GMSC, which then directly establishes the call to the MS.

Let us now turn to the second of the two problems enumerated. In the GSM system there are several permanent channels defined in each cell. Whenever an MS enters a cell it locks into these channels. One of these channels is called the paging and access grant channel (PAGCH). If a call arrives for an MS, and it is determined that the MS may be in a cell, or in a group of cells, then the MS is paged in all these cells. Another such common channel is basically a slotted Aloha random access channel (RACH) (see Chapter 7), and is shared by all the MSs in the cell. When an MS has to respond to an incoming call (i.e., it is paged on the PAGCH) or has to initiate a call, it contends on the RACH in the cell, and conveys a short message to the network. Subsequently, the network allocates a channel to the MS and call set up signaling starts.

4.8 Notes on the Literature

In this chapter we have discussed concepts and techniques that were researched in the 1970s and 1980s, at a time when the first analog cellular telephony systems were being experimented with. Bell System's Advanced Mobile Phone Service (AMPS) and the cellular concept are described in a seminal article by MacDonald in Bell Systems Technical Journal [96]. There are several textbooks devoted to extensive treatments of cellular telephony, including the classic by Lee, and the more recent book by Garg and Wilkes [39]. The widely adopted text by Rappaport [116] discusses cellular mobile systems in conjunction with a detailed coverage of propagation phenomena in cellular mobile communication systems, physical layer techniques, and speech coding. A rigorous derivation of the formula relating the cochannel cell distance $D(i,j)$ and N_{reuse} was carried out by Gamst [38] using group and ring theory. The Fenton-Wilkinson approximation, and other similar techniques have been derived in the text on mobile communications by Stuber [123]. The comparison of fixed channel allocation and maximum packing allocation has been adapted from Kelly [73], which also provides some very useful insights into several problems in networking. A very accessible and extensive coverage of the GSM standard has been provided by Mouly and Pautet [104].

Problems

4.1. The coverage of a cell is first obtained by *ignoring* shadow fading (Rayleigh fading can be assumed to be averaged over). If the shadow fading standard deviation is 8 dB, roughly how much additional power is required so that the outage probability for the same coverage is less than 2%?

4.2. A fade margin of 20 dB is required to combat shadowing and achieve adequate coverage in a cell.

   a. If the shadowing standard deviation is 8 dB, what was the target outage probability?

   b. If the path loss exponent is 4, how much additional coverage would be obtained if there is no shadowing?

4.3. A GSM operator leases 7 MHz of spectrum (i.e., 7 MHz each in the uplink and the downlink), and estimates that a $\frac{D}{R}$ of at least 4 is required. If the cell radius, R, is 2 km (assume hexagonal cells), determine the Erlangs per square kilometer for the network, for a target blocking probability of 1%.

4.4. A GSM operator leases 7 MHz of spectrum. Assuming that the uplink constrains performance, a path loss exponent of 4, and ignoring shadowing and additive noise, and given that an SIR of 14 dB is required, determine the Erlang capacity per cell for a blocking probability of 1%. Do not consider sectorization. Assume a hexagonal cell geometry.

4.5. Consider a highway cellular system. Assume that the highway is exactly linear, the cells are of length 2R, and the cell width (i.e., the width of the highway) can be ignored. Frequencies can be reused in cells whose centers are D units apart. The base station in each cell is at its center, and has two directional antennas, one covering each half of the cell (i.e., the cells are "sectorized" into two sectors).

a. Relate D, R, and the number of reuse groups N.

b. Accounting only for first tier interferers, assuming that Rayleigh fading is averaged out, assuming independent log-normal shadowing for all the received signals, determine the minimum $\frac{D}{R}$ value so that the SIR falls below 12 dB with a probability of 1%. You must analyze both the forward and reverse channels. The standard deviation of log-normal shadowing is 8 dB. Take the power law path loss exponent to be 4.

c. Explain why the SIR analysis is greatly simplified in this problem by assuming directional antennas; that is, by sectorization.

d. Given that there are 200 traffic channels available (assume single channel per carrier) determine the maximum number of Erlangs that each cell can be offered.

4.6. Consider a channelized cellular system with a total of 320 traffic channels. Denote the cell radius (center to apex) by R, and the minimum distance between cochannel cells by D. Assume that we can average over Rayleigh fading. Take the lognormal shadowing to have a standard deviation of $\sigma = 8$ dB, and the path loss component to be 4. Considering only the uplink channel answer the following.

a. Obtain the channel reuse ratio for an uplink channel target SIR of 6 dB and an outage probability of 10%. Use the Fenton-Wilkinson method, and a table of the normal distribution. You may assume that the worst case interferer distance is $D - R$.

b. List two assumptions that this analysis makes. In your solution in (a), where is Rayleigh fading being accounted for (even though it is being averaged over)?

c. For this reuse ratio and the given number of channels, obtain the Erlang capacity per cell assuming a fixed channel allocation, and a call blocking probability of 2%. Use an Erlang blocking table.

4.7. Consider a TDM/TDMA cellular system in which each carrier handles eight calls. Voice activity detection (VAD) is used to reduce cochannel interference; an MS does not transmit when there is no speech activity. The probability of an MS being active is p. Consider a hexagonal cell layout; ignore shadowing and Rayleigh fading; take the path loss exponent to be η. In the following, use the standard approximations for the hexagonal geometry. Use tables of the standard normal distribution and Erlang blocking tables.

a. Considering only the uplink, and accounting for voice activity, determine the minimum D/R ratio required for a SIR γ, if the probability of SIR falling below γ is allowed to be 2.3%. (Hint: consider the total power at the reference BS, and individual powers from each of the interfering MSs.)

b. For $\gamma = 20$ dB, $\eta = 4$ and $p = 0.4$ determine the reuse ratio without and with VAD. Show that the effect of VAD is equivalent roughly to reducing the value of γ by 3 dB.

4.8. In the figure are shown five cochannel cells each with four 90° sectors, oriented as shown.

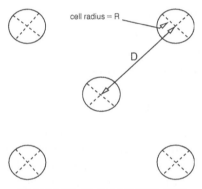

5 cochannel cells, showing the 90 degree vectors

a. Copy the diagram and mark the cochannel *sectors* with g_1, g_2, g_3, and g_4.

b. Ignoring Rayleigh fading and log-normal shadowing, obtain the value of D/R for a reverse channel worst-case S/I of 20 dB. Take the path loss exponent to be 4.

4.9. a. The figure shows seven cells and pairwise reuse constraints between them. Show that for these constraints, and two channels, $\mathcal{X}_{\text{CPA}}$ is strictly larger than $\mathcal{X}_{\text{MPA}}$.

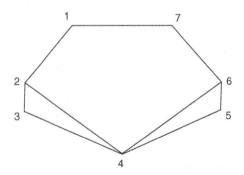

b. Consider three cells with a triangular pairwise reuse constraint graph. There are N channels and calls arrive to each cell at rate λ. The calls have a mean channel holding time of b.

  i. Sketch the set of possible vectors of the numbers of calls that can be present in each of the cells (i.e., $\mathcal{X}$).

  ii. Find the probability that a call is blocked.

4.10. Consider a linear array of K cells, with reuse constraint graph given. Let the $J \times K$ clique incidence matrix be denoted by A. Assume that the maximum number of frequency channels that can be used simultaneously in the j–th maximum clique is given to be $n_j, 1 \leq j \leq J$ ($n_j \leq M$, where M is the total number of frequency channels in the cellular system). Let $N = (n_1, n_2, \ldots, n_J)$.

a. Find the set $S(N)$ of feasible cell occupancy vectors $x = (x_1, x_2, \ldots, x_K)$.

b. Ignoring mobility, assume Poisson call arrivals with traffic intensity ρ_k in cell $k, 1 \leq k \leq K$, and assume that the cell occupancy vector has the steady-state distribution

$$\pi(x) = G(N)\Pi_{k=1}^{K} \frac{\rho_k^{x_k}}{x_k!}, x \in S(N),$$

where

$$G(N) = \left(\sum_{x \in S(N)} \Pi_{k=1}^{K} \frac{\rho_k^{x_k}}{x_k!} \right)^{-1}.$$

Show that the steady-state probability that a call arrival in cell k is blocked is

$$B_k = 1 - \frac{G(N)}{G\left(N - Ae_k^T\right)}$$

where e_k is the length-K vector $(0, \ldots, 0, 1, 0, \ldots, 0)$ with the only non-zero element 1 appearing at the k-th position.

4.11. Consider two neighboring base stations BS 1 and BS 2, a distance $2R$ apart (where $R = 10d_0$), and an MS on the line joining them. Assuming that Rayleigh fading is averaged over, the minimum SINR required for acceptable communication is 10 dB. Let Ψ_0 denote the average SNR at a distance d_0 from a transmitter. Assume that the power law path loss exponent is 4; the shadowing standard deviation is $\sigma = 8\,\text{dB}$.

a. With hand-off (at the cell boundary) obtain the $(\Psi_0)_{\text{dB}}$ required for an outage probability of 10%.

b. With hand-off at 10% beyond the half-way point between the BSs (i.e., $1.1R$), repeat part (a).

CHAPTER 5

Cellular CDMA

We discussed the CDMA concept in Chapter 2, Section 2.4.1. One of the two major technologies for second-generation cellular systems is based on CDMA. Third-generation (3G) cellular access systems that provide high speed data and multimedia access are also based on CDMA. In this chapter we will study various resource allocation problems in cellular CDMA systems, basing our discussion mainly on SINR (signal-to-interference plus noise ratio) analysis.

Overview

Unlike the FDM-TDMA cellular systems discussed in Chapter 4, CDMA cellular systems are based on the principle of *universal frequency reuse*; that is, the same portion of the spectrum is reused at every BS. These systems employ frequency division duplexing; each system is assigned a pair of bands, one for the uplink and the other for the downlink. These two bands then are used at every BS. Thus, every uplink transmission interferes, in principle, with every other uplink transmission in the system; the same holds for the downlink. As discussed in Chapter 2, Section 2.4.1, the performance of an instance of communication between an MS and a BS depends on the SINR achieved at the receiver (see (2.24)). Second-generation CDMA systems were designed mainly for carrying telephone quality voice. CDMA systems have been evolving so as to be able to efficiently carry other guaranteed QoS services, such as interactive video, and also elastic services, such as file transfer and web access. We consider resource allocation for both these types of services.

Each guaranteed QoS connection needs to achieve an SINR target. In Section 5.1, we write down general inequalities that need to be satisfied by the transmission powers used at all the uplink transmitters in the system. An important question that we then ask is about the existence of a set of transmit power levels at all the MSs so that the inequalities are satisfied. This leads to the concept of admission control; arbitrary collections of MSs, each with its SINR target cannot be handled by the system. Hence, some call requests need to be blocked.

Focusing on the uplink problem, we begin by assuming a spatially homogeneous system in which the interference at a BS, from MSs *associated* with other BSs, can be taken to be just a multiple of the total received power at a BS. We develop the case of a single call class (say, voice) in Section 5.2. We find that each call can be characterized by a resource requirement expressed in terms of the target SINR. We find that the resource requirements of the calls just add up, and the admission control ensures that a certain measure of total system resource is not exceeded. This measure of system resource depends on the *other-cell interference*. We show how this is calculated, for *hard handover* and *soft handover* of calls between BSs. In Section 5.3, we expand our discussion to multiclass calls. The Chernoff bound is used to develop an admission control that again treats each call as having a resource requirement, and the requirement due to a set of calls is the sum of the individual resource requirements.

In Section 5.4 we abandon the spatial homogeneity assumption and consider a general configuration of MSs scattered among several BSs. For a given association of MSs and BSs, we develop a necessary and sufficient condition for there to exist a feasible transmit power allocation. The condition is in terms of the Perron-Frobenius eigenvalue of a matrix derived from the channel gains. This also leads to an iterative power control algorithm.

Finally, in Section 5.5, we consider the scheduling of downlink elastic transfers. Depending on the channel power gains between the BSs and the MSs, there is a convex set of transfer rates that can be achieved to the MSs. There is a trade-off between maximizing the total transfer rate over all the MSs (which leads to maximization of operator revenue), and fairness between the rates assigned to the MSs. We use the sum-utility maximization formulation to compare various approaches. One such formulation leads to the idea of proportional fairness, for which we then show how the file transfer delay can be analyzed in terms of the M/G/1 processor sharing model.

A brief overview of 2G and 3G CDMA cellular standards is then provided in Section 5.6.

5.1 The Uplink SINR Inequalities

In CDMA cellular systems, each active mobile station (MS) is associated with one of the base stations (BSs) in its vicinity. When an MS is involved in a conversation, then it is assigned a power level with which it should transmit. As explained in Section 2.4.1, in CDMA access networks the link performance obtained by each mobile station (MS) is governed by the strength of its signal and the interference experienced by the MS's signal at the intended receiver. For each radio link between an MS and a BS, a SINR target needs to be met. Hence it is important to associate MSs with BSs, and to assign them transmit powers in such a way that signal strengths of intended signals are high and interference from unintended signals is low. It is evident that increasing the transmit power to help one MS may not solve the overall problem, as this increase may cause unacceptably high interference at the intended receiver (i.e., a BS) of another MS. We will say that an association of MSs with BSs, and an allocation of transmit powers, is *feasible* if all SINR targets are achieved. In some situations there may be no feasible power allocation. The analysis of CDMA systems is performed via certain SINR inequalities. We will begin our discussion by setting up these inequalities in general.

Consider a CDMA system with multiple interfering cells (see Figure 5.1). The system bandwidth is W (e.g., 1.25 MHz in the IS-95 standard), and the chip rate is $R_c \leq W$ (e.g., 1.2288 Mcps (Mega chips per second)) in IS-95. There are M MSs and N BSs, with $\mathcal{B} = \{1, 2, 3, \ldots, N\}$ denoting the set of BSs. Let $h_{i,j}, 1 \leq i \leq M$, $1 \leq j \leq N$, denote the power "gains" (i.e., attenuations) from MS i to BS j. Let $A = (a_1, a_2, \ldots, a_M), a_i \in \mathcal{B}$, denote an *association* of MSs with the BSs; thus, in the association A, MS i is associated with BS a_i. Let p_i be the transmit signal power used by MS $i, 1 \leq i \leq M$. For the most part of the following discussion, we will assume that the power gains and the association are fixed. With these definitions we can write the uplink received signal power to interference plus noise ratio for MS k as

$$(\text{SINR})_k = \frac{h_{k,a_k} p_k}{\sum_{\{i\,:\,1 \leq i \leq M,\, i \neq k\}} h_{i,a_k} p_i + N_0 W}$$

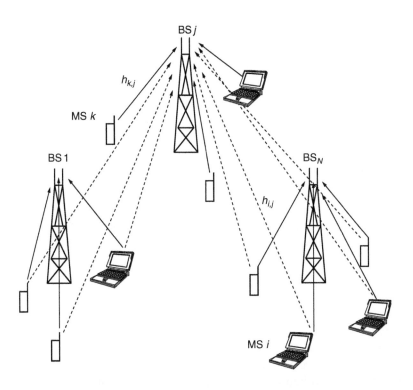

Figure 5.1 **A depiction of the power allocation problem for several MSs in the vicinity of some BSs. The solid lines indicate signals from MSs to the BSs with which they are associated. MS k is associated with BS j, and its signal (solid line labeled $h_{k,j}$) is interfered with by all the other MSs associated with the other BSs (dashed lines), and also by the other MSs associated with BS j. The signal from MS k has a channel "gain" of $h_{k,j}$ to BS j.**

where N_0 is the power spectral density of the additive noise, and W is the radio spectrum bandwidth. Assume that the interference plus noise is well modeled by a white Gaussian noise process.

Various types of calls may be carried on the system; for example, there could be different types of voice telephony calls that use various codecs. Suppose that a call requires a bit rate R_k. In order to ensure a target bit-error-rate (BER) (which is governed by the required QoS for the application being carried; see the discussion to follow later in Section 5.2.2), we need to lower bound the product of the SINR$_k$ and the processing gain $L_k := \frac{R_c}{R_k}$ (see (2.24)). For example, with $R_k = 9.6$ Kbps and $R_c = 1.2288$ Mcps, we obtain $L = 128$. If the desired lower bound is γ_k, then, defining $\Gamma_k := \gamma_k \frac{R_k}{R_c}$, we obtain (see (2.24)), for MS k,

$$\frac{h_{k,a_k} p_k}{\sum_{\{i:1 \leq i \leq M, i \neq k\}} h_{i,a_k} p_i + N_0 W} \geq \Gamma_k \tag{5.1}$$

For a given association and given channel gains, we thus obtain M linear inequalities in the M uplink powers of the M MSs. Suppose $a_k = j$; then if we define

$$I_j := \sum_{\{i:1 \leq i \leq M, a_i = j\}} h_{i,j} p_i$$

the total power received at BS j from MSs associated with it, and define

$$I_{o,j} := \sum_{\{i:1 \leq i \leq M, a_i \neq j\}} h_{i,j} p_i$$

the total interference power at BS j from MSs associated with other BSs, then we can write the SINR inequalities as

$$\frac{h_{k,a_k} p_k}{(I_{a_k} - h_{k,a_k} p_k) + I_{o,a_k} + N_0 W} \geq \Gamma_k \tag{5.2}$$

for each $k, 1 \leq k \leq M$.

Let us understand this derivation by looking at the geometry of the two-user case. Both users are associated with the same BS and there is no interference from any other cell. In Figure 5.2 we depict the analysis for two users. The SINR inequalities are (since there is only one BS we write $h_{i,1}$ as h_i):

$$h_1 p_1 - \Gamma_1 h_2 p_2 \geq \Gamma_1 N_0 W$$

$$-\Gamma_2 h_1 p_1 + h_2 p_2 \geq \Gamma_2 N_0 W$$

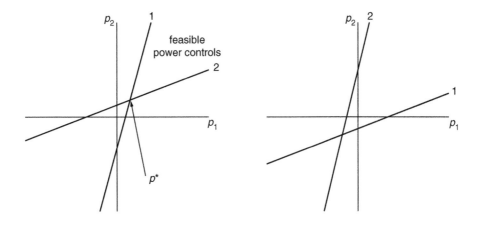

Figure 5.2 Power control feasibility for two users. The left panel shows the situation in which there are feasible power controls; then there is a power control that achieves the SINR targets with equality. The right panel shows a situation in which there is no feasible power control.

with $p_1 \geq 0, p_2 \geq 0$. These inequalities are depicted in Figure 5.2 by the lines labeled 1 and 2, for MS 1 and MS 2, respectively. The region to the right of, and below, the line labeled 1 is feasible for MS 1, and the region to the left of, and above, the line labeled 2 is feasible for MS 2. It is easy to see that there is a nonempty feasible region if

$$\frac{h_1}{\Gamma_1 h_2} > \frac{\Gamma_2 h_1}{h_2}$$

equivalently, if $\Gamma_1 \Gamma_2 < 1$. It can easily be checked that this is equivalent to

$$\frac{\Gamma_1}{1 + \Gamma_1} + \frac{\Gamma_2}{1 + \Gamma_2} < 1$$

A situation in which this holds is depicted in the left panel of Figure 5.2, and an infeasible case is depicted on the right. In the left panel of Figure 5.2 we also show a power vector p^*, which is feasible, and uses the least power in order to satisfy the SINR constraints.

An important step in analyzing cellular CDMA systems is to determine the conditions under which a set of MSs, with given locations and given demands, is admissible in the sense that an association of MSs and BSs, and corresponding power allocations, can be found so that the SINR constraints shown earlier are met. Further, given that a set of users is admissible, distributed algorithms are needed in order to determine which BSs they should associate with, and the transmission powers that should be used. A considerable part of our discussions in this chapter will be devoted to this issue.

5.2   A Simple Case: One Call Class

5.2.1   Example: Two BSs and Collocated MSs

To illustrate several issues, we now limit ourselves to one call class, for example telephony voice, which, in any case, is the single service with which cellular systems are first established. The SINR target for this class is denoted by Γ. Let us further simplify to just $N = 2$; that is, there are just two BSs, as shown in Figure 5.3. Each BS has associated with it M MSs that are situated close together in a group (we say that the MSs are *collocated*) so that the channel power gain from each group of MSs to its associated BS is h, and to the other BS is h'.[1] All the MSs use the same transmit power p, thus yielding the following simplification of (5.1):

$$\frac{hp}{(M-1)hp + Mh'p + N_0 W} \geq \Gamma \tag{5.3}$$

Writing the total received power at a BS from the MSs associated with it as $Q = Mhp$, and $v = \frac{h'}{h}$, we can rewrite the inequality as

$$\frac{hp}{(1+v)Q - hp + N_0 W} \geq \Gamma$$

which, on rearranging, yields

$$hp \geq \frac{\Gamma}{1+\Gamma}((1+v)Q + N_0 W) \tag{5.4}$$

We need to assign a transmit power p to each MS so that this inequality is satisfied. Summing the inequalities for all the M MSs associated with a BS, we obtain the following necessary condition for there to exist a feasible power $p > 0$:

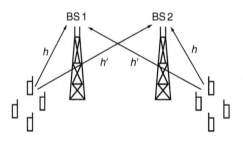

Figure 5.3 **The uplink power control problem for two cells with each of which there are *M* MSs associated. The MSs associated with each BS are collocated, and *h* and *h'* are the channel power gains, as shown.**

---

[1]Note that, for simplicity, we are only considering path loss, and not shadowing, so that, for the geometry shown in the picture, it is plausible that the two groups of MSs have the same gains to the BSs.

$$Q \geq M \frac{\Gamma}{1 + \Gamma}((1 + v)Q + N_0 W)$$

$$> M \frac{\Gamma}{1 + \Gamma}(1 + v)Q$$

where the strict inequality arises because $N_0 W > 0$. Thus, we find that a necessary condition is

$$M \frac{\Gamma}{1 + \Gamma} < \frac{1}{1 + v} \tag{5.5}$$

or, in other words, the number of admitted calls M should satisfy

$$M < \left(\frac{1}{1 + v}\right)\left(\frac{1 + \Gamma}{\Gamma}\right)$$

Notice that v is a spatial property of system, whereas Γ is a property of the calls being carried. We could interpret $\left(\frac{1}{1+v}\right)$ as a quantity of resource, and $\left(\frac{\Gamma}{1+\Gamma}\right)$ as the resource requirement per call.

Assuming the necessary condition in (5.5) to hold, take (5.4) to be an equality, substitute for Q, and then solve for p to obtain

$$p = \frac{N_0 W}{h} \frac{\frac{\Gamma}{1+\Gamma}}{1 - M(1 + v)\frac{\Gamma}{1+\Gamma}}$$

Thus, if the condition (5.5) holds, then we have positive powers that satisfy the power constraints with equality. It follows that the condition in (5.5) is both necessary and sufficient for there to exist $p > 0$. For the power allocation obtained, the value of Q is given by

$$Q = \frac{M\frac{\Gamma}{1+\Gamma}(N_0 W)}{1 - M(1 + v)\frac{\Gamma}{1+\Gamma}}$$

and the interference at a BS from MSs associated with the *other* BS is given by $I_o = vQ$.

5.2.2 Multiple BSs and Uniformly Distributed MSs

We now assume that the MSs are uniformly distributed, and the radio propagation is spatially homogeneous, and, thus, that the BSs are uniformly loaded; that is, each BS receives the same total power Q from the MSs associated with it. Then we can continue to assume that at each BS the interference received from MSs not associated with it is some factor v times Q; that is, $I_o = vQ$ for every BS. The SINR inequalities become

$$h_k p_k \geq \frac{\Gamma}{(1+\Gamma)}((1+v)Q + N_0 W) \qquad (5.6)$$

for each k, $1 \leq k \leq M$, where h_k is the channel gain of MS k to the BS with which it is associated. As before, we sum these inequalities over the MSs associated with a BS to obtain the following necessary condition for a set of powers p_k, $1 \leq k \leq M$, to exist:

$$Q \geq M \left(\frac{\Gamma}{1+\Gamma} \right) ((1+v)Q + N_0 W)$$

But then, noting that all the terms on the right are positive, and hence lower bounding this expression, we see that it is necessary that

$$Q > M \left(\frac{\Gamma}{1+\Gamma} \right) (1+v)Q$$

It follows that a necessary condition for the existence of a set of powers p_k, $1 \leq k \leq M$, that satisfy the SINR inequalities (5.6) is

$$M \left(\frac{\Gamma}{1+\Gamma} \right) < \frac{1}{1+v} \qquad (5.7)$$

Now suppose that this condition holds, by associating new calls with a BS in such a way as to ensure that the condition is not violated. Taking equalities in (5.6) and summing, we obtain the following power allocation. For each k, $1 \leq k \leq M$,

$$p_k = \left(\frac{N_0 W}{h_k} \right) \left(\frac{\frac{\Gamma}{1+\Gamma}}{1 - M(1+v) \left(\frac{\Gamma}{1+\Gamma} \right)} \right) \qquad (5.8)$$

These powers are all positive when (5.7) holds, and, hence, we have a feasible power allocation (which meets the SINR constraints with equality). For this power allocation, by setting $Q = \sum_{k=1}^{M} h_k p_k$, we see that

$$Q = \frac{M \left(\frac{\Gamma}{1+\Gamma} \right) (N_0 W)}{\left(1 - M(1+v) \frac{\Gamma}{1+\Gamma} \right)}$$

Thus, the condition expressed by (5.7) is found to be necessary and sufficient for the existence of a feasible power control, in the present setting (i.e., at a BS, the uplink interference from MSs associated with other BSs can be modeled as a factor v times the total power received at the BS from the MSs associated with it).

Discussion

a. From the previous derivation, we conclude that, in the single class case (with the spatial homogeneity assumptions we made), the following admission control will permit a feasible power allocation. A connection request is characterized by its "effective" resource requirement $\frac{\Gamma}{1+\Gamma}$. The connection is added to the existing calls at a BS if and only if the following inequality is satisfied:

$$\text{Number of existing connections} \times \frac{\Gamma}{1+\Gamma} + \frac{\Gamma}{1+\Gamma} < \frac{1}{1+\nu} \qquad (5.9)$$

where ν is a spatial parameter that captures *other cell interference*. We will discuss how ν can be derived later in this chapter.

b. We notice that a large value of Γ reduces the number of calls we can carry. How is the value of Γ determined? Suppose we wish to carry a new enhanced quality voice call, streaming audio call, or streaming video call using the CDMA access system just described. The source coding scheme that is used will determine the aggregate bit rate R that needs to carried. Also, sophisticated source coders will encode the source into bit streams of varying degrees of importance (called Class A, B, and C bits in some speech coders). When bit errors occur, a radio link layer protocol can recover the CDMA bursts containing the errored bits, but this recovery takes time, which adds to the end-to-end delay for the connection. After some number of attempts, bits may need to be discarded, in the hope that the decoder can reconstruct the speech or audio with some desirable quality using the received bits. It is thus clear that, for each coder, there will be a threshold bit error rate above which the speech (or audio or video) quality will not be acceptable. Finally, the physical layer (PHY) techniques employed (e.g., exploitation of multipath diversity (via a Rake receiver), interference cancellation, multiuser detection) will determine the $\frac{E_b}{N_0}$, γ, required to provide the desired bit error rate to the connection (see the discussions in Chapter 2). More sophisticated PHY techniques will result in a lower value of γ, hence a lower value of $\Gamma = \gamma \frac{R}{R_c}$, and thus a lower resource requirement $\left(\frac{\Gamma}{1+\Gamma}\right)$ for the connection.

c. To get a feel for the numbers, let us consider telephone quality voice over the IS 95 CDMA system. A commonly used speech coder has $R = 9.6$ Kbps. The system bandwidth is 1.25 MHz, and the chip rate is 1.2288 Mcps. Thus the processing gain is $\frac{1.2288 \times 10^6}{9.6 \times 10^3} = 128 \approx 21$ dB (i.e., $10 \log \frac{1.2288 \times 10^6}{9.6 \times 10^3} \approx 21$). It turns out that, for the PHY techniques employed in the IS 95 standard, the target $\frac{E_b}{N_0}$ for this speech coder is 6 dB. It follows that the target SINR, $\Gamma \left(= \frac{\gamma}{\frac{R_c}{R}}\right)$, is $6 - 21 = -15$ dB (in fact, $\Gamma = \frac{1}{32}$). The target SINR of -15 dB should be contrasted with narrowband systems such as FDM-TDMA (see Chapter 4) where the target SINR could be as high as 8 to 10 dB.

d. We can see that the interference $I_o = \nu Q$ can be reduced by exploiting voice activity detection; the voice call transmits only when carrying actual speech, and turns off during silence periods, thereby reducing the other cell interference for a given number of accepted calls. Note that, roughly, this will result in the factor $1 + \nu$ getting multiplied by the voice activity factor (note that even the intracell interference reduces by the voice activity factor, hence we multiply $1 + \nu$ by this factor). The voice activity factor is typically 0.4 to 0.5, and thus this technique results in the capacity being increased by a multiplicative factor of 2 to 2.5.

5.2.3 Other Cell Interference: Hard and Soft Handover

Let us examine the form of the power allocation proposed in (5.8). We will now refer to a BS and the region in which MSs will normally associate with this BS as a *cell*. Notice that, in the one class case, with homogeneous interference at each cell, the received powers $h_k p_k$ are all equal at every BS. Thus, when the entire system carries just one type of call (as is the case in the early deployment of all cellular telephony systems), then the powers of all MSs, in any cell, need to be controlled in such a way that *the received powers at their respective BSs are all equal.*

If the power to be received at each BS from any MS has to be the same, then, in order that an MS uses the least transmit power it should associate with the geographically nearest BS (assuming only deterministic path loss proportional to an inverse power of the distance). For a location with coordinates (x, y) let $r_j(x, y)$ denote the distance of BS j from the location (x, y). We can say that the default *coverage area* of BS j is all (x, y) such that $r_j(x, y) < r_k(x, y)$ for every other BS k. If this is done, then the coverage areas are actually so-called *Voronoi cells*, which are uniquely determined by the BS locations.

We obtained the power allocation shown in (5.8), assuming that the other cell interference factor ν was somehow given. The power allocation actions in one cell, however, affect the other-cell interference seen by other cells. For example, if MS k is at the fringe of the coverage area of the BS with which it is associated then the value of h_k will be small, thus requiring a large value of p_k (see (5.8)). But this large value of p_k will result in a higher level of other-cell interference at neighboring BSs. In fact, it is possible that the MS may have a better channel to a neighboring BS than to the one with which it is associated. If on the basis of this better channel to the neighboring BS the MS is handed over to that BS, then we say that we are performing *soft* handovers. On the other hand if the region is demarcated into coverage areas on the basis of path loss measurements, and MSs are associated with a BS so long as they are in its coverage area, then we say that we are performing *hard* handovers.

We will carry out an interference analysis, assuming that all calls are of the same type, and hence (for a spatially homogeneous system, as assumed in our simple analysis earlier) the target received power from an MS is the same at every BS. This analysis will yield the value of ν for hard hand-off and for soft hand-off.

With this we will have all the ingredients to perform a quantitative evaluation of the system capacity as given by (5.7).

Let us first consider hard hand-off. Let S_r denote the target uplink received power at a BS from any MS associated with it. In Figure 5.4 we show an MS at the location (x, y) in the coverage area of BS 1. The distance of the MS (located at (x, y)) to BS 1 is $r_1(x, y)$, and to BS 0 is $r_0(x, y)$. Modeling the power law path loss and shadowing, it can be seen that the interference power, say, S_0, at BS 0 due to the MS at location (x, y) is given by

$$S_0 = S_r \left(\frac{r_1(x, y)}{r_0(x, y)} \right)^{\eta} \frac{10^{(\xi_1(x, y) + \zeta(x, y))/10}}{10^{(\xi_0(x, y) + \zeta(x, y))/10}}$$

where η is the path loss exponent, $\xi_1(x, y), \xi_0(x, y)$, and $\zeta(x, y)$ are i.i.d. normally distributed random variables with mean 0, and variance $\frac{\sigma^2}{2}$. Here, $(\xi_1(x, y) + \zeta(x, y))$ correspond to the log-normal shadowing on the path to BS 1, and $(\xi_0(x, y) + \zeta(x, y))$ to the log-normal shadowing on the path to BS 2. The shadowing is modeled as being composed of local shadowing around the MS, $\zeta(x, y)$, and the shadowing on the two different paths, $\xi_1(x, y), \xi_2(x, y)$. The total shadowing standard deviation over each path is σ.

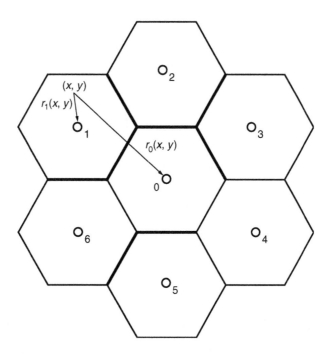

Figure 5.4 Other-cell interference with hard hand-off. An MS at the location (x, y) is power controlled by BS 1, and the power it radiates causes uplink interference at BS 0.

The previous expression can be understood as follows. Starting with the target power S_r at BS 1, we trace back to the MS to obtain its transmission power. This gives the numerator of the expression multiplying S_r. Then we obtain the interference power seen at BS 0. This is obtained by dividing by the channel attenuation along the path from (x, y) to BS 0. Note that the local shadowing terms cancel out, and, further, we assume that the distributions of $\xi_1(x, y)$ and $\xi_0(x, y)$ do not depend on the MS location (x, y). Then denoting these generic random variables by ξ_1 and ξ_0, we get

$$S_0 = S_r \left(\frac{r_1(x, y)}{r_0(x, y)} \right)^\eta \frac{10^{\xi_1/10}}{10^{\xi_0/10}}$$

Then the total expected other-cell interference at BS 0 is obtained by adding up the interference from all the other cell MSs and taking the expectation of this sum. This computation is done by assuming a uniform distribution of MSs over the coverage area, with density d MSs per unit area, and then integrating over the area outside of the cell covered by BS 0. This yields

$$I_o = S_r \, \mathsf{E}\left(10^{\frac{\xi_1 - \xi_0}{10}}\right) \int_{\{(x, y) \notin \text{Cell } 0\}} \left(\frac{r_{BS}(x, y)}{r_0(x, y)} \right)^\eta d \, dxdy \qquad (5.10)$$

where $r_{BS}(x, y)$ denotes the distance of the location (x, y) from the BS in whose cell (x, y) lies. Clearly, the total power received at BS 0 from MSs associated with it is $Q = S_r dA$, where A is the area covered by a BS. It follows that

$$\nu = \frac{I_0}{Q} = \mathsf{E}\left(10^{\frac{\xi_1 - \xi_0}{10}}\right) \int_{\{(x, y) \notin \text{Cell } 0\}} \left(\frac{r_{BS}(x, y)}{r_0(x, y)} \right)^\eta \frac{1}{A} \, dxdy$$

It can be seen that the integral in the right-hand side of this expression does not vary with the cell radius, R. This integral can be numerically evaluated to approximately 0.44 for $\eta = 4$. Further, we observe that

$$\mathsf{E}\left(10^{\frac{\xi_1 - \xi_0}{10}}\right) = \mathsf{E}\left(e^{\frac{\ln 10}{10}(\xi_1 - \xi_0)}\right)$$

$$= e^{\left(\frac{\sigma^2}{2}\left(\frac{\ln 10}{10}\right)^2\right)}$$

where we use the fact that $\xi_1 - \xi_0$ is normally distributed with mean 0 and variance σ^2. For $\sigma = 8$ dB and $\eta = 4$, we then find that $\nu = e^{\left(\frac{\sigma^2}{2}\left(\frac{\ln 10}{10}\right)^2\right)} \times 0.44 = 2.38$. Thus, with an 8 dB standard deviation for the shadowing, and a path loss exponent of 4, the other-cell interference is 2.38 times the power received from MSs within the cell. We notice that with $\sigma = 0$ we have $\nu = 0.44$, for $\eta = 4$.

Let us now turn to the same analysis with soft handovers. Figure 5.5 depicts the concept. An MS at location (x, y) is power controlled by either BS 1 or BS 0. What this means is that the MS will use a transmit power that is the smaller of the two values required to achieve a received signal power of S_r at either of the two BSs. In the situation of random shadowing, this will result in the MS causing less interference than if it was dedicated to the more proximate of the two BSs. Thus, with random shadowing, an MS may get power controlled by a geographically farther away BS.

For two neighboring BSs i and j (e.g., BS 1 and BS 0), and for a location (x, y) in the region where an MS chooses between either of them (e.g., (x, y) in Figure 5.5 is power controlled by BS 1 or BS 0), define

$$\alpha_{i,j}(x, y) = \frac{(r_i(x, y))^\eta \; 10^{\xi_i(x, y)/10}}{(r_j(x, y))^\eta \; 10^{\xi_j(x, y)/10}}$$

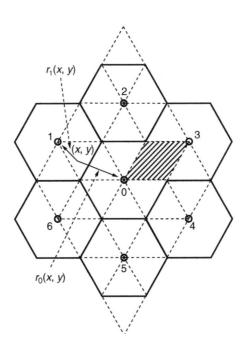

Figure 5.5 In soft hand-off, an MS is power controlled by the best of two or more BSs. This diagram shows an MS located at position (x, y) being power controlled by the best of BS 1 or BS 0. Each diamond shaped area, with a BS at each end of its long diagonal, shows the area in which an MS would be power controlled by either of those two BSs. By $\diamond_{i,j}$ we will mean the diamond between BS i and BS j; as an illustration, $\diamond_{0,3}$ is shown shaded.

where $r_i(x, y)$ and $r_j(x, y)$ are the distances of (x, y) from BS i and BS j, respectively, and $\xi_i(x, y)$ (resp. $\xi_j(x, y)$) corresponds to log-normal shadowing near BS i (resp. BS j). As before, the distributions of these shadowing random variables will be taken to be independent of the location (x, y). Also $\xi_i(x, y)$ and $\xi_j(x, y)$ are assumed statistically independent in the following analysis. We can see that $\alpha_{i,j}(x, y)$ is the relative attenuation from (x, y) to the BSs i and j; $\alpha_{i,j}(x, y) > 1$ implies that the power attenuation from the location (x, y) to BS i is larger (than that to BS j) and hence an MS located at the position (x, y) should be power controlled by j, since this will require the MS to use less transmission power. As before, let d be the density of mobiles per unit of the system coverage area. It can then be seen that the total power received at BS 0 (i.e., intracell power and other-cell interference) is given by

$$6 \int \int_{\Diamond_{0,1}} \left(S_r 1_{\{\alpha_{0,1}(x,y) \le 1\}} + S_r \alpha_{1,0}(x, y)\, 1_{\{\alpha_{1,0}(x, y) < 1\}} \right) d\, dA(x, y)$$

$$+ \sum_{\{\{i,j\}:\, i \ne 0, j \ne 0,\, i,j \text{ are neighbors}\}}$$

$$\times \int \int_{\Diamond_{i,j}} \left(S_r \alpha_{i,0}(x, y)\, 1_{\{\alpha_{i,j}(x,y) \le 1\}} + S_r \alpha_{j,0}(x, y)\, 1_{\{\alpha_{j,i}(x,y) < 1\}} \right) d\, dA(x, y)$$

where $1_{\{\cdot\}}$ denotes an indicator function. This expression can be understood as follows. There are six first tier "diamonds" that include BS 0, the one between BS 1 and BS 0 being denoted by $\Diamond_{0,1}$. Each of these yields a term given by the first integral in the expression. Considering BS 1, if $\alpha_{0,1}(x, y) \le 1$ then an MS at (x, y) is power controlled by BS 0, and hence BS 0 receives power S_r from such an MS. On the other hand, if $\{\alpha_{1,0}(x, y) < 1\}$ then an MS at (x, y) is power controlled by BS 1, and then BS 0 receives interference power $S_r \alpha_{1,0}(x, y)$. This integral is over $\Diamond_{0,1}$, and $dA(x, y)$ denotes the infinitesimal area around (x, y). Then there are terms for every other pair of neighboring BSs, not involving BS 0. Each such neighboring BS pair gives an integral of the form shown in the second term.

Using the fact that $\xi_i(x, y) - \xi_j(x, y)$ is normally distributed with 0 mean and variance σ^2, where σ is the shadowing standard deviation, the expectation of the preceding expression can be numerically evaluated to yield $Q + I_o$ at BS 0. On doing this, for $\sigma = 0$ and $\eta = 4$ we find that $\nu = 0.44$, which is the same as obtained for hard handover, earlier, because, without shadowing, the most proximate BS always power controls an MS. With $\sigma = 8$ dB, we find that $\nu = 0.77$ with soft hand-off, in contrast with the value of 2.38 with hard hand-off. It follows that there is a substantial reduction in intercell interference if soft handovers are employed.

Finally, we note that in this discussion we have assumed that an MS is power controlled by either of two BSs. This idea can be generalized. For example, with reference to Figure 5.5, an MS in the triangle formed by BSs 0, 1, and 2 can be power controlled by the best among the three. Even more general soft hand-off strategies can be considered.

5.2.4 System Capacity for Voice Calls

In the discussion at the end of Section 5.2.2, we calculated the value of Γ for a standard voice coder that is used in the IS 95 cellular CDMA system. We are now in a position to provide an estimate of the call carrying capacity of the system. If power control can be accurately performed, and voice activity is *not* exploited, then we see that the number of calls, M, that can be admitted into a cell is bounded as follows

$$M \le \left\lfloor \frac{1}{1+v} \frac{1+\Gamma}{\Gamma} \right\rfloor$$

With $\Gamma = \frac{1}{32}$ (a value we motivated earlier), $\sigma = 8$ dB, and $\eta = 4$, we use the values of v calculated earlier. For hard handover, we find that M is bounded by nine calls; with soft handover we see that M is bounded by 18 calls. Thus, for the parameters assumed, soft handover doubles the number of calls that can be admitted. In practice, power control is performed in a feedback loop between the MSs and the BSs. Thus, there are inaccuracies due to feedback delay and coarse control. This results in the bound on M being reduced by a *power control inaccuracy factor*, whose calculation is shown in [132]. Let us denote by K the resulting bound on the number of calls that can be admitted.

This analysis ensures that when a call is admitted it obtains the desired bit rate and BER; the *in-call performance* is assured. But we are also interested in the probability of blocking new calls. The *Erlang capacity* is defined as the Erlang load up to which the blocking probability is less than some target, say 0.01.

Now there are two alternatives. We can admit arrivals into the cell until the number of calls is K and then block any additional arrivals. Alternatively, if we want to exploit the fact that voice calls alternate between speech and silence, we can model the activity of the ith admitted call by a 0-1 random variable, V_i, where $V_i = 1$ with probability a (the fraction of time that a voice call is active) and $V_i = 0$, otherwise. When M calls are admitted, we take the random variables $V_1, V_2, \ldots, V_M$, to be i.i.d. with the same Bernoulli distribution. Then, since calls generate power only when they are active, we can admit M_ϵ calls so long as the following criterion is satisfied:

$$\Pr\left(\sum_{i=1}^{M_\epsilon} V_i > K \right) < \epsilon$$

where $\epsilon > 0$ is a suitably chosen *outage probability*. This means that if we admit M_ϵ calls then, over the times during which M_ϵ calls have been admitted, during less than a fraction ϵ of the time the SINR constraints would be violated. If ϵ is small enough then the infrequent outages may be imperceptible to users. Evidently, $M_\epsilon > K$ and hence, the Erlang capacity for a given call blocking probability increases by taking the second approach. The value M_ϵ is called *soft* capacity, as opposed to K being called the *hard* capacity. If only K calls are admitted, the in-call performance has a "hard" assurance, but if M_ϵ calls are admitted then the in-call performance has a "soft" assurance, as the performance can be violated with a small probability.

The SINR analysis assures the QoS within a call (in terms of the voice bit rate, and the quality deterioration due to bit errors). In addition, call blocking probability is also a QoS requirement (e.g., a typical call blocking objective could be 1%), and must be assured by making a good assessment of the expected Erlang load on the system, once it is deployed. Given the admission control limit M_ϵ and the offered Erlang load, if the blocking probability is higher than the objective, then the operator has the following alternatives:

- Deploy more BSs, thus reducing the coverage of each BS and thereby reducing the Erlang load on the cells. There is a limit to how much the cells can be shrunk in this way, while retaining the advantages of CDMA.

- As in the case of FDM-TDMA cellular systems (see Section 4.3.3), cell sectorization and directional antennas can substantially reduce intercell interference, and thereby improve system capacity.

- Better CDMA receiver techniques can be employed, thus reducing the value of Γ, and hence the resource requirement per call.

5.3 Admission Control of Multiclass Calls

We return now to the general SINR inequalities shown in (5.2). Consider BS j, and assume that other-cell interference at BS j is $I_{o,j} = \nu Q$. Then, proceeding in the same way as we did to obtain (5.7) and (5.8), we obtain the following admission control condition for multiclass calls. Calls with SINR requirements $\Gamma_1, \Gamma_2, \ldots, \Gamma_k, \ldots, \Gamma_M$ can be associated with BS j, provided

$$\sum_{k=1}^{M} \frac{\Gamma_k}{1 + \Gamma_k} < \frac{1}{1 + \nu} \tag{5.11}$$

Then the following is a feasible power allocation:

$$p_k = \left(\frac{N_0 W}{h_{k,j}} \right) \left(\frac{\frac{\Gamma_k}{1+\Gamma_k}}{1 - (1 + \nu) \sum_{i=1}^{M} \left(\frac{\Gamma_k}{1+\Gamma_k} \right)} \right) \tag{5.12}$$

These powers are all positive when the condition in (5.11) holds.

5.3.1 Hard and Soft Admission Control

Equation (5.11) shows *hard admission control*. Thus, suppose there are two classes of calls (Class 1 and Class 2) that are being handled by the system. Denote their resource requirements by $g_1 = \frac{\Gamma_1}{1+\Gamma_1}$ and $g_2 = \frac{\Gamma_2}{1+\Gamma_2}$. Then the admission control in (5.11) becomes the following. If there are already n_1 calls of Class 1 and n_2 calls of Class 2 associated with a BS, then admit a call of Class $i \in \{1, 2\}$, if and only if

$$n_1 g_1 + n_2 g_2 + g_i < \frac{1}{1+v} \tag{5.13}$$

Define $S = \{(n_1, n_2) : n_1 g_1 + n_2 g_2 < \frac{1}{1+v}\}$. If calls of each class arrive in independent Poisson processes of rates λ_1 and λ_2, and the times for which calls stay in the system are exponentially distributed, with rates μ_1 and μ_2, and independent from connection to connection, then $(X_1(t), X_2(t)), t \geq 0$, is a Markov chain on S. As shown at length in [89], analysis of this Markov chain yields the probability that a connection of each class is blocked. If the resulting blocking probability is not acceptable to the customers of the system then the arrival rates will need to be reduced. This can be achieved, to some extent, by reducing the area covered by each cell.

   Another approach for capacity enhancement (already discussed for the single class voice case) is to employ *soft admission control*; that is, when a connection is not active (as, for example, when a party is listening in a speech telephony connection) then the term corresponding to that flow is set to 0 in the left-hand side of (5.11). Define, for a connection of type k, the random process $Z_k(t) = g_k$ when the call is active at instant t, and $Z_k(t) = 0$ when the call is inactive. Assume that this is a stationary random process, and let Z_k denote a random variable with the marginal distribution of $Z_k(t)$. With p_k denoting the fraction of time Connection k is active, we have $Z_k = 1$ with probability p_k, and $Z_k = 0$ otherwise. We may then say that a set of connections $(1, 2, 3, \ldots, n)$ is admissible if

$$\Pr\left(\sum_{k=1}^{n} Z_k \geq \frac{1}{1+v}\right) \leq \epsilon \tag{5.14}$$

where ϵ is the probability of *outage*, the fraction of time that the system violates the connection QoS requirements. During such times the SINR targets of calls will not be met and users will experience poor in-call QoS.

5.3.2 Soft Admission Control Using Chernoff's Bound

Let us limit our discussion now to the situation in which there are K classes of calls, with resource requirements $g_k, 1 \leq k \leq K$. For example, two of the classes could be speech telephony with two different types of coders, one class could be streaming audio, and the other could be streaming video. If n_k calls of Class k are to be admitted, (5.14) is equivalent to

$$\Pr\left(\sum_{k=1}^{K}\sum_{i=1}^{n_k} Z_{k,i} \geq a\right) \leq \epsilon$$

where (1) $Z_{k,i}$ is the resource requirement random variable of Connection i of Class k, and (2) we have defined $a := \frac{1}{1+v}$, for notational convenience. Although it is possible to implement such an admission control by storing a look-up table in the admission controller, such a table needs to be updated each time that a new call type is included. The following approximate calculation yields an additive call admission rule, which has been found to be quite efficient.

We assume that, for each $k, 1 \leq k \leq K$, $Z_{k,i}$, $1 \leq i \leq n_k$, are independent and identically distributed, and all the $Z_{k,i}$ are mutually independent. Now, for any $\theta \geq 0$, let us define $M_k(\theta) = \ln \mathsf{E}(e^{\theta Z_{k,1}})$, and then use Chernoff's Bound (see Appendix B) to obtain

$$\Pr\left(\sum_{k=1}^{K}\sum_{i=1}^{n_k} Z_{k,i}(t) \geq a\right) \leq e^{-\theta a}e^{\sum_{k=1}^{K} n_k M_k(\theta)}$$

$$= e^{-(\theta a - \sum_{k=1}^{K} n_k M_k(\theta))}$$

Because this is true for each $\theta \geq 0$, and e^x is increasing in x, we can further write

$$\Pr\left(\sum_{k=1}^{K}\sum_{i=1}^{n_k} Z_{k,i}(t) \geq a\right) \leq e^{\inf_{\theta \geq 0}(-(\theta a - \sum_{k=1}^{K} n_k M_k(\theta)))}$$

$$= e^{-\sup_{\theta \geq 0}(\theta a - \sum_{k=1}^{K} n_k M_k(\theta))}$$

Exercise 5.1

For the two valued Z_k random variables shown earlier, show that

$$\sum_{k=1}^{K} n_k M_k(\theta) = \sum_{k=1}^{K} n_k \ln\left(1 - p_k\left(1 - e^{\theta g_k}\right)\right)$$

Hence observe the following (see Figure 5.6 for a graphical depiction)

(1) $\sum_{k=1}^{K} n_k M_k(\theta) \sim_{\theta \to \infty} \sum_{k=1}^{K}(n_k g_k \theta + n_k \ln p_k)$; that is, as $\theta \to \infty$, $\sum_{k=1}^{K} n_k M_k(\theta)$ is an affine function of θ with slope $\sum_{k=1}^{K} n_k g_k$.

(2) $\lim_{\theta \to -\infty} \sum_{k=1}^{K} n_k M_k(\theta) = \sum_{k=1}^{K} n_k \ln(1 - p_k)$.

(3) The derivative of $\sum_{k=1}^{K} n_k M_k(\theta)$ with respect to θ at $\theta = 0$ is given by $\sum_{k=1}^{K} n_k p_k g_k$ the mean resource requirement from the sources.

Let us assume that $\sum_{k=1}^{K} n_k E(Z_{k,1}) < a$, which implies that the mean resource requirement is less than a, or, in other words, the system in *not overloaded*. We can then observe that

$$\theta a - \sum_{k=1}^{K} n_k M_k(\theta) = \theta a - \sum_{k=1}^{K} n_k \ln E\left(e^{\theta Z_{k,1}}\right)$$

$$\leq \theta a - \sum_{k=1}^{K} n_k \ln e^{\theta E(Z_{k,1})}$$

$$= \theta\left(a - \sum_{k=1}^{K} n_k E(Z_{k,1})\right)$$

where we have used Jensen's Inequality (see Appendix B) in the second step (the exponential being a convex function). Under the "not overloaded" assumption, the right-hand side of the last equality is ≤ 0 whenever $\theta \leq 0$. It follows that $\sup_{\theta \in \mathbb{R}}(\theta a - \sum_{k=1}^{K} n_k M_k(\theta)) = \sup_{\theta \geq 0}(\theta a - \sum_{k=1}^{K} n_k M_k(\theta))$, and we can write

$$\Pr\left(\sum_{k=1}^{K}\sum_{i=1}^{n_k} Z_{k,i}(t) \geq a\right) \leq e^{-\sup_{\theta \in \mathbb{R}}\left(\theta a - \sum_{k=1}^{K} n_k M_k(\theta)\right)}$$

Let us write the outage probability as $\epsilon = e^{-\delta}$, for a suitably chosen δ. Then the outage probability will be met provided we ensure that the vector of the numbers admitted calls, $\mathbf{n} = (n_1, n_2, \ldots, n_K)$, is in the region defined by

$$\left\{\mathbf{n} : n_k \geq 0, 0 \leq k \leq K, \sup_{\theta \in \mathbb{R}}\left(\theta a - \sum_{k=1}^{K} n_k M_k(\theta)\right) \geq \delta\right\} \tag{5.15}$$

As shown in Figure 5.6 this admission control basically means that the vector $\mathbf{n}$ should be such that the maximum vertical gap between the line θa and the curve $\sum_{k=1}^{K} n_k M_k(\theta)$ should be greater than δ. Notice that if this is done with $\delta > 0$ then the "not overloaded" condition assumed earlier is also met.

Exercise 5.2
Show that the function $f(\mathbf{x}) = \sup_{\theta \in \mathbb{R}}\left(\theta a - \sum_{k=1}^{K} x_k M_k(\theta)\right)$ is convex, and, hence, that the boundary

$$\left\{\mathbf{x} : x_k \geq 0, 0 \leq k \leq K, \sup_{\theta \in \mathbb{R}}\left(\theta a - \sum_{k=1}^{K} x_k M_k(\theta)\right) = \delta\right\}$$

is convex.

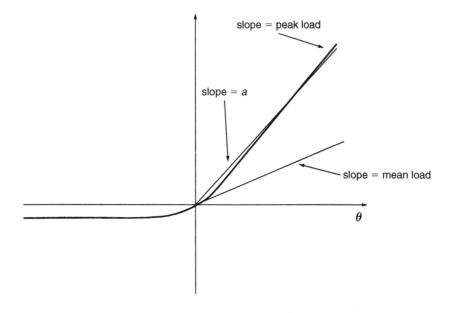

Figure 5.6 Depiction of the Chernoff's bound based admission control. The thick curve is a sketch of $\sum_{k=1}^{K} n_k M_k(\theta)$, which depends on the vector n of calls admitted. The slope of this curve at $\theta = 0$ is the mean resource requirement, the asymptotic slope as $\theta \to \infty$ is the peak resource requirement, and the line with slope a corresponds to the resource.

Consider a point n^* on the boundary of the set $\{n : n_k \geq 0, 1 \leq k \leq K,$ $\sup_{\theta \in \mathbb{R}} (\theta a - \sum_{k=1}^{K} n_k M_k(\theta)) \geq \delta\}$, and let $\theta_{n^*}^*$ be the value of θ that achieves the sup. Thus $(\theta_{n^*}^* a - \sum_{k=1}^{K} n_k^* M_k(\theta_{n^*}^*)) = \delta$. Now consider the admission control region

$$\left\{ n : \left(a - \frac{\delta}{\theta_{n^*}^*} \right) \geq \sum_{k=1}^{K} n_k \frac{M_k(\theta_{n^*}^*)}{\theta_{n^*}^*} \right\} \tag{5.16}$$

Exercise 5.3
Show that if admission control is done according to (5.16) then the QoS requirements of all the admitted calls are satisfied. (Hint: Show that an n in the set defined by (5.16) is also in the set defined by (5.15).)

Observe that the complex admission control region defined by (5.15) has been simplified to a linear region defined by (5.16). The admission control defined by (5.16) has the following interpretation. Each call of Class k requires an amount of resource $\frac{M_k(\theta^*_{n*})}{\theta^*_{n*}}$, and the total resource available is $(a - \frac{\delta}{\theta^*_{n*}})$. The reason why this construction works is that the boundary of the region defined by (5.15) is convex, and hence any hyperplane at a point on the boundary is below the boundary. It follows that the polytope below this hyperplane (and in the positive orthant) lies inside the set defined by (5.15). It has been found that the boundary of the admission region is almost a straight line, hence, the approximation is quite accurate.

5.4 Association and Power Control for Guaranteed QoS Calls

In the previous analysis we made the simplification of assuming spatial homogeneity so that we could model the interference received at a BS as a scaled total intracell received power. In this section we dispense with this assumption and consider the joint problem of power allocation and MS to BS association in a multicell system. We start with the SINR constraints in (5.1). Before we proceed, let us write these inequalities in matrix form as follows:

$$\mathbf{p} \geq \mathbf{F}\mathbf{p} + \mathbf{g}$$

where $\mathbf{p}$ is the (column) vector of powers, $\mathbf{F}$ is an $m \times m$ matrix with a zero diagonal, $F_{k,i} = \frac{\Gamma_k h_{i,a_k}}{h_{k,a_k}}$, and $g_k = \frac{N_0 W \Gamma_k}{h_{k,a_k}}$, $1 \leq k, i \leq m$. Notice that g_k is the uplink power required at MS k if there was no interference from other users. For fixed $\mathbf{F}$ and $\mathbf{g}$, we need to know if the power allocation problem is *feasible*; that is, if the set

$$\{\mathbf{p} : \mathbf{p} \geq \mathbf{F}\mathbf{p} + \mathbf{g}\}$$

is nonempty. The answer to this question has been shown to depend on the matrix $\mathbf{F}$ and hence on the association and on the channel gains. First, we need some definitions and results from the Perron-Frobenius theory of nonnegative matrices (see this chapter's appendix). Applying these to the problem at hand, we can first establish the following property of the matrix $\mathbf{F}$.

Exercise 5.4

Show that the matrix $\mathbf{F}$ is primitive for $m \geq 3$.

Then, let ρ be the eigenvalue provided by the Perron-Frobenius Theorem 5.1. Suppose also that $\rho < 1$. Consider the matrix $\mathbf{I} - \mathbf{F}$, where $\mathbf{I}$ denotes, as usual, the $m \times m$ identity matrix. Suppose this matrix is singular, or, equivalently, its columns are linearly dependent. Then there will exist a column vector $\mathbf{x}$ (which could have positive, negative, or even zero elements, but at least one element will be nonzero) such that $(I-F)x=0$. But then 1 becomes an eigenvalue of F, contradicting the third conclusion of Theorem 5.1 that ρ (< 1) is the largest eigenvalue. It follows that $\rho < 1$ implies that $I - F$ is invertible, and, thus, there exists p^* such that

$$p^* = (I - F)^{-1}g$$

Thus, $p^* \in \{p : p \geq Fp + g\}$, and the power allocation problem for the given association and channel gains is feasible.

Our aim should be to obtain feasible power allocations that require the MSs to use as little power as possible. A power allocation p is called *Pareto* if there does not exist another power allocation p' with $p' \leq p$ with $p'_j < p_j$ for some j. Thus, a feasible power allocation is Pareto if there is no other feasible power allocation in which the power of some MS is strictly smaller, without increasing the power of some other MS. In other words, a Pareto power allocation is on the "lower left boundary" of the set of feasible power vectors.

Lemma 5.1
The power control p^* is Pareto.

Proof: Suppose that p^* is not Pareto. Then there exists a feasible power vector p such that $p \leq p^*$, with $p_j < p^*_j$ for some j, $1 \leq j \leq m$. Let $k = \arg\max_{1 \leq j \leq m} \frac{p^*_j}{p_j}$, and let $\alpha := \frac{p^*_k}{p_k}$. By the choice of p, we have $\alpha > 1$. Then, $\alpha p_k = p^*_k$ and $\alpha p \geq p^*$. Now, we can see that

$$p^* = Fp^* + g$$

$$\leq \alpha Fp + g$$

$$< \alpha(Fp + g)$$

$$\leq \alpha p$$

where the first inequality follows from $\alpha p \geq p^*$ and the fact that F is nonnegative, the strict inequality follows since $\alpha > 1$ and $g > 0$, and the last inequality follows because p is feasible. We find that $p^* < \alpha p$ (i.e., $p^*_j < \alpha p_j$, for all j, $1 \leq j \leq m$), which is a contradiction, since $\alpha p_k = p^*_k$. Thus, p^* is Pareto. ∎

At this point, we have learned that (1) if the Perron-Frobenius eigenvalue of F is less than 1, then there exists a feasible power allocation p^* such that $p^* = Fp^* + g$, and (2) that such a power allocation (i.e., p^* for which $p^* = Fp^* + g$) is Pareto.

Exercise 5.5

Show that, if the Perron-Frobenius eigenvalue of F is less than 1, then there is a unique solution of $p = Fp + g$.

We need to answer one more question. Can there be a feasible Pareto power allocation that does not satisfy $p \geq Fp + g$ with an equality? Let p be such that $p \geq Fp + g$, with strict inequality for some $j, 1 \leq j \leq m$. Consider, $p' = Fp + g$. Now, $p' \leq p$, with strict inequality for j. Further, $p' = Fp + g \geq Fp' + g$, since $p \geq p'$. Hence, even p' is feasible. It follows that p is not Pareto, since it can be *improved*. Hence, every Pareto power allocation must satisfy $p = Fp + g$.

We conclude that when the Perron-Frobenius eigenvalue of F is less than 1, then (1) there is a unique solution p^* of $p = Fp + g$, and (2) p^* is the unique Pareto power allocation.

In order to minimize their battery drain, the MSs should operate with the power vector p^*. The following is an iterative algorithm that converges to p^* starting from an initial feasible power allocation $p^{(0)}$. Since $p^{(0)}$ is feasible, we have

$$p^{(0)} \geq Fp^{(0)} + g$$

Define, for $i \geq 1$,

$$p^{(i)} = Fp^{(i-1)} + g \tag{5.17}$$

Note that this iteration is exactly the "improvement" step that was done earlier when we were showing that p^* was uniquely Pareto. Hence, if $p^{(i-1)}$ is feasible, it follows that

$$p^{(i)} \leq p^{(i-1)}$$

that is, $p^{(i)}$ is a nonnegative and nonincreasing sequence, and hence converges. Since (by nonnegativity of terms) $Fp^{(i-1)} + g \geq Fp^{(i)} + g$, it follows that $p^{(i)}$ is also feasible. Further, taking limits as $i \to \infty$ in both sides of (5.17) it follows that the iterations converge to p^*. There remains the question of how the computation on the right side of (5.17) is carried out without knowledge of F and g. It is easily seen from the derivation that led to this matrix expression that this computation can be done if, with the power vector set at $p^{(i-1)}$, the total interference plus noise experienced by the signal from each MS can be measured, as also the channel attenuation h_{k,a_k} for each k. Note that this yields the following iterative power control algorithm:

$$p_k^{(i)} = \frac{\Gamma_k}{\dfrac{h_{k,a_k}}{\sum_{\{i:1 \leq j \leq M, j \neq k\}} h_{j,a_k} p_j^{(i-1)} + N_0 W}} \tag{5.18}$$

Thus, if the channel gain h_{k,a_k} and the total interference power for MS k at BS a_k can be measured at each iteration then we obtain a distributed power control update. Observe that the iteration in (5.18) can also be written in the following form

$$p_k^{(i)} = \frac{\Gamma_k}{\text{SINR}_k^{(i-1)}} p_k^{(i-1)} \tag{5.19}$$

where $\text{SINR}_k^{(i-1)}$ is the uplink SINR for MS k in the $(i-1)$-th iteration. Thus, each MS increases or decreases its transmit power accordingly as its measured SINR is below or above target in the previous iteration.

We note that these algorithms are *synchronous* and distributed; synchronous because it is assumed that all MSs update their powers at the same instants. An asynchronous version would be more practical; indeed asynchronous distributed power control algorithms have also been studied in the literature.

Discussion

a. In the previous development we have seen that, for a given association of MSs to BSs, there is a feasible power vector (for which the SINR constraints in (5.1) are met) if the matrix $\mathbf{F}$ has its Perron-Frobenius eigenvalue less than 1. On the other hand, if there is a feasible power allocation $\mathbf{p}$ then the iteration $\mathbf{p}^{(0)} = \mathbf{p}, \mathbf{p}^{(i)} = \mathbf{F}\mathbf{p}^{(i-1)} + \mathbf{g}$ yields a sequence of feasible power controls that converges to a $\mathbf{p}^*$ that satisfies the SINR constraints with equality. Thus, the existence of a feasible power allocation implies the existence of one that achieves the SINR constraints with equality. It can be shown that such a power control is also unique (see the proof of Lemma 5.1, and Problem 5.4). This further implies that $\mathbf{I} - \mathbf{F}$ is nonsingular. For if not, then there exists a nonzero vector $\mathbf{x}$ such that $(\mathbf{I} - \mathbf{F})\mathbf{x} = 0$, and $\mathbf{x}$ can be taken to be arbitrarily small by scaling it. It follows that, if $\mathbf{p}$ is a positive solution of $(\mathbf{I} - \mathbf{F})\mathbf{p} = \mathbf{g}$, then so is $\mathbf{p} + \mathbf{x}$ for a suitably chosen $\mathbf{x}$, leading to a contradiction of the uniqueness of the equalizing power control. Further, it can be shown that, for $\mathbf{I} - \mathbf{F}$ to be nonsingular, the Perron-Frobenius eigenvalue of $\mathbf{F}$ must be less than 1 (see this chapter's appendix). Thus, this is a necessary and sufficient condition for feasibility of the SINR constraints in (5.1).

b. From the previous point we conclude that, given an association of MSs with BSs, admission control should be exercised so that the desired property of the matrix $\mathbf{F}$ is obtained. Thus, in general, the admission control should depend on the channel power gains between the MSs and the BSs. In Problem 5.6, we show that in the simple case of a single cell, the admission control criterion does not depend on the channel gains, and indeed reduces to the form shown in (5.9).

c. Note that the power control update (e.g., the one in (5.19)) requires the knowledge of the target SINR, Γ. We recall from our discussion in Chapter 2, Section 2.1.4, that, for a given bit error rate, the target SINR depends on the channel statistics (e.g., AWGN, or Rayleigh, etc.). The channel statistics in turn depend on the propagation characteristics of the environment, which cannot be accurately predicted. Hence, in practice, Γ (or, γ) is really not known in advance. The approach is to use a nominal value of target SINR, and at the BS determine if the target BER is being violated. This is then used to increase or decrease the target SINR, at a slow time scale (e.g., a couple of times a second). Thus, this slow *outer loop* provides the SINR target for the fast iterative power control discussed earlier. ∎

This discussion showed how the optimal power allocation can be achieved for a given association. However, there are several alternative associations and the complete problem is to find the association that yields the smallest power allocation vector. Let $\mathcal{A}$ denote the set of all feasible associations—those for which the set of feasible power allocations is nonempty. Let $\mathbf{p}^*(A)$ denote the optimal power allocation for the association A. It can be shown that among all the feasible associations $A \in \mathcal{A}$, there is an association A^* such that $\mathbf{p}^*(A^*) \leq \mathbf{p}^*(A)$. There is an iterative distributed algorithm for achieving this optimal association and the corresponding optimal power allocation.

We have discussed uplink power allocation. Let us now turn to a discussion of some issues in downlink power allocation. The approach is to allocate to each BS a certain amount of average power budget, which the BS then allocates to the MSs that associate with it. Since transmissions from a BS to all of its MSs are chip synchronous, intracell interference is less of a problem. However, multipath propagation does result in some intracell interference. This is because, even though the BS transmits the user symbols synchronously, multipath propagation causes multiple phase shifted copies of the transmitted signal to arrive at the receiver. Thus, since cross-correlations of the spreading sequences are not perfectly zero, some residual intracell interference is obtained at the correlation receiver. Turning to intercell interference, since there is universal spectrum reuse, the power radiated by all other BSs potentially interferes with the transmission from a BS to one of its MSs. However, now there are a few large interferers rather than several small ones, and hence the interference levels can be more variable than in the uplink, and also the white Gaussian interference assumption is less valid. Yet, the analysis of downlink power allocation usually is done using the same modeling approach as discussed earlier for the uplink.

5.5 Scheduling Elastic Transfers

In the previous section we discussed the optimal association and power control problem for calls that require a guaranteed bit rate and bit error rate, such as a voice call. Cellular networks originally were designed for mobile telephony, and have been used primarily for this since their inception over two decades ago. With the rapid developments in digital communication over fading wireless channels, the most eagerly awaited service is ubiquitous wireless access to the Internet. Hence, considerable attention is being paid to high speed wireless Internet access in the next generation cellular systems. In this section we will briefly consider the problem of power control and scheduling in high speed downlink elastic data transfers in a CDMA system. Such techniques have been adopted in third-generation CDMA cellular systems. See Chapter 3 for a discussion of the concept of elastic traffic, and of the TCP protocol, which controls the elastic traffic in the Internet.

The downlink power allocation problem was discussed briefly at the end of Section 5.4. Each BS is assigned a certain total average power, P_d, which it allocates among the ongoing downlink transmissions. We use the notation defined in Section 5.4. Further, define S to be the set of all MSs, and, for $1 \leq j \leq n$, let S_j denote the set of MSs associated with BS j. The sets $S_j, 1 \leq j \leq n$, constitute a partition of S, and such a partition is equivalent to an association A. Let $m_j = |S_j|, 1 \leq j \leq n$. For $i \in S_j$, let p_i be the power assigned by BS j to MS i. Thus

$$\sum_{i \in S_j} p_i \leq P_d$$

Now, ignoring the intracell interference (see the discussion on downlink power allocation at the end of the previous section), the downlink received signal power to interference plus noise power ratio is given by

$$(SINR)_i = \frac{h_{i,a_i} p_i}{\sum_{j:1 \leq j \leq n, j \neq a_i} h_{i,j} P_d + N_0 W}$$

where the first term in the denominator is the total interference power received at MS i from the other BSs, assuming that they are all transmitting at their maximum downlink power P_d. Note that $h_{i,j}$ now denotes the power gain from BS j to MS i. For a given association, define, for $1 \leq i \leq m$,

$$\eta_i := \frac{\sum_{j:1 \leq j \leq n, j \neq a_i} h_{i,j} P_d + N_0 W}{h_{i,a_i}}$$

If MS i has to be provided the rate R_i, then, as explained in the previous sections of this chapter, the SINR target $\Gamma_i = \gamma_i \frac{R_i}{R_c}$, for User i. The power allocation needs to satisfy the following inequality:

$$\Gamma_i \eta_i \leq p_i \tag{5.20}$$

We will assume that the CDMA access link is the bottleneck on the path of the elastic transfer connection from an MS (see Section 3.4.3 for a discussion of this assumption). This permits us to assume that the queue of packets at the BS (that serves the MS) is *backlogged*. Thus, when the system allocates a rate R_i to MS i, the data actually get transferred at rate R_i. Further, since γ_i and the chip rate R_c are fixed, Γ_i is proportional to R_i, and, hence, we can think of Γ_i as equivalent to the rate performance provided to MS i. Note that the value of γ_i (i.e., the target $\frac{E_b}{N_0}$) relates to the BER. The BER in turn relates to packet error probability, which in turn affects the performance of TCP controlled transfers (see Section 3.4.3).

Note that, in (5.20), the term η_i captures the downlink interference from the other BSs. If η_i is larger, then more power will be required to obtain the same connection throughput. Under our assumptions, the value of η_i is constant, as long as the channel power gains are constant.

For a given association, we now need to obtain the power allocation that is optimal in some sense. Allocating all the downlink power from a BS to the best MS (i.e., in BS j, to the MS with index $\arg\min_{\{i \in S_j\}} \eta_i$) in that cell will maximize the overall throughput carried by the network but will provide zero throughput to several MSs. One approach is to evaluate the *utility* obtained by an MS when a certain rate is allocated to it, and then optimize the total network utility. The utility function can be chosen to capture the desired trade-off between network throughput and fairness between users.

Let $U(\cdot)$ be the utility function, so that the utility to user i is evaluated as $U(\Gamma_i)$. Let us fix an association and ask for a power allocation in each cell so that the constraints $\Gamma_i \eta_i \leq p_i$ are met for the users, and the network utility is maximized. This leads to the following optimization problem.

$$\max \sum_{i \in S_j} U(\Gamma_i) \tag{5.21}$$

subject to

$$\sum_{i \in S_j} \Gamma_i \eta_i \leq P_d$$

$$\Gamma_i \geq 0 \quad \text{for} \quad i \in S_j \tag{5.22}$$

Let us consider the specific utility $U(\Gamma_i) = \ln(\Gamma_i)$. We then have a problem of maximizing a concave function over a set of linear constraints. The KKT Theorem (see Appendix C) can be applied to obtain the following solution

$$\Gamma_i = \frac{P_d}{m_j \eta_i} \tag{5.23}$$

where we recall that $m_j = |S_j|$.[2] We see that this formulation yields a *proportionally fair* solution. Each MS obtains a throughput proportional to the best it can obtain, $\frac{P_d}{\eta_i}$.

Note that in the problem defined by (5.21) and (5.22), in each cell, some power may be allocated to every user. In order to avoid the problem of intracell interference (which we have ignored in the previous formulation) an alternative is to allocate the entire power in each cell (i.e., P_d) to a user at a time, and obtain a power allocation over the users by *time sharing*. Let ϕ_i be the fraction of time power is allocated to MS i, by the BS a_i; then, of course, $\sum_{i \in S_j} \phi_i = 1$, and we obtain the following optimization problem.

$$\max \sum_{i \in S_j} U\left(\phi_i \frac{P_d}{\eta_i}\right) \tag{5.24}$$

subject to

$$\sum_{i \in S_j} \phi_i = 1$$

$$0 \le \phi_i \le 1 \quad \text{for } i \in S_j \tag{5.25}$$

For the utility function $U(\cdot) = \ln(\cdot)$, it can be shown that this problem, too, yields the same proportionally fair solution, $\Gamma_i = \frac{P_d}{m_j \eta_i}$. This solution is implemented in the framework of (5.21) and (5.22) by always allocating to an MS $i \in S_j$ the power $p_i = \frac{P_d}{m_j}$, and, in the framework of (5.24) and (5.25), by allocating power P_d to each station associated with BS j a fraction $\frac{1}{m_j}$ of the time. The latter solution also avoids intracell interference and is the one that is preferred in practice.

One might expect that the rate allocation provided by the log-utility function yields a smaller total rate, while providing some fairness between users. On the other hand, serving the best user in each cell provides the maximum total throughput, but is very unfair. This issue is explored in Problem 5.8. For additional discussion on utility functions and fairness, see [89].

Let us now examine this time sharing solution and obtain the mean file transfer delay under a certain traffic model. Assume that the same value of $\gamma_i = \gamma$

---

[2]To see this, first note that we are maximizing a concave function subject to linear constraints. Hence, the KKT conditions are necessary and sufficient. Then consider the function $L(\lambda, \Gamma_i, i \in S_j) := \sum_{i \in S_j} \ln(\Gamma_i) - \lambda \sum_{i \in S_j} \Gamma_i \eta_i$, obtained by relaxing the constraint $\sum_{i \in S_j} \Gamma_i \eta_i \le P_d$. Now optimize this function subject to the remaining constraints $\Gamma_i \ge 0, i \in S_j$. Take partial derivatives with respect to each $\Gamma_i, i \in S_j$, and set these derivatives equal to zero. These equations will yield $\lambda = \frac{1}{\Gamma_i \eta_i}, i \in S_j$. Then taking an equality in the constraint $\sum_{i \in S_j} \Gamma_i \eta_i \le P_d$, obtain $\lambda = \frac{P_d}{m_j}$. Finally, verify that this value of λ, along with $\Gamma_i = \frac{\lambda}{\eta_i}, i \in S_j$, satisfy the KKT conditions.

is required for all users. When a user is being served (and is therefore allocated the full downlink power P_d in its cell), the user receives a downlink physical bit rate of $R_i = \frac{R_c P_d}{\gamma \eta_i}$. If the γ is appropriately chosen, then the TCP packet loss probability will be small and the TCP throughput will be close to R_i (see Section 3.4.3); let us assume this to be the case. Thus if a file of size V bits has to be downloaded by MS i, and if the MS receives the transfer rate R_i (bps), then it will take $\frac{V}{R_i}$ seconds to download the file. Here again we are assuming that the transfer is backlogged at the BS.

In order to obtain a simple analytical model, we assume that there is a large population of MSs associated with BS j. These MSs can be partitioned into sets, with each set containing a large number of MSs, such that all MSs in a set obtain the same downlink rate R_k, for each $k, 1 \leq k \leq K$. For example, such sets could be obtained by partitioning the coverage area of BS j into K concentric rings, such that all MSs in a ring obtain the same value of R_k (this would be true if they all have the same path loss $h_{i,j}$, and the interference from the other BSs is the same for all MSs). Let the aggregate arrivals of transfer requests from all the MSs in the k-th set constitute a Poisson process of rate λ_k, $1 \leq k \leq K$. The transfer requests are of random sizes (in bits), denoted by the random variable V, which has the cumulative distribution function $F(v)$. Let, $F_k(x) := F(R_k x), x \geq 0$, and $F_k(x) = 0, x < 0$. Notice that $F_k(x)$ is the *distribution of time* taken to transfer a file requested by an MS in the k-th set, if the BS power is dedicated to this MS. Let T_k denote the expectation of the distribution $F_k(x), 1 \leq k \leq K$; T_k is the mean time taken to complete a transfer to an MS in the k-th set if the BS power is dedicated to it, where the mean is taken over successive file transfers.

Let $\lambda = \sum_{k=1}^{K} \lambda_k$. Then transfer requests arrive at rate λ, and a request is from an MS in the k-th set with probability $\frac{\lambda_k}{\lambda}$. Let

$$T = \sum_{k=1}^{K} \frac{\lambda_k}{\lambda} T_k$$

T is the average time to transfer a file if the BS power is dedicated to this transfer, where the average is over MSs and over transfers.

Now the file transfer requests from the MSs in the k-th set can be viewed as bringing an amount of time distributed as $F_k(t)$ to be served by the downlink CDMA "server." This is depicted in Figure 5.7, where the horizontal bars represent the amount of time remaining for each transfer. The CDMA server then serves these time requirements in a round-robin fashion (as suggested by the solution to the optimization problem defined by (5.24) and (5.25)). Since the number of MSs is large, we assume that from each MS there is never more than one ongoing transfer. With this assumption, each ongoing transfer is served in a round-robin manner for an equal fraction of the time (if the MSs are served for equal fractions of the time). As the service quantum goes to 0, we obtain the standard M/G/1 Processor Sharing (PS) model (see Appendix D, Section D.5.2). If $N(t), t \geq 0$, is

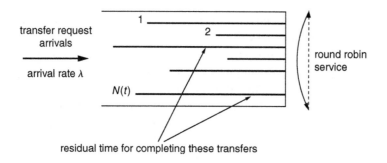

transfer request arrivals

arrival rate λ

round robin service

$N(t)$

residual time for completing these transfers

Figure 5.7 The model for downlink file transfers. The bars depict the remaining transfer times of the $N(t)$ files that are currently in transfer. The transfer times are served in a round-robin manner. As the round-robin quantum goes to 0 we obtain a processor sharing model.

the number of ongoing transfers at time t, then it can be shown that this process is stable when $\lambda T < 1$; the average amount of transfer time brought in per second by the users is less than 1 second. Then the mean transfer delay can be seen to be

$$\frac{T}{1 - \lambda T} \qquad (5.26)$$

where we have used Little's Theorem (see, Appendix D, Theorem D.10), and the fact that the stationary distribution of the number of customers in an M/G/1 PS queue is the same as that of the M/M/1 PS queue, which in turn is the same as that of the M/M/1 queue (see Appendix D, Section D.5.2).

Notice, from (5.26), that the mean transfer delay becomes very large as λT becomes close to 1 (though less than 1). Thus, from the point of view of traffic engineering, we will need to limit how close λT is to 1 in order to provide some mean transfer delay guarantee to the users. This can be done by improved physical layer techniques that result in a reduction of γ, for then, for a given power P_d, the MSs get larger download rates. For a given system, the physical layer techniques are fixed, and load control, or reduction of cell coverage, remain the only alternatives for managing large transfer delays.

5.6   CDMA-Based 2G and 3G Cellular Systems

The first generation of cellular networks were based on analog FDM. The second-generation systems used digital modulation, and there have been two competing physical layer technologies: FDM-TDMA (e.g., GSM systems) and CDMA (e.g., the IS-95 standard, now also called cdmaOne). The IS-95 standard was adopted in 1993, and was deployed commercially from 1994. We have already seen examples based on IS-95 parameters in the course of this chapter. The signals are spread over a bandwidth of 1.25 MHz, and the chip rate is 1.2288 Mcps (mega chips per second). The uplink and downlink are frequency division duplexed, with two 1.25 MHz allocations that are 45 MHz apart. The uplink band is 824–849 MHz, and the downlink band is 869–894 MHz. We notice that $1.2288 \times 10^6 = 128 \times 9600$. Indeed, 9600 bps is the maximum data rate that is provided, and this yields a processing gain of 128. Other data rates that can be provided are 1200, 2400, and 4800 at processing gains of 1024, 512, and 256. Higher data rates were defined later so as to accommodate higher quality voice coders.

Although, there were both FDM-TDMA and CDMA systems in the second generation, much of the standardization activity in third-generation systems has been in CDMA systems. The rate flexibility provided by the CDMA physical layer is useful when carrying a variety of multirate services. WCDMA (wideband CDMA) has emerged as the most widely adopted third generation air interface. The first full standard was released in 1999, and services began in some countries in 2001. Europe and many countries in East Asia have standardized on a common band for WCDMA deployment: 1920–1980 MHz for the uplink and 2110–2170 MHz for the downlink, for frequency division duplex operation. In the standardized WCDMA system, the user signals are spread over 5 MHz (i.e., 4 times that of IS-95). The chip rate is 3.84 Mcps. Because of the higher chip rate, the time resolution becomes finer, and hence Rake receivers can resolve multiple paths even in small cells. A variety of speech coders have been defined to be carried over WCDMA systems, with source rates ranging from 4.75 Kbps to 12.2 Kbps. A variety of other services are defined, and bit rates up to 2 Mbps can be assigned to a connection. Thus WCDMA systems bring cellular networks closer to being able to provide shared mobile broadband services.

5.7 Notes on the Literature

The other-cell interference analysis in this chapter has been taken from [132], where Viterbi has used various mathematical models to analyze several technical issues in cellular CDMA, such as Erlang capacity, hard and soft handovers, other-cell interference, and imperfect power control. Holma and Toskala [58] have provided an updated account of the emergence of CDMA cellular standards from the second generation to the third generation.

The analytical approach to soft admission control of multiclass calls, based on Chernoff's bound, was developed by Kelly [74]. Earlier it was reported by Hui [60] that the admission region has an almost straight line boundary; hence, Kelly's approach would work well. Evans and Everitt [32] have also reported various analytical approximations for call admission control.

The iterative power control for multiple MSs and BSs was proposed by Foschini and Miljanic [35]. A review of the theory and the algorithm were provided by Holliday et al. [57], who also develop a power control algorithm for the realistic situation in which the channel gains vary over time. The joint problem of BS selection and power control was studied by Hanly [52]. Yates [138] has developed a generalized framework for power control algorithms. Other power controls were proposed by Zander [142] and Mitra and Morrison [101]. A survey of the area was provided by Bambos [4].

In [8], Bender et al. have explained the ideas behind high data rate downlink Internet access in cellular CDMA systems (called CDMA/HDR technology). The processor-sharing queue based model for downlink high-speed TCP transfers over systems such as CDMA/HDR was studied by Bonald and Proutiere [15].

5.8   Appendix: Perron-Frobenius Theory

The standard source for the following material is Seneta's book [120]. A square $(m \times m)$ nonnegative matrix $\mathbf{M}$ is *primitive* if there exists a $k \geq 1$ such that $\mathbf{M}^k > 0$, where 0 is the $m \times m$ matrix of zeros, i.e., $[\mathbf{M}^k]_{i,j} > 0$ for every i, j.

Obviously, $\mathbf{M} = \begin{bmatrix} 1 & 0 \\ 0 & 1 \end{bmatrix}$ is not primitive. Check that even $\mathbf{M} = \begin{bmatrix} 0 & 1 \\ 1 & 0 \end{bmatrix}$ is not primitive. One way to think about this concept is to view the positive elements of $\mathbf{M}$ as defining a directed graph on the nodes $(1, 2, \ldots, m)$. There is a directed edge (i, j) if $M_{i,j} > 0$. Observe that $[\mathbf{M}^k]_{i,j} > 0$ if and only if there is a k hop directed path from i to j. So $\mathbf{M}$ is primitive if and only if, for some $k \geq 1$, there is a k hop directed walk from each node i to each node j (including from each node i to itself). Consider now $\mathbf{M} = \begin{bmatrix} 0 & 1 & 1 \\ 1 & 0 & 1 \\ 1 & 1 & 0 \end{bmatrix}$. We see that there is a directed walk $1 \to 2 \to 3 \to 1$, and $1 \to 2 \to 1 \to 2$, both of which take three hops. We can conclude that $\mathbf{M}^3 > 0$; thus, $\mathbf{M}$ is primitive.

Theorem 5.1

Suppose M is a square nonnegative matrix that is also primitive. Then $\mathbf{M}$ has an eigenvalue ρ such that

a. ρ is real, simple, and $\rho > 0$,

b. ρ has strictly positive right and left eigenvectors, that are unique up to constant multiples (i.e., the vector spaces of the right and left eigenvectors are each of rank 1), and

c. for all eigenvalues $\lambda \neq \rho$, $\rho > |\lambda|$.

Often, ρ is called the *Perron-Frobenius eigenvalue* of $\mathbf{M}$, and, by virtue of the last conclusion of this theorem, ρ is also called the *spectral radius* of $\mathbf{M}$. This term becomes clear if we think of all the eigenvalues as being plotted on the complex plane, and a circle of radius ρ encircling them all. Let the eigenvalues be indexed from 1 to m so that $\lambda_1 = \rho > |\lambda_2| \geq |\lambda_3| \geq \cdots \geq |\lambda_m|$.

Theorem 5.2

M is an $m \times m$ primitive nonnegative matrix, and ρ is the Perron-Frobenius eigenvalue of M. Corresponding to the eigenvalue ρ, let w be a right eigenvector (column vector) and v be a left eigenvector (column vector) such that $v^T w = 1$. Then

a. If $\lambda_2 \neq 0$, for $k \to \infty$,

$$M^k = \rho^k w v^T + O\left(k^{m_2-1}|\lambda_2|^k\right)$$

where m_2 is the multiplicity of the eigenvalue λ_2.

b. If $\lambda_2 = 0$, then for $k \geq n - 1$,

$$M^k = \rho^k w v^T$$

■

An important consequence of Theorem 5.1 and Theorem 5.2 is that, if M is primitive, then $\sum_{k=0}^{\infty} M^k$ is finite if and only if $\rho < 1$, where ρ is the spectral radius of M. Consider the matrix $I - M$, where I is the identity matrix. Suppose $I - M$ has the inverse A. It follows that

$$(I - M)A = I$$

that is,

$$A = MA + I$$

$$= M^2 A + M + I$$

$$= \cdots$$

$$= M^k A + \sum_{i=0}^{k-1} M^i$$

Evidently, if M is primitive, then $I - M$ is invertible if and only if $\rho < 1$, and in that case the inverse is $\sum_{i=0}^{\infty} M^i$.

Problems

5.1 M MSs communicate with a BS using CDMA. The processing gain is $L = 200$. We consider only the uplink.

a. $(M-1)$ MSs are at distance d from the BS and one MS is at distance $\frac{d}{2}$ from the BS. All MSs are transmitting at power P. Considering only path loss ($\eta = 4$) determine the maximum value of M so that the uplink $\frac{E_b}{N_0}$ between any MS and the BS is at least 5.

b. The M MSs are now power controlled to a target BS received power P_r, but there is power control inaccuracy, which can be modeled as $(S_r)_{\text{dB}} + \phi$ where ϕ is $N(0, \sigma^2)$. The power control inaccuracies for the M MSs are i.i.d. The performance target is to achieve $\frac{E_b}{N_0} = \Gamma$ dB with outage probability P_{out}. Show how you will go about determining the maximum value of M. No numerical computation required.

5.2 The following figure depicts a linear array of cells, say on a highway.

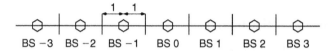

Mobiles are confined to lie on the doubly infinite line shown. Each BS covers a cell of width 2 units, 1 unit on either side. Cellular CDMA is used for spectrum sharing. Assuming an MS density of δ, and log-normal shadowing with MS and BS components both having a variance of $\sigma^2/2$, obtain an expression for reverse channel other cell interference at BS 0 due to BS -2, BS -1, BS 1, BS 2, and so on. You may leave your answer in integral form.

5.3 There are two MSs being served by a cellular system with two BSs. The following is the uplink channel gain matrix

	BS_1	BS_2
MS_1	$h_{11} = 1$	$h_{12} = 0.5$
MS_2	$h_{21} = 0.5$	$h_{22} = 1$

The uplink SIR target is -20dB. Consider two associations $(MS_1 \longrightarrow BS_1, MS_2 \longrightarrow BS_2)$, and $(MS_1 \longrightarrow BS_2, MS_2 \longrightarrow BS_1)$.

a. In each case sketch the feasible power vector region, and show the Pareto optimal power assignment. Leave your answers in terms of the receiver noise power $N_0 W$.

b. Which is the better association of the two? Illustrate your answer with a diagram of the feasible power vectors.

5.4 In a CDMA cellular network there are n BSs and m MSs. The uplink channel gain from MS i to BS j is denoted by $H_{i,j}$. The target SINR for

MS i is Γ_i. The system bandwidth is W and the one-sided noise spectral density is N_o. The uplink transmit power of MS i is P_i. Let **P** denote the vector of transmit powers. Fix an association of MSs to BSs so that MS i is assigned to BS a_i.

a. Show that the set of feasible power vectors can be expressed in the form

$$\mathcal{P} = \{\mathbf{p} : (\mathbf{I} - \mathbf{F})\mathbf{p} \geq \mathbf{g}\}$$

for appropriate matrices $\mathbf{I}, \mathbf{F}$ and $\mathbf{g}$.

b. Show that if there exists $\mathbf{p}^{(0)} \in \mathcal{P}$ then there is a $\mathbf{p}^* \in \mathcal{P}$ that satisfies $(\mathbf{I} - \mathbf{F})\mathbf{p}^* = \mathbf{g}$ and such a $\mathbf{p}^*$ is the unique Pareto power vector.

5.5 Consider the uplink of a single CDMA cell (i.e., no other cell interference) carrying several multimedia calls. User i, $1 \leq i \leq m$, has SIR target Γ_i. Let h_i be the uplink channel gain for user i, and let p_i be the power allocated to i, $1 \leq i \leq m$. Show that the power allocation problem is feasible if and only if $\sum_{k=1}^{m} \frac{\Gamma_k}{1+\Gamma_k} < 1$. (Hint: Use the iterative power control idea.)

5.6 Consider the uplink power control problem in a single cell CDMA system with system bandwidth W and m MSs. The power gain from MS i to the BS is h_i and the target SINR Γ_i. Let $\mathbf{p} = (p_1, \ldots, p_m)$ denote vector of transmit powers at the MSs. Show that the criterion that the Perron-Frobenius eigenvalue, ρ, of **F** is < 1 does not depend on the channel gains, but only the SINR requirements Γ_i, $1 \leq i \leq m$.

5.7 Consider downlink Internet access over the WDCMA–HSDPA system with proportional fairness. Focus on users in a single cell, which is surrounded by several other cells. Assume that the downlink interference power from the other cells does not depend on user location within this cell. There are two sets of users; those who are close to the *BS* and can obtain a download rate (at full dedicated power) of 512 Kbps. The other set of users have a 9 dB lower receive power from this *BS*. The mean file transfer size is 100 KB. Half the transfers are due to each class of users.

a. Obtain the maximum transfer arrival rate below which the system is stable.

b. Obtain the mean file transfer delay at 90% of this rate, using the processor sharing (PS) model.

c. Do you need the PS model for (a)?

5.8 For downlink elastic traffic in a CDMA cellular system, consider the framework developed in Section 5.5 and focus on BS_j. Suppose the average rate Γ_i is allocated to user i. The set of MSs associated with BS_j is S_j, with $m_j :=| S_j |$. We have three allocation objectives:

(i) max $\sum_{i \in S_j} \Gamma_i$ (max sum rate)

(ii) max $\sum_{i \in S_j} \ln(\Gamma_i)$ (proportional fair rates)

(iii) max $\min_{i \in S_j} \Gamma_i$ (max $-$ min fair rates)

a. Obtain the optimum power control and optimum fair rates in each case.

b. Obtain the sum average rate (over all users) in each case and obtain an ordering between them.

c. Discuss the results in (a) and (b).

CHAPTER 6

Cellular OFDMA-TDMA

Wmark{W}e discussed the basic concept of OFDM in Chapter 2, Section 2.4.2. This has become the PHY layer in the latest commercial systems based on the IEEE 802.16 series of standards for broadband wireless access networks. Since, in general, each OFDM carrier can be allocated to different users over time, these systems can be said to employ OFDMA-TDMA. In this chapter we study resource allocation problems in OFDMA-TDMA cellular systems by formulating various constrained optimization problems, the decision variables being the carriers assigned to the various users, the powers used on each carrier by each user, how much data each user sends in a slot, and so on. The constraint could be on the average total power.

Overview

We focus on the case of a single isolated cell. The basic model is of a sequence of OFDMA-TDMA frames, each with a certain number of symbols. Given the user requirements, the problem in each frame is to determine how much of the frame to allocate to the uplink and downlink data, and how much data of each user to carry in each direction. Since the subcarriers have time varying fading there is also the problem of determining the transmission powers. We begin with the simplest problem of one user transmitting over a single carrier with time varying channel gains that stay constant over frames. We consider the problem of maximizing the time average capacity, subject to a time average power constraint, and derive the *water pouring* power allocation. We perform the analysis by Lagrangian relaxation and provide insights about the Lagrange multiplier, or the *power price*. We then show that the water pouring power allocation stabilizes any arrival rate smaller than the water pouring capacity, and no arrival rate larger than the water pouring capacity can be stabilized. We then study the trade-off between power and delay. Next we show how a delay optimal, power constrained scheduler can be derived. We then turn to the multicarrier case. The single user case is considered, first with constant channel gains on all the carriers, and then with a stationary and ergodic channel gain process on each carrier. Finally, for multiple users over multiple carriers, the sum rate maximizing power control is derived. A brief overview of the WiMAX standard is also provided.

6.1 The General Model

Figure 6.1 shows a schematic of an OFDMA-TDMA system. The unit of time over which the OFDMA-TDMA scheduling takes place is the *frame*, whose duration is denoted by τ. Each frame has K OFDMA symbols. In the *time division duplex* version, each frame is partitioned into a *downlink subframe* and an *uplink subframe*. We denote the number of MSs by M and the number of carriers by N. The downlink part of each frame is used to send data to the MSs; such data would be queued in per MS queues at the BS. The uplink part of the frame is used to send data from the MSs to the destinations, via the BS. The uplink-downlink boundary can be adapted over time, depending on the ratio of the uplink-downlink load on the system. If Internet access is a major application in a system, then the downlink part of the frame would be substantially larger than the uplink part, due to the asymmetry of Internet access traffic.

Resource allocation decisions are made only at frame boundaries, as the information required for resource allocation (i.e., the channel gains, and the backlogs of the various MSs) is updated only at this slow time scale. The nth frame is the interval $((n-1)\tau, n\tau)$, $n \in 1, 2, 3, \ldots$ (see Figure 6.2). Let us consider one direction of transmission; say, the downlink. The power gain for MS i on the j-th carrier is $H_{j,n-1}^{(i)}$ during the nth frame, $n \geq 1$. Thus, we are assuming that the channel coherence time $T_c > \tau$ in order that the channel gains can be assumed to be constant over a frame. The data buffer occupancy at the downlink queue for MS i, $1 \leq i \leq M$, at the beginning of frame n, $n \geq 1$, is denoted by $Q_{n-1}^{(i)}$.

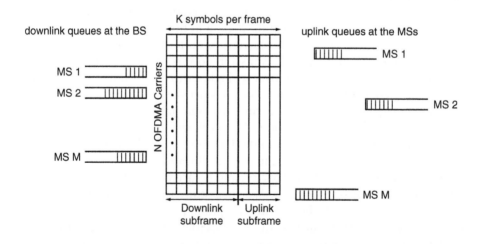

Figure 6.1 A conceptual model of uplink and downlink queues corresponding to *M* MSs associated with an OFDMA-TDMA system. The downlink queues are at the base station subsystem, while the uplink queues are in the MSs, which could be located anywhere in the coverage area of the base station.

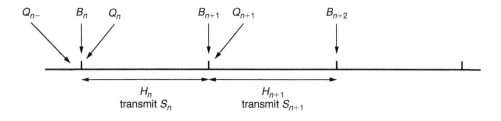

Figure 6.2 The processes in the model of joint buffer and power control at one MS. The superscript (i) indicating the MS index, and the subscript (j) indicating the carrier index are not shown.

If MS i is downloading (or uploading) a file under the control of TCP then, as explained in Chapter 3, Section 3.4.3, we assume that the wireless network is the bottleneck, and hence all the packets in the TCP window are backlogged either in MS i's buffer at the BS, or in the buffer at MS i. On the other hand, if MS i is using an application such as packet voice telephony then open-loop arrivals will occur for MS i, at either end of the wireless access link. In this case, we assume that new data arrive at the end of each frame (see Figure 6.2). For the i-th MS, we define the following downlink processes. The initial queue length just before time 0 is denoted by $Q_{0-}^{(i)}$, taken to be 1. The arrivals, if any, at 0 are denoted by $B_0^{(i)}$, and the arrivals in the nth frame, $n \geq 1$ are denoted by $B_n^{(i)}$, which are assumed to arrive at the frame boundary between Frame n and Frame $n + 1$ (see Figure 6.2). $Q_{n-}^{(i)}$ denotes the amount of data queued in the downlink for MS i at the end of Frame n, before any arrivals in the frame are taken into account. The amount of data that is transmitted in the downlink to MS i in Frame n is denoted by $S_{n-1}^{(i)}$, with $S_{n-1}^{(i)} \leq Q_{n-1}^{(i)}$. Thus, the evolution equation for the downlink buffer of MS i is (for $n = 0, 1, 2, \ldots$),

$$Q_{n+1}^{(i)} = Q_n^{(i)} - S_n^{(i)} + B_{n+1}^{(i)} \tag{6.1}$$

with $Q_0^{(i)} = B_0^{(i)}$.

6.2 Resource Allocation over a Single Carrier

In this section, we consider the situation in which a single channel is being used to transport data for a single MS; this could model the uplink or the downlink. Consider again the model shown in (2.8). We retain the multiplicative fading term, but remove the interference term. The removal of the interference term is an idealization that ignores issues such as the interference between carriers due to inaccuracies in timing, and due to carrier frequency offsets. We obtain the symbol level model

$$Y_k = G_k X_k + Z_k \tag{6.2}$$

where k indexes the symbol, and, as before, we will denote by $H_k(= |G_k|^2)$ the sequence of power gains. Given the power constraint in (2.14), the question naturally arises as to how to choose the energy to use in the k-th symbol, for each k, so as to stabilize the process $Q_n, n \geq 0$. Although there are several notions of stability of a random process, here by "stabilize" we mean formally that this random process should converge to a random variable that is finite with probability 1, as $n \to \infty$.

In order to address this question, we first seek to maximize the bit carrying capacity of the channel subject to the power constraint on the transmitted symbols. Such an objective can arise in various situations. Suppose the MS is performing a TCP controlled file download (i.e., a closed loop controlled transfer), and the MS's packets are backlogged at the MS's queue at the BS; then maximizing the bit carrying capacity will maximize the file transfer throughput. This assumes, of course, that TCP can adapt the window so as to utilize the entire capacity of the channel. This may not be possible in a situation in which the packet loss rate on the link is high, as was shown in Section 3.4.3. On the other hand, if the MS's application is generating open-loop traffic (such as streaming packet video) then maximizing the bit carrying capacity will maximize the streaming rate that can be handled. The latter observation will be formally proved after we derive the capacity maximizing power control (see Section 6.2.2, later in this chapter).

As assumed earlier, the channel power gain remains constant over all the symbols in a frame time, and $H_n, n \geq 0$, is the sequence of channel power gains (over the successive frames) in this single-channel–single-MS setting. Suppose that the transmitter knows the channel attenuation at the beginning of each frame. This is practically possible (by using *pilots*) if the channel is varying slowly. If the channel happens to be highly attenuating in a frame, we can try to combat the attenuation by boosting the power, or, alternatively, we can wait out the fade and use the power only when the channel is good again. If the random process $H_n, n \geq 0$, is ergodic (see Appendix B) then the fraction of frames in which this process will be in each state will be the same as the probability of that state. Hence, if the distribution of $H_n, n \geq 0$, gives positive mass to good states, then eventually we will get a good period to transmit in. Of course, this argument ignores the fact that there could be urgent information waiting to be sent out, but we will come back to this issue later in this chapter (in Section 6.2.2). For the moment, we seek a *power control* that will maximize the rate at which the channel can send bits.

Now consider those frames in which $H_n = h$, for some nonnegative number h. Suppose, in these frames we use the transmit power $P(h)$

(i.e., over the symbols in such frames, $\mathsf{E}(|X|^2) = P(h)$, or, more precisely, $\lim_{t\to\infty} \frac{\sum_{n=0}^{t-1} I_{\{H_n=h\}} \sum_{\{\text{symbol } k \in \text{frame } n\}} |X_k|^2}{\sum_{n=0}^{t-1} I_{\{H_n=h\}} K} = P(h)$, where we recall that there are K symbols in each frame). Then, over such frames, using the model in (6.2), the channel capacity achievable is

$$C = \frac{K}{\ln 2} \ln\left(1 + \frac{hP(h)}{\sigma^2}\right)$$

bits per frame, where we write σ^2 for the noise power, $N_0 B$, B being the subcarrier bandwidth.[1] In this expression, the division by $\ln 2$ converts nats into bits, and the multiplication by K (symbols per frame) converts bits per symbol, into bits per frame. For notational simplicity, let us denote $\frac{K}{\ln 2}$ by κ, thus yielding

$$C = \kappa \ln\left(1 + \frac{hP(h)}{\sigma^2}\right) \tag{6.3}$$

6.2.1 Power Control for Optimal Service Rate

Let the channel power gain process H_k, $k \geq 1$, be stationary and ergodic, taking values in the finite set $\mathcal{H} = \{h_1, h_2, \ldots, h_J\}$, where $J := |\mathcal{H}|$. Note that, although we introduced a continuous model for channel power gains in Chapter 2, for simplicity, here we will assume that the channel power gain is discretized into a few levels. Let us denote by $P_j = P(h_j)$, the power used when the channel power gain is h_j, and write the power control as the J vector $\mathbf{P}$. Then, for this power control, the average channel capacity over the first t frames (for large t) is given by

$$C_t(\mathbf{P}) = \sum_{h_j \in \mathcal{H}} \left(\frac{1}{t} \sum_{n=0}^{t-1} I_{\{H_n=h_j\}}\right) \kappa \ln\left(1 + \frac{h_j P(h_j)}{\sigma^2}\right)$$

bits per frame. This is because $\frac{1}{t} \sum_{n=0}^{t-1} I_{\{H_n=h_j\}}$ is the fraction of frames in which the channel power gain is h_j, and over these frames the capacity $\kappa \ln\left(1 + \frac{h_j P(h_j)}{\sigma^2}\right)$ bits per frame is achieved. But this is exactly the same as writing

$$C_t(\mathbf{P}) = \frac{1}{t} \sum_{n=0}^{t-1} \kappa \ln\left(1 + \frac{H_n P(H_n)}{\sigma^2}\right) \tag{6.4}$$

Now, taking t to ∞, using the Ergodic Theorem (see Theorem B.5), and defining the resulting limit as $C(\mathbf{P})$, we find that $C_t(\mathbf{P})$ converges with probability 1 to

---

[1]We note that to achieve this rate, we will need to *code across* symbols with the same power gains h, thus adding even more delay.

$$C(\mathbf{P}) := \kappa \; \mathsf{E}\left(\ln\left(1 + \frac{HP(H)}{\sigma^2} \right) \right) \tag{6.5}$$

where H denotes the marginal random variable of the process H_n, $n \geq 0$. It can be shown that this capacity is achievable by a coding scheme, albeit with large coding delays.

We wish to ask the question: "What is the best power control?" First, as explained earlier we impose a power constraint on the input symbols,

$$\lim_{k' \to \infty} \frac{1}{k'} \sum_{k=1}^{k'} |X_k|^2 \leq \overline{P}$$

To calculate the left-hand side of this expression, we consider the average symbol power over the first t frames, and let $t \to \infty$.

$$\lim_{t \to \infty} \frac{1}{Kt} \sum_{n=0}^{t-1} \sum_{\{\text{symbol } k \in \text{frame } n\}} |X_k|^2$$

$$= \lim_{t \to \infty} \frac{1}{Kt} \sum_{h_j \in \mathcal{H}} \sum_{n=0}^{t-1} I_{\{H_n = h_j\}} \sum_{\{\text{symbol } k \in \text{frame } n\}} |X_k|^2$$

$$= \lim_{t \to \infty} \sum_{h_j \in \mathcal{H}} \left(\frac{\sum_{n=0}^{t-1} I_{\{H_n = h_j\}} \sum_{\{\text{symbol } k \in \text{frame } n\}} |X_k|^2}{K \sum_{n=0}^{t-1} I_{\{H_n = h_j\}}} \right) \frac{1}{t} \sum_{n=0}^{t-1} I_{\{H_n = h_j\}}$$

$$= \sum_{h_j \in \mathcal{H}} P_j g_j$$

where g_j is the fraction of symbols that find the channel in the power attenuation h_j, i.e., $\lim_{t \to \infty} \frac{1}{t} \sum_{n=0}^{t-1} I_{\{H_n = h_j\}} = g_j$. We also have used the fact, discussed earlier, that in those symbols in which the power gain is h_j, the average transmitter power used is P_j. Notice that, in terms of g_j we can now write (6.5) as follows:

$$C(\mathbf{P}) = \kappa \sum_{h_j \in \mathcal{H}} g_j \ln\left(1 + \frac{h_j P_j}{\sigma^2} \right)$$

Since κ is a constant it plays no role in the optimization of $C(\mathbf{P})$; hence it can be suppressed. Equivalently, we work with capacity in units of nats per symbol. This leads to the following optimization problem.

$$\max \sum_{h_j \in \mathcal{H}} g_j \ln\left(1 + \frac{h_j P_j}{\sigma^2} \right) \tag{6.6}$$

subject to

$$\sum_{h_j \in \mathcal{H}} P_j g_j \leq \overline{P}$$

$$P_j \geq 0 \qquad \text{for every } h_j \in \mathcal{H} \tag{6.7}$$

This is a nonlinear optimization problem with a concave objective function and linear constraints (see Appendix C). We will solve it from first principles. For each power control **P**, and a number $\lambda \geq 0$, consider the function defined as follows:

$$L(\mathbf{P}, \lambda) := \sum_{h_j \in \mathcal{H}} g_j \ln \left(1 + \frac{h_j P_j}{\sigma^2}\right) - \lambda \sum_{h_j \in \mathcal{H}} g_j P_j$$

It is as if we are penalizing ourselves for the use of power, and λ is the *price* per unit power.[2] The function $L(\mathbf{P}, \lambda)$ can be viewed as a *net payoff*, which is the difference between the throughput "reward" and power cost.

Exercise 6.1
Show that, for fixed λ, $L(\mathbf{P}, \lambda)$ is a strictly concave function of the vector argument **P**.

Let us maximize the function $L(\mathbf{P}, \lambda)$, for a given λ, over the power controls **P**, while requiring only the nonnegativity of these powers.[3] The strict concavity of $L(\mathbf{P}, \lambda)$ in **P** implies that a locally maximizing power vector will also provide a global maximum over $\mathbb{R}^+$ (see Appendix C). Rewriting,

$$L(\mathbf{P}, \lambda) := \sum_{h_j \in \mathcal{H}} g_j \left(\ln \left(1 + \frac{h_j P_j}{\sigma^2}\right) - \lambda P_j\right)$$

we make the simple observation that, since the power constraint is no longer imposed, we can maximize this expression term by term for each channel gain h_j.

Exercise 6.2
Consider the maximization of $\ln(1 + \alpha p) - \lambda p$ over $p \geq 0$, given that $\alpha > 0$. Show that the unique optimizer is $p = \left(\frac{1}{\lambda} - \frac{1}{\alpha}\right)^+$.

---

[2]The dimension of the function $L(\cdot)$ could be monetary, in which case we should multiply the capacity term by a monetary value per unit capacity. But on dividing across by this value we again get the same form as displayed.

[3]We say that the power constraint has been *relaxed*, leaving only the nonnegativity constraint.

It then follows that the power vector $\mathbf{P}_\lambda$ that maximizes $L(\mathbf{P}, \lambda)$ has the form

$$P_{\lambda,j} = \left(\frac{1}{\lambda} - \frac{\sigma^2}{h_j} \right)^+ \tag{6.8}$$

Thus, for the chosen λ, if we maximize the function $L(\mathbf{P}, \lambda)$ the capacity with the maximizing power control, $P_{\lambda,j}$, is given by

$$C(\mathbf{P}_\lambda) = \kappa \sum_{h_j \in \mathcal{H}} g_j \ln \left(1 + \frac{h_j P_{\lambda,j}}{\sigma^2} \right) \tag{6.9}$$

bits per frame, and the average power is given by

$$\overline{P}_\lambda := \sum_{h_j \in \mathcal{H}} P_{\lambda,j} g_j \tag{6.10}$$

Let us understand the structure of $P_{\lambda, j}$ and of the resulting capacity and power cost. Given the set of values taken by the channel power gain process (i.e., $\mathcal{H}$), let $h^{(1)}, h^{(2)}, \ldots, h^{(J)}$, be these values in descending order. Thus $h^{(1)}$ is the best

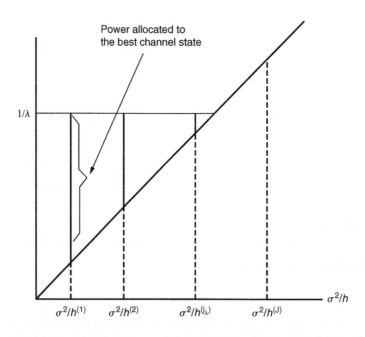

Figure 6.3 The "water pouring" structure of the optimal power allocation for a given power price λ.

channel gain and $b^{(J)}$ the worst. In Figure 6.3 the horizontal axis represents $\frac{\sigma^2}{b}$, and the various channel states also are shown along this axis. The slant line is at 45°. A horizontal line is drawn at the level $\frac{1}{\lambda}$. Now observe the form of the power control $P_{\lambda,j}$ shown in (6.8). Positive power is allocated to those states for which $\frac{1}{\lambda} > \frac{\sigma^2}{b_j}$. These power allocations are shown as solid vertical lines drawn from the slant line to the horizontal line at the level $\frac{1}{\lambda}$. Thus, the best channel state is allotted the most power, and the power allocation successively decreases, until, for channel state indices larger than j_λ, no power is allotted.

If we imagine vertical glass tubes standing on the slant line at the positions corresponding to each of the channel states, then the power allocation is as if water is poured into these glass tubes until the level corresponding to the inverse of the power price is reached. There is no water in any tube whose bottom is above this level.

For a given λ, the total average power allocated is the power allocated to each channel state, averaged over the state probabilities; this is what (6.10) says. Further, the optimal capacity is the capacity achievable with the allocated power at each channel state, averaged over the channel states; this is what (6.9) says.

Now let us observe what happens if we vary the power price, λ. Reducing the price to $\lambda' < \lambda$ causes the horizontal line to be raised, thus permitting more power to be poured into the tubes, and perhaps making channel states worse than j_λ worthy of assigning positive power. This additional power permits a larger capacity. Thus, at each power price, this solution provides the power that should be used, and how it should be allocated over the channel states so that the overall payoff is the best possible.

We return to the problem of obtaining the optimal power control for a *given* average power constraint, $\overline{P}$. Having solved the problem for a given λ, consider another vector $\mathbf{P}$, with

$$\sum_{b_j \in \mathcal{H}} P_j g_j \leq \overline{P}_\lambda$$

We are asking for power controls whose average power is no more than that of the power control that maximizes $L(\mathbf{P}, \lambda)$, for a given power price λ. Since $\mathbf{P}_\lambda$ maximizes $L(\mathbf{P}, \lambda)$ for the given λ, over all nonnegative power controls, we have, for the particular $\mathbf{P}$ just chosen,

$$C(\mathbf{P}_\lambda) - \lambda \overline{P}_\lambda \geq C(\mathbf{P}) - \lambda \sum_{b_j \in \mathcal{H}} P_j g_j$$

which implies that

$$C(\mathbf{P}_\lambda) \geq C(\mathbf{P}) + \lambda \left(\overline{P}_\lambda - \sum_{b_j \in \mathcal{H}} P_j g_j \right)$$

$$\geq C(\mathbf{P})$$

since **P** was chosen to make the second term on the right nonnegative. We conclude that $\mathbf{P}_\lambda$ is optimal for the optimization problem (6.6, 6.7) among all power controls that satisfy the power constraint $\overline{P}_\lambda$. It follows that if we can choose a λ such that $\overline{P}_\lambda = \overline{P}$, then the resulting power control will be optimal with the power constraint $\overline{P}$. That is, we need λ so that

$$\sum_{h_j \in \mathcal{H}} \left(\frac{1}{\lambda} - \frac{\sigma^2}{h_j} \right)^+ g_j = \overline{P} \tag{6.11}$$

We notice that the expression on the left is 0 for $\lambda \geq \frac{h^{(1)}}{\sigma^2}$, and increases continuously and strictly monotonically to ∞ with decreasing λ, for $\lambda < \frac{h^{(1)}}{\sigma^2}$. Hence, for each $\overline{P} > 0$, there exists a unique λ that solves (6.11). Let us denote the resulting power price by $\lambda(\overline{P})$. Then the optimal power control becomes $\mathbf{P}_{\lambda(\overline{P})}$, with

$$P_{\lambda(\overline{P}),j} = \left(\frac{1}{\lambda(\overline{P})} - \frac{\sigma^2}{h_j} \right)^+ \tag{6.12}$$

Looking again at Figure 6.3, we see that, for a given power constraint $\overline{P}$, the optimal power control can be thought of in the following way. Start with a large value of λ and decrease it (i.e., raise the horizontal line) until the average power is equal to the available power $\overline{P}$.

Let us write the optimal capacity with power constraint $\overline{P}$ as $C_{opt}(\overline{P})$. Substitution of the optimal power control $\mathbf{P}_{\lambda(\overline{P})}$ into (6.9) and simplification shows that

$$C_{opt}(\overline{P}) = \kappa \sum_{\{h_j \in \mathcal{H}, h_j > \sigma^2 \lambda(\overline{P})\}} g_j \ln\left(\frac{h_j}{\sigma^2 \lambda(\overline{P})} \right) \tag{6.13}$$

Exercise 6.3
Show that

$$\frac{\partial C_{opt}(\overline{P})}{\partial \overline{P}} = \lambda(\overline{P})$$

(Hint: First use (6.11) to obtain $\frac{\partial \frac{1}{\lambda(\overline{P})}}{\partial \overline{P}}$, and then differentiate the right-hand side of (6.13) and substitute.)

This exercise demonstrates the useful result that the rate at which optimal capacity increases with additional power is just the power price. We have solved the optimization problem from first principles, but it is instructive to apply the KKT Theorem to the problem and obtain the same results.

Exercise 6.4
Solve the optimization problem (6.6, 6.7) using the KKT Theorem (Theorem C.3 in Appendix C) and obtain all the preceding results.

6.2.2 Power Control for Optimal Power Constrained Delay

In the previous power control problem the concern was just with optimizing the physical channel capacity subject to an average power constraint. The optimal policy assigned no power to certain channel states, and low power (hence a low rate) to some other channel states. Although this might optimize the channel capacity, it might not be completely satisfactory from the point of view of the application that is sending data on the channel. This application may have a delay constraint for its data. Deferring transmission during a poor channel state would make the data wait longer for transmission. Thus, there would be a trade-off between data delay and satisfying the power constraint. Before we consider the question of delays, we should consider the question of stability of the MS's buffers under open-loop traffic.

Stability of the Buffer Process
Consider the single MS version of the buffer evolution shown in (6.1); for $n \geq 0$,

$$Q_{n+1} = Q_n - S_n + B_{n+1} \tag{6.14}$$

where $Q_0 = B_0$, and $S_n (\leq Q_n)$ is the number of bits served from the buffer in frame n. Later we will see how S_n may be chosen depending on the history of the queue length process and the channel state process. For now, let us take it to be an arbitrary number of bits served in the nth frame, with $S_n \leq Q_n$, and such that the power constraint $\overline{P}$ is respected.

Let $A_t, t \geq 0$, be defined by

$$A_t = \sum_{n=0}^{t} B_n$$

A_t is the cumulative arrivals until the end of the t-th frame. We will assume that there is a well-defined arrival rate of a bits per frame. Thus, with probability one,

$$\lim_{t \to \infty} \frac{A_t}{t} = a$$

Given an average power constraint $\overline{P}$, we wish to show that the buffer process Q_n converges in distribution to a random variable that is finite with probability 1 (and, hence, is *stable* in this sense) if $a < C_{opt}(\overline{P})$, and will "blow up" if $a > C_{opt}(\overline{P})$.[4] We can show this as follows. Consider first a service policy that, when $H_n = h_j$, uses the power control $P_{\lambda(\overline{P}),j}$ shown in (6.12). Let C_n denote the corresponding number of bits that can be served, which we assume is given by the Shannon capacity formula, even though the frame is of finite length. The evolution of the buffer process, $Q_n, n \geq 0$, is seen to be

$$Q_{n+1} = (Q_n - C_n)^+ + B_{n+1}$$

Exercise 6.5

Show, by unraveling the preceding recursion that we can write Q_n in terms of the sequences $B_n, n \geq 0$, and $C_n, n \geq 0$, as follows.

$$Q_n = B_n + \max_{0 \leq k \leq n} \sum_{j=k}^{n-1} (B_j - C_j)$$

Using this equation, it can be argued [89] that Q_n converges in distribution to a proper random variable if $a < \lim_{t \to \infty} \frac{1}{t} \sum_{n=1}^{t} C_n = C_{opt}(\overline{P})$. This is also quite intuitive, as it just says that the arrival rate is less than the service rate that can be applied. We have thus demonstrated a service policy that stabilizes the buffer.

On the other hand, suppose $a > C_{opt}(\overline{P})$, and consider any service sequence $S_n, n \geq 0$, in (6.14). It is clear that (assuming that the limit on the left exists)

$$\lim_{t \to \infty} \frac{1}{t} \sum_{n=0}^{t-1} S_n \leq C_{opt}(\overline{P})$$

since $C_{opt}(\overline{P})$ is the optimal time average rate we can get with the power constraint $\overline{P}$. Recursing (6.14), it is easily seen that, for $t \geq 0$,

$$Q_t = \sum_{j=0}^{t} B_j - \sum_{j=0}^{t-1} S_j$$

Now, if $a > C_{opt}(\overline{P})$, we have

$$a = \lim_{t \to \infty} \frac{1}{t} \sum_{n=0}^{t} B_n > C_{opt}(\overline{P}) \geq \lim_{t \to \infty} \frac{1}{t} \sum_{n=0}^{t-1} S_n$$

---

[4]The case $a = C_{opt}(\overline{P})$ is more complicated, in general, and is beyond the scope of this text. Recall that in a D/D/1 queue in which customers arrive periodically at intervals of T, and each customer requires exactly the time T to serve, the queue never exceeds 1, although the arrival rate and the service rate are equal. On the other hand, in the M/M/1 queue the arrival rate being equal to service rate leads to instability.

(where the left-hand inequality actually holds with probability 1). The previous equation implies that (with probability 1)

$$\lim_{t \to \infty} \frac{1}{t} \left(\sum_{n=0}^{t} B_n - \sum_{n=0}^{t-1} S_n \right) > 0$$

from which it follows that, as $t \to \infty$, $\left(\sum_{n=0}^{t} B_n - \sum_{n=0}^{t-1} S_n \right) \to \infty$, that is, Q_n goes to ∞.

Discussion

We have seen that, with a given power constraint $\overline{P}$, for a fading channel the capacity $C_{opt}(\overline{P})$ (obtained by water filling over the channel states) can be viewed as an average service rate. If the arrival rate of an open-loop traffic source (such as voice or streaming video) is less than this average service rate then the link buffer is stable. The service rate, $C_{opt}(\overline{P})$, increases with the power constraint $\overline{P}$. It follows that, for a given arrival rate a, there is a minimum power ($\overline{P}_a = \inf\{\overline{P} : C_{opt}(\overline{P}) > a\}$) that is needed to ensure stability of the link buffers. However, if $\overline{P}$ is greater than but very close to $\overline{P}_a$ then the delays will be large (see Figure 6.4). Hence, more

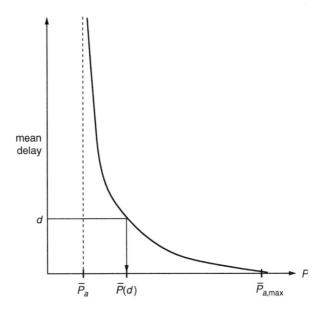

Figure 6.4   A sketch of mean delay in the queue vs. the average power $\overline{P}$, for an arrival rate a. $\overline{P}_a$ is the minimum power required to stabilize the queue, and $\overline{P}_{a,\max}$ is the power required so that the arrivals are completely served in the next frame. If the desired mean delay is d, then an average power $\overline{P}(d)$ is required.

power than just $\overline{P}_a$ will be needed in order to achieve reasonable delays. In fact, if it is required that all data are transmitted in the next frame after which they arrive, the rate required in the n-th frame will be $S_n = B_n, n \geq 0$. Then the power, P_n, required in the n-th frame will be obtained by solving

$$B_n = \kappa \ln \left(1 + \frac{H_n P_n}{\sigma^2} \right)$$

from which we see that

$$P_n = \frac{\sigma^2}{H_n} \left(e^{\frac{B_n}{\kappa}} - 1 \right)$$

It follows that the average power required if all data need to be sent in the frame after the one in which they arrive is given by

$$\overline{P}_{a,\max} := \lim_{n \to \infty} \frac{1}{n} \sum_{j=0}^{n-1} \frac{\sigma^2}{H_j} \left(e^{\frac{B_j}{\kappa}} - 1 \right) \tag{6.15}$$

$$= \mathsf{E} \left(\frac{\sigma^2}{H} \left(e^{\frac{B}{\kappa}} - 1 \right) \right) \tag{6.16}$$

where H and B denote random variables that have the ergodic distributions of the fading process and the arrival process.

Exercise 6.6

Show that with Rayleigh fading, for which H is exponentially distributed, we get $\overline{P}_{a,\max} = \infty$.

We notice that the solution to the optimization problem (6.6, 6.7) depended only on the stationary probability distribution $g_j, h_j \in \mathcal{H}$, and not on the correlations, if any, in the power gain process $H_k, k \geq 0$. Consider an H_k process that takes two states $h_0 << h_1$, where h_0 corresponds to a very poor channel. Suppose that H_k is a Markov chain with transition probabilities $p_{0,0}$ and $p_{1,1}$. Consider two cases (1) $p_{0,0} = 0.9 = p_{1,1}$, and (2) $p_{0,0} = 0.5 = p_{1,1}$. In both cases, the stationary probability distribution gives equal probabilities to the two states. This then determines the value of $C_{opt}(\overline{P})$, and hence the stability condition. But, in the first case the server spends an average of 10 frames in the "bad" state before returning to the "good" state, where it then spends an average of 10 frames. In the second case, the server spends an average of two frames in each state, thus providing "smoother" service. Thus, for each $\overline{P}$ and $g_j, h_j \in \mathcal{H}$, for which $a < C_{opt}(\overline{P})$, the correlations in the H_k process determine the queue build up in the buffer and hence the delays. It follows that, in Figure 6.4, although $\overline{P}_a$ and

$\overline{P}_{a,\max}$ depend only on the stationary distribution of the H_k process, the curve itself depends on the correlations in the H_k process.

The question then arises that if we are given a finite power constraint $\overline{P}$, with $\overline{P}_{a,\max} > \overline{P} > \overline{P}_a$, how should this power be allocated to frames over time in order to get the best possible delay performance. This is our next topic.

But before we address that topic it is opportune to point out that although we have considered only mean delay as the performance objective, applications such as interactive voice or streaming video would require delay or packet loss bounds. Such requirements would be expressed as a stochastic QoS objective such as $\Pr(Q > x) < \epsilon$, where Q is the stationary queue length random variable. For a given arrival process of rate a, and a given fading process, it would be possible to determine an average power $\overline{P}_a$ such that this stochastic QoS requirement is met. Some approaches for carrying out such calculations are discussed in [89].

Delay Minimizing Power Control

For the queue evolution

$$Q_{n+1} = Q_n - S_n + B_{n+1}$$

and given a power constraint, $\overline{P}$, we wish to determine a sequence of power allocations, $P_n, n \geq 0$, so that the mean buffer delay is minimized in some sense (to be elaborated later). We will assume that the number of bits transmitted, $S_n \, (\leq Q_n)$, and the power required, P_n, are related by Shannon's capacity formula,

$$S_n = \kappa \ln\left(1 + \frac{H_n P_n}{\sigma^2}\right)$$

If $S_n \, (\leq Q_n)$ bits need to be transmitted in a frame in which the channel gain is H_n then the power required in that frame is

$$P_n = \frac{\sigma^2}{H_n}\left(e^{\frac{S_n}{\kappa}} - 1\right)$$

We will also assume that the channel gain in each frame is known at the transmitter at the beginning of the frame. Also, the sequence of arrivals, $B_i, 0 \leq i \leq n$, until the beginning of Frame $n + 1$ is, of course, known to the transmitter. A *control policy*, π, prescribes for each n, and each possible *history* up to the beginning of Frame $n + 1$ (i.e., $((Q_0, B_0, H_0, P_0), (Q_1, B_1, H_1, P_1), \ldots, (Q_n, B_n, H_n)))$ the power control P_n to be used during Frame $n + 1$. Thus, each control π produces a *controlled stochastic process* $((Q_0, B_0, H_0, P_0), (Q_1, B_1, H_1, P_1), \ldots,)$. Let $\mathsf{E}^\pi(\cdot)$ denote the expectations when the policy π is used. Define[5]

$$\hat{Q}(\pi) = \limsup_{n\to\infty} \frac{1}{n} \mathsf{E}^\pi\left(\sum_{k=0}^{n-1} Q_k\right)$$

---

[5]We use lim sup since, in general, a control policy need not guarantee the existence of the limit.

and

$$\hat{P}(\pi) = \limsup_{n \to \infty} \frac{1}{n} \mathsf{E}^{\pi} \left(\sum_{k=0}^{n-1} P_k \right)$$

The problem is

$$\min_{\{\pi : \hat{P}(\pi) \leq \overline{P}\}} \hat{Q}(\pi) \qquad (6.17)$$

Proceeding in a manner similar to how we handled such problems before in this chapter, let us define, with a *power price* $\beta \geq 0$,

$$L(\pi, \beta) = \limsup_{n \to \infty} \frac{1}{n} \mathsf{E}^{\pi} \left(\sum_{k=0}^{n-1} (Q_k + \beta P_k) \right) \qquad (6.18)$$

and consider the minimization of $L(\pi, \beta)$ over all policies π. We notice (from the basic definition of lim sup; see Appendix B) that, for each policy π,

$$L(\pi, \beta) \leq \hat{Q}(\pi) + \beta \hat{P}(\pi)$$

with equality if each lim sup in the right-hand side is actually a limit.

Suppose now that π^* minimizes $L(\pi, \beta)$, for a given β, and further $L(\pi^*, \beta)$, $\hat{Q}(\pi^*)$, and $\hat{P}(\pi^*)$ are achieved as limits, then the following calculation follows:

$$L(\pi^*, \beta) = \hat{Q}(\pi^*) + \beta \hat{P}(\pi^*)$$

$$\leq L(\pi, \beta) \quad \text{(by the optimality of } \pi^*)$$

$$\leq \hat{Q}(\pi) + \beta \hat{P}(\pi)$$

Suppose further that $\hat{P}(\pi^*) = \overline{P}$, then using the inequality just derived

$$\hat{Q}(\pi^*) \leq \hat{Q}(\pi) - \beta(\overline{P} - \hat{P}(\pi))$$

$$\leq \hat{Q}(\pi)$$

whenever $\hat{P}(\pi) \leq \overline{P}$. It follows that π^* is optimal for the optimization problem (6.17). Hence, this approach says that we should look for a policy π^* that minimizes the right-hand side of (6.18), and has the stated properties (i.e., $L(\pi^*, \beta), \hat{Q}(\pi^*)$, and $\hat{P}(\pi^*)$ are achieved as limits, and $\hat{P}(\pi^*) = \overline{P}$). When the arrival process and the channel power gain process are Markov then this is an *average cost Markov decision problem*. The solution of such problems is beyond the scope of this text. But suppose the minimization of the function defined by (6.18) was

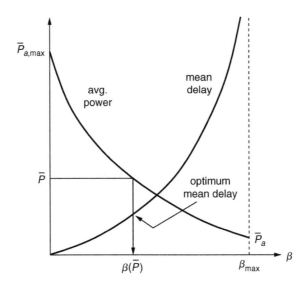

Figure 6.5 A sketch of optimum mean delay and time average power vs. the "power price" β for the problem of optimum dynamic control of the buffer subject to a power constraint.

to be done, and suppose the optimal policy yielded the time average queue length (or, equivalently, the mean delay) and the time average power as limits. If the optimum mean delay and time average power were to be plotted against β, we would obtain the trade-off sketched in Figure 6.5. As the "power price" β increases, power becomes more expensive, we use less of it, and mean delay increases. There is a β_{max} beyond which there is insufficient power (less than $\overline{P}_a$), and the buffer is unstable. When $\beta = 0$, power is free, and there is zero delay, which requires the power $\overline{P}_{a,max}$. For a given $\overline{P}$, the optimal policy is the one that is obtained by finding the policy that minimizes the objective in (6.18) using $\beta(\overline{P})$, which is obtained as shown in Figure 6.5.

In the special case in which B_n and H_n are independent, and each is an i.i.d. process, we can obtain some characterization of the structure of the optimal policy. The form of the optimum policy is depicted in Figure 6.6. The decision in each frame is based on the buffer state q, and the channel gain h at the beginning of the frame. We see, as expected, that when the channel is good ($\frac{1}{h}$ is small) and the buffer is not very large, the entire contents of the buffer are served. Even for a good channel, if the buffer is large not all of it is served. This happens because the power required increases exponentially with the amount to be transmitted. Because of the average power constraint, it is better to serve large amounts of data in two or more transmissions rather than all at once. When the channel gain is small ($\frac{1}{h}$ is large), nothing is served until the buffer crosses a threshold (the upper curve in the figure). Beyond this threshold some of the data are served. It can be

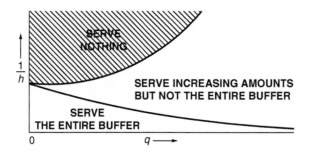

Figure 6.6 The form of the optimum policy for the joint buffer and power control problem.

shown that for any value of h the amount of data served increases monotonically with the buffer occupancy s.

6.3 Multicarrier Resource Allocation: Downlink

6.3.1 Single MS Case

We consider first the situation in which all the channels are being used to transmit data for a single MS. Recall the parallel channel model that we introduced in Section 2.4.2. This could be in the uplink or in the downlink. We work with the model in (2.25); for $1 \leq j \leq N$ (the carriers), and $k \geq 1$ (the OFDM symbols),

$$Y_{j,k} = G_{j,k} X_{j,k} + Z_{j,k}$$

where $Z_{j,k}$ are Gaussian random variables that are i.i.d. in j and in k, with variance σ^2. As before, define $H_{j,k} = |G_{j,k}|^2$, the power gain on the j-th carrier during the k-th OFDM block. There is a total average power constraint

$$\lim_{\ell \to \infty} \frac{1}{\ell} \sum_{k=1}^{\ell} \sum_{j=1}^{N} |X_{j,k}|^2 \leq \overline{P} \qquad (6.19)$$

meaning that the average power transmitted over all the carriers over time is no more than $\overline{P}$. We wish to maximize the total bit rate achieved by the MS, subject to this power constraint.

Let us assume that $H_{j,k} = h_j, 1 \leq j \leq N$, so that the power gain on each carrier is constant over time. This might be a good approximation to use when the fading is very slow compared to an OFDMA frame size. The problem is now to split the power $\overline{P}$ over the N carriers. Let $\mathbf{P} = (P_1, P_2, \ldots, P_N)$ be the vector of powers assigned to the carriers, such that $\sum_{j=1}^{n} P_j \leq \overline{P}$. The total capacity achievable with this power allocation is

$$C(\mathbf{P}) = \kappa \sum_{j=1}^{N} \ln \left(1 + \frac{h_j P_j}{\sigma^2} \right) \tag{6.20}$$

bits per frame. The optimal power splitting problem becomes

$$\max \sum_{j=1}^{N} \ln \left(1 + \frac{h_j P_j}{\sigma^2} \right) \tag{6.21}$$

subject to

$$\sum_{j=1}^{N} P_j \leq \overline{P}$$

$$P_j \geq 0 \quad \text{for every } j, 1 \leq j \leq N \tag{6.22}$$

It is easy to see that the approach for solving this problem, and the solution that is obtained, are the same as for the optimization problem (6.6, 6.7). There is a power price $\lambda(\overline{P})$ defined by

$$\sum_{j=1}^{N} \left(\frac{1}{\lambda(\overline{P})} - \frac{\sigma^2}{h_j} \right)^{+} = \overline{P} \tag{6.23}$$

Then the optimal power allocation becomes $\mathbf{P}_{\lambda(\overline{P})}$, with

$$P_{\lambda(\overline{P}),j} = \left(\frac{1}{\lambda(\overline{P})} - \frac{\sigma^2}{h_j} \right)^{+}$$

This is of the "water pouring" form *across the carriers* (rather than over time, as was the case in problems 6.6, 6.7). If we write $C_{opt}(\overline{P})$ as the optimal total capacity over the carriers then, as before, it also follows that

$$\frac{\partial C_{opt}(\overline{P})}{\partial \overline{P}} = \lambda(\overline{P}) \tag{6.24}$$

Suppose that now we include a time varying fading process, $H_{j,k}$, in the model, this process taking values in the finite set $\mathcal{H}$, as before. The stationary and ergodic process $H_{j,k}, k \geq 1$, has marginal probabilities $g_j(h), h \in \mathcal{H}, 1 \leq j \leq N$. Let us take the approach of first splitting the average power $\overline{P}$ over the carriers, and allocating power to each carrier over time, irrespective of the actions on the other carriers. Let $\overline{\mathbf{P}} = (\overline{P}_1, \overline{P}_2, \ldots, \overline{P}_N)$, with $\sum_{j=1}^{N} \overline{P}_j \leq \overline{P}$. If the power $\overline{P}_j$ is utilized optimally on Carrier j, we have the optimization problem defined by

(6.6, 6.7). Let the optimal rate on Carrier j, for the chosen power split be denoted by $C_{opt,j}(\overline{P}_j), 1 \leq j \leq N$. The problem then becomes

$$\max \sum_{j=1}^{N} C_{opt,j}(\overline{P}_j) \tag{6.25}$$

subject to

$$\sum_{j=1}^{N} \overline{P}_j \leq \overline{P}, \quad \overline{P}_j \geq 0, 1 \leq j \leq N \tag{6.26}$$

Again, defining

$$L(\overline{\mathbf{P}}, \lambda) = \sum_{j=1}^{N} C_{opt,j}(\overline{P}_j) - \lambda \sum_{j=1}^{N} \overline{P}_j$$

differentiating with respect to $\overline{P}_j$, and setting these derivatives to 0, we obtain, for $j, 1 \leq j \leq N$,

$$\frac{\partial C_{opt,j}(\overline{P}_j)}{\partial \overline{P}_j} = \lambda$$

Denoting the optimal power "price" for the problem on Carrier j by $\lambda_j(\overline{P}_j)$, and using the result of Exercise 6.3, we see that, for each $j, 1 \leq j \leq n$,

$$\lambda_j(\overline{P}_j) = \lambda$$

Thus the solution is to use the same power price on each carrier, and vary this common price until all the power is utilized; this is depicted in Figure 6.7.

Discussion

    a. At this point the following question can arise with respect to the previous derivation. We have chosen to split the power across the carriers and then

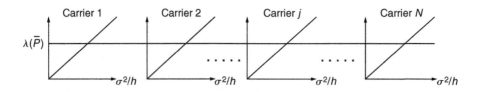

Figure 6.7 The structure of "water pouring" across carriers obtained for the problem of serving a single MS by several fading carriers, under a total power constraint.

use a carrier-by-carrier power control, that looks only at the state of each carrier separately. Suppose that, instead, we use a power vector $\mathbf{P}(\mathbf{h}) = (P_1(\mathbf{h}), P_2(\mathbf{h}), \ldots, P_N(\mathbf{h}))$, where $\mathbf{h}$ is the joint state of the N carriers in the next frame; the power $P_j(\mathbf{h})$ is used in the next frame on Carrier j. Let $g(\mathbf{h})$ denote the joint distribution across carriers of the vector fading process $\mathbf{H}_k = (H_{1,k}, H_{2,k}, \ldots, H_{N,k}) \in \mathcal{H}^N$. Then, the time average rate provided to the buffer becomes

$$\sum_{\mathbf{h} \in \mathcal{H}^N} g(\mathbf{h}) \sum_{j=1}^{N} \kappa \ln\left(1 + \frac{h_j P_j(\mathbf{h})}{\sigma^2}\right)$$

with the power constraint

$$\sum_{\mathbf{h} \in \mathcal{H}^N} g(\mathbf{h}) \sum_{j=1}^{N} P_j(\mathbf{h}) \leq \overline{P}$$

It turns out that the problem of maximizing the rate under this setup yields the same power splitting solution that we have derived (see Problem 6.6).

b. The problem of dynamic power control so as to minimize the time average queue length subject to a time average power constraint can also be formulated in the multicarrier context. See Problem 6.5.

6.3.2 Multiple MSs

Let us now turn to the problem of resource allocation for multiple MSs in the downlink of an OFDMA system. There are now m MSs, indexed $1 \leq i \leq M$, and N carriers, indexed $1 \leq j \leq N$. Time is divided into OFDMA frames, with several OFDM symbols in each frame. The MSs share the carriers in a TDMA fashion over time. Since the channel to each MS can be different, the model between the downlink OFDM transmitter and the i-th MS, over the j-th carrier becomes

$$Y_{i,j,k} = G_{i,j,k} X_{j,k} + Z_{i,j,k}$$

This is the model for the signal received at the i-th MS on the j-th carrier, if the symbol $X_{j,k}$ is sent on the j-th carrier, in the k-th OFDM symbol. Suppose a fraction $\alpha_{i,j}$ of the OFDM blocks carry data for the i-th MS on the j-th carrier. For simplicity, let us assume that the channel power gain for MS i on Carrier j (i.e., $H_{i,j,k}$), does not vary with k, and can thus be written as $h_{i,j}$. Then it suffices to assume that the transmissions to the i-th MS on Carrier j are done at the constant power $P_{i,j}$. Thus, we have, for each Carrier $j, 1 \leq j \leq N$,

$$\sum_{i=1}^{M} \alpha_{i,j} = 1$$

and the power constraint becomes

$$\sum_{j=1}^{N} \sum_{i=1}^{M} \alpha_{i,j} P_{i,j} \leq \overline{P}$$

The power constraint is of this form since the average power utilized for the i-th MS on Carrier j is just $\alpha_{i,j} P_{i,j}$. Subject to these constraints, suppose we wish to maximize the total capacity of the system,

$$\sum_{j=1}^{N} \sum_{i=1}^{M} \alpha_{i,j} \ln \left(1 + \frac{h_{i,j} P_{i,j}}{\sigma^2} \right)$$

To understand this expression, note that, on Carrier j, during times when MS i is being served, this MS obtains the rate $\ln \left(1 + \frac{h_{i,j} P_{i,j}}{\sigma^2} \right)$. Thus, the sum over i yields the total rate obtained by all the MSs over Carrier j. First, let us use Jensen's inequality to observe that

$$\sum_{i=1}^{M} \alpha_{i,j} \ln \left(1 + \frac{h_{i,j} P_{i,j}}{\sigma^2} \right) \leq \ln \left(1 + \sum_{i=1}^{M} \alpha_{i,j} \frac{h_{i,j} P_{i,j}}{\sigma^2} \right)$$

$$\leq \ln \left(1 + \frac{h_{i,j} P_j}{\sigma^2} \right)$$

where $i_j := \arg\max_{1 \leq i \leq M} h_{i,j}$, and where we have defined $P_j = \sum_{i=1}^{M} \alpha_{i,j} P_{i,j}$, the average power utilized on Carrier j. Thus, to maximize the total capacity, the approach should be to split the power $\overline{P}$ over the N carriers and then use all the power allocated to each channel to serve the MS with the best power gain on that channel. Then we may as well think of a single "MS" with power gains $h_j = h_{i_j,j}$, and allocate the power $\overline{P}$ so as to maximize this "MS's" capacity. We are back to the single MS problem discussed earlier in this section.

These arguments have yielded the best partition of the carriers over the MSs, and the optimal power split over the carriers, so as to maximize the *sum capacity*. Note that, at this point, this approach completely ignores the characteristics of the actual applications that the MSs wish to carry. For example, if a particular MS has poor channel conditions, but has delay sensitive data to be delivered, then we may wish break this optimal rule and serve this MS. Also, in the context of elastic traffic we will wish to impose some sort of a fairness requirement when formulating the earlier resource allocation problems. Recent literature on such topics has been listed in the "Notes on the Literature" section.

6.4 WiMAX: The IEEE 802.16 Broadband Wireless Access Standard

OFDMA has been adopted by the IEEE 802.16 series of standards for broadband wireless access. The much talked about WiMAX system, which is being commercially deployed in several countries, is a subset of the IEEE 802.16 standard. This system provides broadband wireless access to fixed stations. On the other hand, Mobile WiMAX is a subset of the IEEE 802.16e standard that is designed to support broadband access for mobile stations. Mobile WiMAX systems can occupy system bandwidths (i.e., W) from 1.25 MHz to 20 MHz, in the 2.3 GHz, 2.5 GHz, 3.3 GHz, and 3.5 GHz bands. In the initial releases of WiMAX, the downlink traffic and uplink traffic will share the same bandwidth in a time-division-duplex (TDD) fashion. In Figure 6.8 we provide a rough sketch of the WiMAX frame structure in the TDD mode. As an example, one of the WiMAX profiles is $W = 5$ MHz, number of carriers $N = 512$, number of symbols per frame $K = 48$ symbols, with a frame time of 5 ms. Various modulation and coding schemes (such as QPSK, 16QAM, and 64QAM, along with various rate convolutional codes) are available for putting the MSs' bits onto the OFDM symbols.

Note that our discussions earlier in this chapter can easily be adapted to model the fact that a part of the channel time is used for the downlink and the remaining time is used for the uplink. As discussed earlier in this chapter, the scheduling decisions for each frame rely on channel power gain estimates. These are provided by means of the mobile stations making measurements on downlink *pilots* and then feeding the measurements back in the channel state feedback overhead channel in the uplink part of the frame (see Figure 6.8). The scheduling decisions in each frame are made known to the MSs in the "MAP" overhead part of the downlink part of the frame.

6.5 Notes on the Literature

The antecedents of the present wireless OFDM systems lie in the so-called multicarrier modulation schemes that were developed for digital communication over telephone channels; see Kalet [65] and Bingham [13] where some history of this technique also is provided. In our discussions we have focused on multiple access packet communication over FDMA-TDMA systems. The single user case that we began with is essentially the same as reported in the seminal work of Goldsmith and Varaiya [44]. In this paper, the authors have derived the

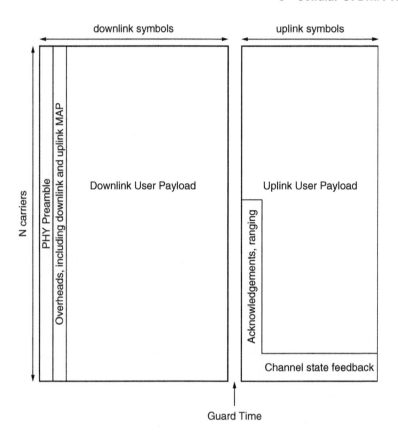

Figure 6.8 A rough sketch of the WiMAX system's TDD frame structure. A guard time is allowed for turn around of the link direction.

ergodic capacity of the fading channel with complete channel information at the transmitter and the corresponding water-pouring-in-time form of the power control, and have proved the related channel coding theorem. The optimal power control problem for a single-user, single-carrier buffered system and the related power-delay trade-off were studied by Berry and Gallager [9]. The approach that we have presented and the form of the optimal control for i.i.d. fading and arrivals were obtained by Goyal et al. [45]. For the multiuser and multicarrier case, we have discussed only some simple problem formulations. Resource allocation in OFDMA-TDMA cellular networks remains an active area of research. Some recent results are reported in the following sampling of articles: Agrawal et al. [2], Huang et al. [59], Kittipiyakul and Javidi [76] (where the authors study optimal downlink scheduling for multiple users, based on queue occupancies), Ergen et al. [31], and Yoon et al. [140], [141].

Problems

6.1 For the single-user, single-carrier problem, described in the text, consider the problem

$$\min \sum_{h_j \in \mathcal{H}} P_j \, g_j$$

subject to

$$\sum_{h_j \in \mathcal{H}} \gamma \ln \left(1 + \frac{h_j P_j}{\sigma^2} \right) g_j \geq a$$

where $a > 0$.

    a. Characterize the optimal solution of this problem.

    b. Let the optimum value of this problem be denoted by $\tilde{P}_a$. Show that $\tilde{P}_a = \overline{P}_a$ (defined in the text).

6.2 Consider the optimal power control problem over a single carrier (in the OFDMA context) and recall the notation $C_{opt}(\overline{P})$ for the optimal capacity with power constraint $\overline{P}$. Show that $C_{opt}(\overline{P})$ has the following properties: (1) $C_{opt}(0) = 0$, (2) $C_{opt}(\overline{P})$ is monotonically increasing with $\overline{P}$, (3) $C_{opt}(\overline{P})$ is concave in $\overline{P}$.

6.3 Consider the problem of m users sharing n OFDMA carriers in the downlink. Given a particular $(C_1, C_2, \ldots, C_m)$ of the set of carriers $C = \{1, 2, \ldots, n\}$, and total power constraint $\overline{P}$, characterize the solution of the problem of weighted sum throughput maximization subject to a power constraint:

$$\max \sum_{i=1}^{m} w_i \, C_{opt}(\overline{P}_i)$$

subject to

$$\sum_{i=1}^{m} P_i \leq \overline{P}$$

where $\overline{P} = (\overline{P}_1, \overline{P}_2, \ldots \overline{P}_m)$ is the power allocation to user i.

6.4 Repeat Problem 6.3 for sum log rate maximization subject to a power constraint.

6.5 Formulate the delay minimizing dynamic power control problem for the case of one user being served by n block fading carriers.

6.6   Consider the power constrained sum rate optimizing power allocation problem for a single user and n OFDMA carriers. The carriers have a joint block fading process $\{\mathbf{H}_k, k \geq 0\}$ taking values in $\mathcal{H}^n$ (where $\mathcal{H}$ is a finite set of channel power gain values) with joint probabilities $g(\mathbf{h}), \mathbf{h} \in \mathcal{H}^n$. Show that optimizing over joint power allocations across the carriers is equivalent to first splitting the power across the carriers and then optimal water pouring over time on each carrier.

CHAPTER 7

Random Access and Wireless LANs

In Chapter 4 we considered stream traffic over circuit multiplexed cellular networks. A centrally coordinated mechanism for sharing the channels provides capacity on demand to a call. The call is blocked if the requisite resource is not available. In Chapter 5 we first considered allocating resources to stream traffic to satisfy in-call QoS requirements like BER. Here, an arriving call is blocked if the in-call QoS cannot be met. We also considered resource allocation for packet multiplexed elastic traffic. In Chapter 6 we considered packet multiplexing with a centralized resource allocation of the OFDMA system. The carrier and timeslots for packet transmissions are the resources and, if a packet cannot be transmitted on arrival, it is queued and not dropped or blocked. In this chapter, we continue to consider packet multiplexed wireless networks with queueing of packets, but with distributed resource sharing mechanisms. Several nodes share a wireless medium with possibly no central coordination. We will discuss distributed multiplexing using random access techniques.

Overview

We consider networks in which all the nodes use the same part of the spectrum. Our interest is in the use of *random access* based medium access control (MAC) protocols for distributed access control. Such networks originated in the *Aloha* experiments in the early 1970s. Transport of "bursty" packetized data was the primary objective of these networks. After discussing basic issues and some terminologies in Section 7.1, the Aloha MAC protocol is described and analyzed in Section 7.2. The analysis is based on elementary probability theory and Markov chains. Some current applications of the Aloha protocol (e.g., signaling and control channels in cellular networks and in VSAT networks) will also be described. In Section 7.3 we consider MAC protocols for wireless *local area networks* (WLANs). We describe the *hidden* and *exposed* nodes in multihop WLANs and the handshake mechanism for *collision avoidance*. We then describe the popular *carrier sense multiple access with collision avoidance (CSMA/CA)* protocol of the IEEE 802.11 WLAN MAC standard. A brief overview of the *ETSI HIPERLAN* is also provided. In Section 7.4 we develop a simplified version of a popular *saturation throughput* analysis of the IEEE 802.11 MAC protocol. Here saturation implies that all the nodes always have a packet to transmit. This analysis will use the *renewal reward theorem* and a *fixed point theorem*. In Section 7.6, *service differentiation* mechanisms in IEEE 802.11 networks are described and an overview of the extension of the fixed point analysis of Section 7.4 is also provided. The performance of TCP-based data and voice traffic over WLANs is discussed in Section 7.6. Optimal *association* of an 802.11 node to an *access point* is discussed in Section 7.7.

7.1 Preliminaries

We consider two aspects of the channel usage model and then describe some commonly used terms.

Recall from Chapter 2 that a minimum SINR of, say, θ is required by a receiver to decode a packet that is being received. Like in CDMA networks, in WLANs also all nodes use the entire allocated spectrum when they are transmitting. However, it is difficult to work with the general approach of calculating the SINRs as in Chapter 5 and we use the following simplified model of channel usage. Whenever two or more transmissions in the same frequency band and of sufficient strength arrive simultaneously at a receiver, neither can be detected; this event is called a *collision*. A collision can occur even after a receiver has successfully decoded a part of a transmission. This can happen if the receiver is "hit" by one or more interfering signals from other nodes in the network. In practice, it can happen that if a receiver is simultaneously receiving signals from one or more transmitters, one of them is strong enough for the resulting SINR to be above the threshold. If this happens, we say that a *capture* has occurred. See Problem 2.2 for an illustration of capture. The throughput and capacity results obtained by not accounting for capture essentially provide a lower bound on what is achievable.

Another aspect of the channel usage model is the following. Around a receiver, we define an *interference* region (also called the *carrier sense region*) and a subset of it called the *decode* region. Consider a transmitter and its receiver in the network. If the transmitter is in the interference region of the receiver, the received power at the receiver is significant and causes a collision at the receiver if another transmission is being received at the same time. If the transmitter is also in the decode region, then the SNR at the receiver is greater than the prescribed threshold, say θ, and the receiver can decode the transmission. The received power from transmitters outside of the interference region is assumed insignificant. The interference and decode regions for a receiver at A are shown in Figure 7.1. The solid line is the boundary of the decode region and the dashed line is that of the interference region. When no other node is transmitting, A can decode transmissions from B (or C). Transmissions from D or E (or both) when receiving from B (or C) can cause a collision at A. However, the transmissions from D and E cannot be decoded at A. Transmissions from G or H do not cause a collision at A.

In general, the decode and interference regions depend on the locations of the other nodes that are transmitting. In this description we have ignored this dependency. We can also define the interference and decode regions for a

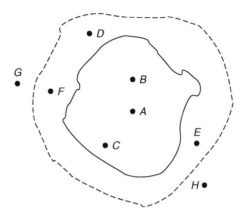

Figure 7.1 Decode and interference regions for a receiver at A. These regions depend on the topography around the transmitter and the receiver and may not have any regular shape.

transmitter. In general, the interference regions for transmission and reception may not be equal because the topography around the transmitter and the receiver may not be the same. Likewise for decode regions.

In *single hop* networks, every node is within the decode region of every other node; that is, all nodes are one hop away from each other. Such networks are also called *broadcast* networks because a transmission from a node can be decoded by every other node in the network. They are also known as *colocated* networks. In this chapter we will consider only single hop networks.

The channel is also called the *medium* and a medium access control (MAC) protocol regulates use of the medium by prescribing the rules to initiate a transmission and continue with it. In random access networks, collisions may occur and the MAC protocol has to resolve collisions; it has to arbitrate among the nodes contending to use the medium. The arbitration is a distributed algorithm that typically prescribes forced silences on the nodes. Thus some amount of transmission time is lost to collisions and arbitration. The fraction of time so lost is a measure of the efficiency of the protocol. Simple protocols, even if of low efficiency, are useful if the per node throughput that the protocol obtains is significant compared to the throughput required by the nodes in the network. In this chapter we will study two key wireless MAC protocols—Aloha and CSMA/CA.

7.2 Random Access: From Aloha to CSMA

Random access protocols can be motivated by the following simple example. Consider a 100-node multiple access network using a 10 Mbps channel. Each node requires an average throughput of 1 Kbps, but in bursts. As an example, a node may have to transmit 1000-bit packets, on an average once every second. In a TDM scheme, each node would be statically allocated every 100-th slot. If each slot corresponds to a packet transmission time, the waiting time before the packet transmission is completed could be as high as 10.1 ms and the expected delay would be 5.1 ms. This is assuming that the queue is empty when this packet arrives at the node. In a polling scheme, there would be polling overheads and a corresponding delay. In a random access scheme, the transmission would be at 10 Mbps and the transmission time is just 100 μs. Also since the offered load to the network is low (1% of the maximum possible throughput), the *access delay*, the delay between the packet being ready and the beginning of transmission, will also be low.

An additional issue in TDM and polling schemes is that the exact number of nodes in the network needs to be known. In most wireless networks, at any time, this is typically a random number. Thus, in such networks, random access is possibly the only option. And, as we have seen in the simple example earlier, it is not a bad option.

In this section we analyze the simplest of the random access protocols—Aloha and slotted Aloha. We will also briefly discuss the carrier sensing CSMA protocol as a precursor to the discussion of CSMA/CA in the next section.

7.2.1   Protocols without Carrier Sensing: Aloha and Slotted Aloha

The Aloha, also called the *pure Aloha*, is the earliest random access MAC protocol. The idea is as simple as it can be: If a node has a packet to transmit, it just transmits! To see why this is not such a bad idea, consider the example of a satellite network in which every node transmits to a satellite, which then reflects the transmission back for every node to receive it. Let the data rate on the network be 1 Mbps and the packets be 1000 bits long so that the packet transmission time is 1 ms. Now consider the propagation delay in the network. Every packet has to travel approximately 75,000 Kms to reach the receivers resulting in a propagation delay of approximately 250 ms. Thus, what a node that wants to transmit is hearing on the channel at time t is actually a transmission from $(t-250)$ ms. Hence, there is no use deferring to a carrier and it is best just to transmit the packet and hope that no other node is transmitting at the same time. Of course, if there indeed was another transmission at the same time, there would be a collision and neither packet can be decoded correctly by the corresponding receivers. The nodes will have to use additional mechanisms to determine if the packet was successfully received. If the packet is not correctly received, the packets will have to be retransmitted using a suitable retransmission algorithm. Figure 7.2 shows a space-time diagram of a transmission and reception in a network with large propagation delays and the futility of deferring to a carrier.

We now analyze the performance of the Aloha protocol. To keep the analysis simple, we consider the following simple model. We will assume fixed length

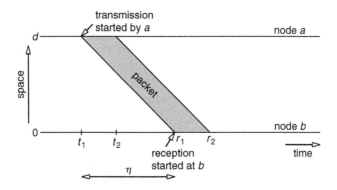

Figure 7.2 Space-time schematic of a transmission and reception in Aloha. Nodes are arranged on a line. Node a is transmitting during $[t_1, t_2]$, which is being received by Node b during $[r_1, r_2]$, after a delay of η. If Node b is not the receiver, it need not defer transmission during $[r_1, r_2]$ and cannot defer when Node a is transmitting (during $[t_1, t_2]$) because it does not know that a is transmitting at that time.

packets and time will be measured in terms of the packet transmission time; the packet transmission times are of unit duration. The nodes are located along a straight line of length η. We will measure distances in terms of the propagation delay; the signal travels a unit length in unit time. Thus the maximum propagation delay in the network is also η.

The packet transmission attempts in the network are assumed to form a Poisson process of rate G attempts per second. The location of the transmitting node is chosen uniformly in $[0, \eta]$ and independently of the other transmissions. Thus, each packet transmission attempt is characterized by an ordered pair (t, y), where $t \geq 0$ is the time at which the transmission started, and $0 \leq y \leq \eta$ is the location of the transmitting node. A sample realization of this *space-time attempt process* in the region $([0, \infty) \times [0, \eta])$ is shown in Figure 7.3. Now consider two nonoverlapping areas in this region, say A and B in Figure 7.3. It is easy to see that the number of attempts in A and B are independent. Further, the number of attempts in A has a Poisson distribution with mean $((G/\eta) \times (\text{Area of } A))$. Thus the space-time attempt process in the region $([0, \infty) \times [0, \eta])$ is a two-dimensional Poisson point process of rate $\frac{G}{\eta}$ attempts per meter-second.

This model means that there is an infinite number of nodes in the network and that at any instant, each node has at most one packet to transmit. This is a good model for a network with a large number of nodes and a low packet arrival rate per node.

Consider a node at location T transmitting a packet to a node at location R. For this transmission, we can define a *collision window* in time at each location in $[0, \eta]$. If a transmission is begun at that location in the collision window, then it will arrive at R when it is receiving the packet from T, thus causing a collision. The

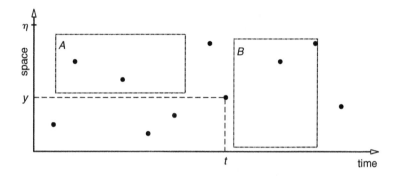

Figure 7.3 A sample realization of packet attempt instants in space and time. The points indicate the times and locations of packet arrivals; for example, a node at location y attempted a packet at time t.

union of the collision window at all points in the network gives us a *collision cone*. Any transmission attempt in this collision cone will cause a collision at the intended receiver, R in the example. See Figure 7.4 for an illustration. A transmission is begun at space-time point A (corresponding to the transmitter at T) whose receiver is at space-time point a (corresponding to receiver R). Transmissions initiated in the interval (b, c) at η and in the interval (f, e) at 0 will cause a collision at a. Note that these intervals are of length 2 time units. The space-time area covered by the collision cone is clearly 2η.

Since $\frac{G}{\eta}$ is the space-time arrival rate, multiplying it by 2η gives the mean of the Poisson distribution of the number of arrivals in the collision cone. Thus, the probability that a reception is successful, P_s, is given by

$$P_s = \Pr(\text{No transmission attempt in collision cone}) = e^{-(\frac{G}{\eta} \times 2\eta)} = e^{-2G}$$

Defining the throughput, S, as the mean number of successful attempts per unit time, we get $S = GP_s = Ge^{-2G}$. The maximum value of S is achieved for $G = 0.5$ and $S_{\max} = 1/(2e) \approx 0.18$.

Exercise 7.1

To simplify the analysis of random access protocols it is sometimes assumed that all nodes are at the same distance from one another. In this case, we do not need the space-time model just described. An example of a network where this is valid is a satellite network. Show that the maximum throughput is still $\frac{1}{2e}$ under this simplifying assumption.

There is a simple way to make pure Aloha more efficient. Instead of allowing a node to begin transmission at any time, let time be slotted and the nodes be allowed

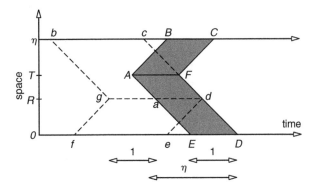

Figure 7.4 Space-time diagram showing a packet transmission and its collision cone in Aloha. A packet of unit duration starts transmission at space-time point *A* and propagates along *ABCFDEA*. For successful reception at space-time point *a* then no other transmission should be initiated in the "collision cone" *bcdefgb*. This collision cone has area *2η*, line segment *ga* has unit length, and the horizontal distance from *c* to *D* is *η*.

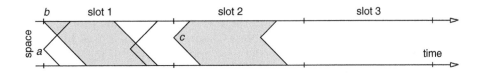

Figure 7.5 Time slotting in slotted Aloha. In slot 1, there are two transmissions, from *a* and *b*, and they collide everywhere. In slot 2 the one transmission from *c* is successfully received everywhere. There is no transmission in slot 3.

to begin transmission only at the beginning of a slot. The slot length is made equal to the sum of the packet transmission time (unity) and the maximum propagation delay in the network. This is the *slotted Aloha* (S-Aloha) MAC protocol and it has found wide applications in practice.

As before, assume that nodes are arranged on a line of length η. Nodes begin transmission only at slot boundaries and the transmission and reception of a packet is completed in one slot. Packets that arrive in a slot are transmitted at the beginning of the next slot. Thus, for a collision at a receiver, a second packet should begin transmission in the same slot; that is, another packet should have arrived in the previous slot. This means that the collision cone is now a rectangle of sides $(1 + \eta)$ and η. Figure 7.5 illustrates this.

Let us now analyze this slotted Aloha protocol. Assume as before that transmission attempts arrive according to a Poisson process of rate G and the source node is uniformly distributed in $[0, \eta]$. The Poisson rate of packet arrivals that can cause a collision is thus the expected number of Poisson arrivals in a slot, $G(1 + \eta)$. The probability that a transmission is successfully received is $P_s = e^{-G(1+\eta)}$. We can obtain the throughput as before and it is $S = GP_s = Ge^{-G(1+\eta)}$. The maximum achievable throughput, S_{max}, is $S_{max} = 1/(e(1 + \eta))$.

The slot length was made equal to $(1 + \eta)$, rather than 1, to absorb the variations in the propagation delays between the nodes. This is not always necessary. In fact it is rarely necessary. In a satellite network the propagation delays between any pair of nodes is very nearly the same, approximately 250 ms, and we can use a slot length of one unit. In terrestrial networks, like in the cellular and the cable networks, the nodes usually transmit to a central node. The nodes use ranging to determine the propagation delay to the central node and advance or delay their transmission times to approximate a time slotted link and absorb the differences in the propagation delays.

Example Applications of Slotted Aloha

An example use of the slotted Aloha protocol is in the GSM cellular networks that we discussed in Chapter 4 in Section 4.7. Recall from Chapter 4 that in the GSM network the TDM channels are either traffic channels carrying voice or are control channels carrying control and signaling information to or from the mobile stations. A control channel on the reverse link from the mobile node to the base station, called the Random Access Channel (RACH), is used by the GSM mobile stations to send messages to the network. The types of messages include those to initiate new calls, register locations of the mobile stations, and reply to paging queries. The messages are small and are generated at a very low rate compared to the capacity of the RACH channel. The number of mobile nodes in a cell is not fixed and also quite large and signaling bandwidth cannot be allocated statically to these nodes. Hence, slotted Aloha is used on this channel. After transmitting on the RACH using the slotted Aloha protocol, the mobile station waits for a fixed duration to know if the transmission was successful. If an acknowledgment is not received before this duration, a retransmission is attempted.

Another application of slotted Aloha is in *very small aperture terminal (VSAT)* networks. A VSAT network is a satellite network in which there are several geographically widespread, small terminals. Figure 7.6 illustrates such a network. These terminals are attached to individual computers or to the local area networks of small organizations through the digital interface unit (DIU). The terminals share a satellite link to a large hub. The nodes can communicate only with the hub and all internode communications are over two hops via the hub. Hence the *inbound* channel from the terminals to the hub needs to be shared. The terminals request for reservations of time on the inbound channel. When a remote node wants to transmit, it first requests for a reservation on the slots in the inbound channel to the hub. This reservation request is made using the slotted Aloha protocol on the uplink from the remote station to the hub. This reservation scheme can be very efficient if the bandwidth allocated for the reservation requests is small and the amount of reserved bandwidth is large.

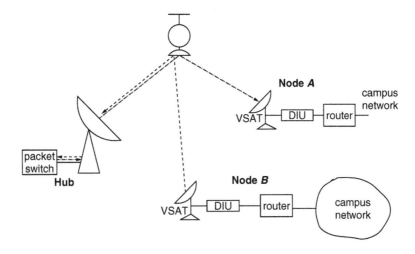

Figure 7.6   A VSAT satellite network. Data packets from *A* are first sent to the Hub, which then transmits them to *B*. *A* first requests slots to be reserved on the "inbound channel" to the hub using S-Aloha.

Slotted Aloha is also used in wireline networks. The DOCSIS (Data Over Cable Service Interface Specifications) standard for digital data transmission over cable TV networks is similar to the VSAT network. All communication is controlled by the head-end of the cable TV network. Separate parts of the spectrum over the cable are reserved for upstream (node to head-end) and downstream (head-end to node) traffic and the control of the channels in both directions is with the head-end. The upstream spectrum, in which the nodes transmit data to the head-end, is available only by reservation and is obtained as follows. This spectrum is divided into contention slots, in which the nodes request bandwidth on the upstream channel, and transmission slots, during which the nodes transmit actual data. A node with data to transmit requests the head-end to reserve some of the upstream transmission slots for it. The requests are made using a slotted Aloha protocol.

Instability of Aloha

In the discussion on the throughput of Aloha and S-Aloha, we had implicitly assumed that packets involved in collisions are lost. We now analyze the S-Aloha protocol under the more realistic assumption that a packet that has suffered a collision stays in the network and makes retransmission attempts until it is successful. The number of such packets is called the backlog.

Let A_k be the number of new packet arrivals into the network and $D_k \in \{0, 1\}$ the number of successful transmissions (departures from the network) in slot k. We assume that all fresh arrivals during a slot will attempt a transmission at the beginning of the next slot. Let B_k denote the backlog at the beginning of slot k. It is easy to see that B_k evolves as

$$B_{k+1} = B_k + A_k - D_k$$

We assume that new packet arrivals form a Poisson process of rate λ independent of everything else in the network. A_k will then be i.i.d. Poisson random variables. We also assume that the backlogged nodes attempt retransmission independently in each slot with probability r. Under these assumptions, $\{B_k\}$ is a discrete time Markov chain. Our interest is in the stability of this Markov chain.

For stability analysis of $\{B_k\}$, consider $d(n)$ defined by

$$d(n) := \mathsf{E}\big(B_{k+1} - B_k | B_k = n\big) = \mathsf{E}\big((A_k - D_k) | B_k = n\big)$$

$d(n)$ is the expected change in the backlog in one slot when the backlog is n and is called the *drift* from state n. If the number of new arrivals is more than one, the backlog increases by that amount because all of them will transmit in the slot causing a collision irrespective of the backlogged packets attempting a transmission. The backlog increases by one if there is exactly one arrival in the slot and at least one of the backlogs attempts a retransmission causing a collision. The backlog decreases by one if no new arrivals occur and only one of the backlogged nodes attempts a transmission in the slot resulting in a successful transmission. For other combinations of retransmission attempts and new arrivals, the backlog does not change. Thus we can write

$$\mathsf{Pr}(A_k - D_k = +m | B_k = n) = \frac{\lambda^m}{m!}e^{-\lambda} \quad \text{for } m \geq 2$$

$$\mathsf{Pr}(A_k - D_k = +1 | B_k = n) = \lambda e^{-\lambda}\left(1 - (1-r)^n\right)$$

$$\mathsf{Pr}(A_k - D_k = -1 | B_k = n) = e^{-\lambda}nr(1-r)^{n-1}$$

Using this and simplifying, we get

$$d(n) = \left(-e^{-\lambda}nr(1-r)^{n-1}\right) + \left(\lambda e^{-\lambda}\left(1 - (1-r)^n\right)\right) + \left(\sum_{m=2}^{\infty} m\frac{\lambda^m}{m!}e^{-\lambda}\right)$$

$$= \sum_{m=0}^{\infty} m\frac{\lambda^m}{m!}e^{-\lambda} - \lambda e^{-\lambda}(1-r)^n - e^{-\lambda}nr(1-r)^{n-1}$$

$$= \lambda - e^{-\lambda}(1-r)^n\left(\lambda + \frac{nr}{1-r}\right)$$

For any $\lambda, r > 0$, for large n, the second term becomes very small and $d(n)$ will be positive; we can find an $n^*_{\lambda,r}$ such that $d(n) > 0$ for all $n > n^*(\lambda, r)$. Since at most one packet can depart from the system in each slot, we can apply Theorem D.4 to conclude that the Markov chain $\{B_k\}$ is not positive recurrent for any $\lambda, r > 0$. What this means is that when the backlog becomes large, the network has a tendency to increase the backlog rather than decrease it. This in turn implies that the S-Aloha protocol can eventually develop a large backlog that will never be cleared.

The assumption of an infinite number of nodes is for analytical convenience. It is also a worst-case analysis because with finite nodes, packets from the same node do not compete with one another and in that case the performance can only improve. However, it can be shown that even when the number of nodes in the network is finite, the behavior is qualitatively similar to the infinite node case. In this case, if the backlog becomes large, the network has a tendency to operate with a large backlog for very long times.

Stabilizing Aloha

An obvious issue now is to design mechanisms to make the network stable for some $\lambda > 0$ so that the network can support new packet arrivals at that rate. This is done by making the retransmission probabilities adaptive. To see how this can be done, assume that all the nodes know the size of the backlog at the beginning of every slot and also the stationary packet arrival rate. A packet is successfully transmitted in a slot if either of the two conditions is satisfied: (1) Exactly one new packet arrives and none of the backlogs attempts a retransmission or (2) no new packet arrives and exactly one of the backlogs attempts a retransmission. Thus, the probability of a successful transmission when the backlog is n, $P_s(n)$, is given by

$$P_s(n) = \lambda e^{-\lambda}(1-r)^n + e^{-\lambda}nr(1-r)^{n-1} \tag{7.1}$$

The retransmission probability that will maximize $P_s(n)$, say $r(n)$, is obtained by differentiating the RHS of (7.1) with respect to r, equating it to 0 and solving for r. We obtain $r(n)$ to be

$$r(n) = \frac{1-\lambda}{n-\lambda} \tag{7.2}$$

With adaptive retransmission probability using $r(n)$ obtained in (7.2), the drift will become

$$d(n) = \lambda - e^{-\lambda}\left(\frac{n-1}{n-\lambda}\right)^{n-1}$$

Note that now $d(n) \to \lambda - e^{-1}$ as $n \to \infty$. Hence, using Theorem D.3 with $f(i) = i$, we conclude that for $\lambda < \frac{1}{e}$, the Markov chain $\{B_k\}$ is positive recurrent; the S-Aloha network with adaptive retransmission of (7.2) is stable for packet arrival rates less than $\frac{1}{e}$.

It is not practical for the nodes to know B_k, and a node should learn the network state from the events that it can observe. Let Z_k be the event observed in slot k. Z_k takes the following values:

$$Z_k = \begin{cases} 0 & \text{idle slot; no transmission is attempted} \\ 1 & \text{successful transmission; exactly one transmission is attempted} \\ e & \text{error (collision); more than one transmission is attempted} \end{cases}$$

To learn the state, a learning variable, S_k, is used. S_k is updated based on Z_k. A node transmits with probability $\frac{1}{S_k}$ in slot k if it has either a fresh packet or a backlogged packet.

Typically, two kinds of updates are used. S_k could be updated by an additive constant depending on the event in slot k as follows:

$$S_{k+1} = \max\{1, S_k + aI_{\{Z_k=0\}} + bI_{\{Z_k=1\}} + cI_{\{Z_k=e\}}\}$$

Here a, b, and c are fixed constants. $a = -1$, $b = 0$, and $c = 1$ is a common choice for these constants. An idle slot indicates a possible overestimate of the backlog and S_k is decremented in this case. Similarly, a collision in a slot indicates a possible underestimate and S_k is incremented in this case. An alternative adaptation is to update S_k multiplicatively. For example, we could have constants $a(Z_k)$ and adapt S_k as follows:

$$S_{k+1} = \max\{1, a(Z_k) \cdot S_k\}$$

It has been shown that there exist $a(Z_k)$ that achieve the maximum possible throughput of $\frac{1}{e}$.

Using the feedback from the network—adapting the retransmission times based on Z_k—requires that all nodes be active all the time. The Z_k observed by all of them should also be the same for all k. This is clearly a very strong requirement and avoidable. To make the protocol more robust, in many random access protocol standards, a node uses its own transmission attempt history to adapt the retransmission times. Usually, the history is reset after every successful transmission. A node will make the m-th transmission attempt after a backoff period of x_m units of time. Here x_m is a uniformly distributed random integer in the interval $[0, B_m - 1]$. B_m is updated by the node at every event (collision or a success). A typical update equation has the form

$$B_m = \begin{cases} \min(a \times B_{m-1}, B_{max}) & \text{if } (m-1)\text{-th transmission collides} \\ \max(B_{m-1} - b, B_{min}) & \text{if } (m-1)\text{-th transmission is successful} \end{cases} \tag{7.3}$$

where a, b, B_{min}, and B_{max} are predefined.

Maximum Throughput in S-Aloha

In this discussion, S-Aloha was stabilized to provide a maximum throughput of $\frac{1}{e}$. Clearly this maximum throughput is a bit too low and leads to the natural question, how much more throughput can be achieved? Toward answering that question, observe that the randomized retransmission strategy is a mechanism to resolve collisions; that is, to assign transmission times to the nodes when there are multiple contenders for the multiaccess channel. The key to improving the throughput is to resolve collisions quickly. A large number of collision resolution algorithms have been proposed, their stability proved, and the corresponding maximum throughput calculated. An algorithm that leads to the best known stable throughput is described next.

The basic idea of this collision resolution algorithm is to define a time interval, $\mathcal{I}_k := (T_k, T_k + \alpha_k)$ for slot k, and prescribe that all the packets that have arrived during this enabled interval transmit in the slot. Here T_k is the left boundary of the enabled interval and α_k is its length. If slot k has a successful transmission or is idle, then it implies that all the arrivals in $\mathcal{I}_k$ have been transmitted successfully by the end of the slot k. If there is a collision in the slot, then there are more than one arrivals in $\mathcal{I}_k$ and the enabled interval for slot $(k+1)$ will be half of $\mathcal{I}_k$. The left half of $\mathcal{I}_k$ is chosen for slot $(k+1)$, so, if there is a collision in slot k, arrivals in $(T_k, T_k + \alpha_k/2)$ are enabled to transmit in slot $(k+1)$. After resolving the left half, the right half of the interval is resolved. A variable called σ keeps track of the half that is being resolved. Some optimizations are possible. The following is a formal description of the algorithm.

For each time slot k, we define the following three parameters: T_k is the left boundary of the interval, α_k is the duration of the interval, and σ_k is used to indicate the part of the starting enabled interval (left or right) that needs to be resolved. These parameters are updated as follows.

$$
\begin{array}{lllll}
\text{If } Z_{k-1} = e & \text{then} & T_k = T_{k-1} & \alpha_k = \frac{\alpha_{k-1}}{2} & \sigma_k = L \\
\text{If } Z_{k-1} = 1 \text{ \& } \sigma_{k-1} = L & \text{then} & T_k = T_{k-1} + \alpha_{k-1} & \alpha_k = \alpha_{k-1} & \sigma_k = R \\
\text{If } Z_{k-1} = 0 \text{ \& } \sigma_{k-1} = L & \text{then} & T_k = T_{k-1} + \alpha_{k-1} & \alpha_k = \frac{\alpha_{k-1}}{2} & \sigma_k = L \\
\text{If } Z_{k-1} = 0/1 \text{ \& } \sigma_{k-1} = R & \text{then} & T_k = T_{k-1} + \alpha_{k-1} & \alpha_k = \min(\alpha_0, k - T_k) & \sigma_k = L
\end{array}
$$

$$(7.4)$$

Observe that the enabled intervals in successive slots are such that the packets are made to depart in the order in which they arrived. Hence this is also called the first-come-first-served (FCFS) collision resolution algorithm. A numerical example is considered in Problem 7.5. By appropriately choosing α_0 and with minor modifications to the basic idea of (7.4), a maximum throughput of 0.487760 is known to be achievable.

7.2.2   Carrier Sensing Protocols

In networks in which the propagation delay is small compared to the packet transmission time, it is possible to infer channel state (busy or idle) through carrier

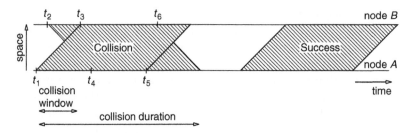

(a) A collision and a successful transmission in a CSMA network.

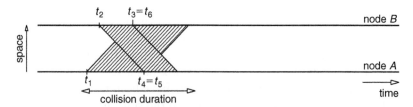

(b) A collision in a CSMA/CD network.

Figure 7.7   Collisions in CSMA and CSMA/CD networks. *A* starts transmission at t_1 and *B* at t_2. The transmission from *A* reaches *B* at t_3 and that from *B* reaches *A* at t_4. Node *A* stops transmitting at t_5 and *B* stops at t_6. Collision ends when there is no carrier on the network. During a collision in a CSMA network, $t_5 = t_1 + 1$ and $t_6 = t_2 + 1$ while in CSMA/CD $t_5 = t_4$ and $t_6 = t_3$.

sensing and thereby obtain a random access protocol that is more efficient than the pure random access strategy of Aloha. In such networks, if a node senses the channel to be busy and yet transmits, it can cause a collision at the receiver of the ongoing transmission. Further, it is likely that the ongoing transmission is being heard at the receiver of the new transmission and a collision will occur there as well. Thus both transmissions are lost. Hence, a node should listen to the channel before beginning to transmit and defer to an ongoing transmission. This is the principle of the *carrier sense multiple access* (CSMA) protocol. In this protocol, once a node begins transmitting, it transmits the complete packet.

The collision window for the CSMA is as follows. It is the time since the beginning of a transmission during which another node (not having heard the ongoing transmission) can begin its own transmission, and hence collide with the first transmission. The maximum collision window is equal to the maximum propagation delay in the network. This is because, after this interval, the carrier would have reached every node in the network and all nodes will defer a transmission attempt until the end of the packet transmission that is in progress. Two or more nodes can begin transmission within a short time of each other (less than the collision window) and collide. In this case, all the colliding transmissions will be lost. Note that during a collision in CSMA, the entire packet is transmitted by all the colliding nodes. The duration of a collision in the network is the time from the beginning of the first transmission in the collision until the earliest time at which a fresh transmission can begin. Clearly, the maximum duration of a collision will be $(t_{trans} + 2t_{propgn})$, where t_{trans} is the packet transmission time and t_{propgn} is the maximum propagation delay. Figure 7.7(a) shows a time-space representation of a collision and a successful transmission.

A further improvement over CSMA is possible by the node continuing to monitor the channel after beginning transmission. If it senses a collision on the channel, then the node can immediately stop transmission and minimize the loss of channel capacity. This is called CSMA with collision detect or CSMA/CD. The maximum collision duration in the network is reduced to $3t_{propgn}$, which is also the maximum collision duration seen at a node. Figure 7.7(b) illustrates this. The CSMA/CD protocol is used in the popular Ethernet local area network.

7.3 CSMA/CA and WLAN Protocols

Recall from our discussions in Chapters 2, 4, and 5 that in wireless networks, spatial reuse allows the spectrum to be simultaneously used in different parts of the network and significantly increases the traffic carrying capacity. Spatial reuse requires that the interference region of the transmitters be much smaller than the geographical spread of the network. This allows different transmitter-receiver pairs to be active in geographically different parts of the network. We will see that using a carrier sensing protocol in such networks can be inefficient because of *hidden* and *exposed* nodes. Further, in wireless networks, the received signal energies are very low compared to the transmitted signal energy, and it is very difficult to design reliable collision detection hardware. Hence in wireless LANs, the emphasis is on avoiding collisions, rather than detecting them. In this section, we first discuss the development of the collision avoidance mechanisms and then describe a carrier sense multiple access with collision avoidance (CSMA/CA) protocol.

Our goal in this book has been to primarily discuss wireless networking independent of technology. However, certain technologies are so widely accepted and deployed that it is imperative to understand their detailed mechanisms, performance, and other related issues. Such is the case with the CSMA/CA based wireless LAN technologies standardized by IEEE under the IEEE 802.11 series of standards, and by ETSI as the HIPERLAN 1 and HIPERLAN 2 standards. Here, we will limit ourselves to a detailed description of the IEEE 802.11 standard, by far the most widely used and ubiquitous of WLANs. We will also provide a short overview of the HIPERLAN protocol.

7.3.1 Principles of Collision Avoidance

Consider the arrangement of nodes shown in Figure 7.8. *A* and *B* are the interference regions of transmitters *a* and *b*, respectively, and *X* is the intersection

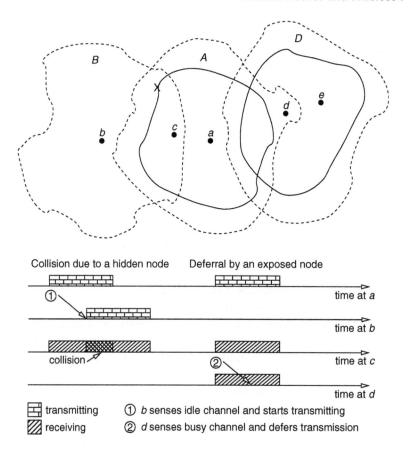

Figure 7.8   Hidden and exposed nodes in a wireless network. Propagation delays are assumed to be zero. Interference regions of nodes *a*, *b*, and *d* are regions *A*, *B*, and *D* respectively. The decode regions of *a* and *d* are shown using solid lines.

of *A* and *B*. Consider an ongoing transmission from *a* to *c*. Since *b* is outside the interference region of *a*, it cannot sense the carrier from this transmission and can decide to transmit. If *b* transmits at the same time as *a*, there will be a collision at the receivers in the region *X* including at *c*. However, *a* will not know of the collision at *c* and will continue to transmit. In this scenario, we say that *b* is *hidden* from *a* with reference to a transmission to *c*.

Now consider the transmitter *d* whose interference region is shown as *D*. *d* has to send a packet to *e* when *a* is transmitting to *c*. Node *d* is in the interference region of *a*, and can therefore sense the carrier from *a*. The two transmissions, *d-e* and *a-c*, can coexist because *c* is outside the interference region of *d*, and *e* is outside the interference region of *a*. Yet, on sensing the carrier from *a*, *d* will be forced to defer transmission. Here we say that node *d* is *exposed* to a transmission from *a*.

The bottom part of Figure 7.8 illustrates a collision due to a hidden node and an unnecessary deferral by an exposed node. In summary, in a wireless network, hidden nodes reduce the capacity by causing collisions at receivers without the transmitter knowing about it, and exposed nodes force a node to unnecessarily defer in its transmission attempts, thus reducing spatial reuse.

Collision avoidance (CA) mechanisms prevent collisions due to transmissions by hidden node. These mechanisms assume that the interference regions, and also the decode regions, for transmission and reception are identical. A simple CA mechanism is to have a narrowband auxiliary signaling channel in addition to the data channel. A node actively *receiving* data on the data channel transmits a *busy tone* on the signaling channel to enable the hidden nodes to defer to receiving nodes in their interference regions.

Dividing the available spectrum into two parts with sufficient spectral gap between them to enable the busy tone signaling mechanism to work properly is both cumbersome and inefficient. The effect of the busy tone is achieved by preceding the actual data transfer by a handshake between the transmitter and the receiver. This handshake is used to convey an imminent reception to the hidden nodes. Before transmitting a data packet, a source node transmits a (short) *request to send* (RTS) packet to the destination. If the destination receives the RTS correctly, it means that it is not receiving any other packet and it acknowledges the RTS with a *clear to send* (CTS) packet. CTS informs the neighborhood of a receiver about an impending packet reception. The source then begins the packet transmission. If CTS is not received within a specified timeout period, the source assumes that the RTS had a collision at the receiver (most likely with another RTS packet) and a retransmission is attempted after a random backoff period.

The RTS serves to inform nodes in the decode region of the transmitter about the imminent transmission of a packet, and the CTS serves the same purpose for nodes in the decode region of the receiver. Thus nodes that are not in the interference region of the transmitter, but are in the decode region of the receiver (i.e., the hidden nodes), are informed of the imminent packet transmission. If the transmission duration information is included in the RTS and CTS packets, then the nodes in the decode region of both the transmitter and the receiver can maintain a network allocation vector (NAV) that indicates the remaining time in the current transmission and schedule their own transmissions to avoid collision. Thus this handshake is a *collision avoidance* scheme and the protocol is the *carrier sense, multiple access with collision avoidance* (CSMA/CA).

Observe that after completion of the RTS/CTS exchange, the medium is reserved in the region that is the union of the decode regions of the transmitter and the receiver. Hence this basic channel access mechanism was called *multiple access with channel acquisition* (MACA) when it was first proposed. It is an adaptation of the handshake protocols used in RS-232-C (between terminal equipment like personal computers and peripheral equipment like modems and printers) and Appletalk (between communicating terminal equipment).

Collision avoidance as just described helps to reduce the inefficiency that is introduced by not being able to do collision detection in wireless networks. In principle, only RTS packets collide; these are short packets, and hence the time lost to collisions is small.

The RTS/CTS scheme helps ameliorate the hidden terminal problem but does not eliminate it. Note that only the nodes in the decode region of the receiver have been alerted by the CTS. Those in the interference region but not in the decode region of the receiver have just sensed a carrier but they do not know of an impending packet transmission. During the packet transmission, they do not sense the carrier and can still cause a collision.

Note that we have not yet addressed the exposed node problem. In fact, in the CSMA/CA scheme, it seems difficult to be able to allow an exposed node to transmit. Any node in the interference region of the transmitter of the ongoing packet is exposed. Even if such a node were allowed to transmit a RTS to a node (outside the interference region of the transmitter of the ongoing packet), it will itself not be able to receive the subsequent CTS and hence it will not know if it can transmit.

The CSMA/CA protocol has been adopted for wireless LANs in the IEEE 802.11 series of standards, which we discuss in detail later in this section.

Exercise 7.2
If all receivers transmit a busy tone, can a combination of the busy tone and carrier sensing solve the exposed node problem? Explain.

Many improvements to this basic protocol have been suggested with the most popular being called MACAW or MACA for Wireless (MACAW). An additional feature in MACAW is the use of an acknowledgment from the receiver after the successful reception of a packet.

MACAW also specifies the transmission of a short *Data Sending* (DS) packet preceding the actual data transfer. This is because it is possible that an exposed node has heard the RTS and not the CTS. If such a node is not sensing the channel, it will not know if the RTS-CTS handshake was successful and may attempt to transmit an RTS. This in turn could collide with the ACK at the transmitter of the packet. To avoid this situation, the DS packet provides information to the exposed nodes about the beginning and end of transmission times. This is also useful in networks where the nodes do not have carrier sensing capability. It has since been decided that DS is not necessary and the IEEE 802.11 standard does not include this message in its handshake protocol. The handshake and the data exchange sequence of the MACAW protocol are shown in Figure 7.9.

7.3.2 The IEEE 802.11 WLAN Standards

As was mentioned earlier, the basic ideas of the CSMA/CA protocol of MACA and MACAW have been formalized in IEEE 802.11 (Wi-Fi) wireless LAN standards.

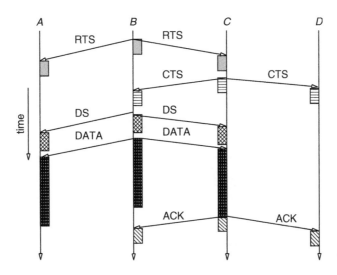

Figure 7.9   Handshake and data exchange sequence in MACAW. *A* and *B* are in each other's decode range; *B* and *C* are in each other's decode range, and *C* and *D* are in each other's decode range. *B* wants to transmit data to *C*. RTS from *B* is heard by *A* and it defers transmission till CTS may be received at *B*. CTS from *C* is heard by *D* and it defers transmission till data exchange is complete. DS is heard by *A* and it knows that the RTS/CTS handshake was successful and data exchange is in progress. It also knows the duration of the data exchange. The ACK from *C* is heard by *D* and it can then infer that data exchange between *C* and *B* is complete and *D* can now enable its transmitter.

According to this standard, the network configuration could be either of the following two modes.

- *Independent* or *ad hoc network* mode. In this mode, the nodes form an independent multihop wireless network and they communicate directly with one another. A routing protocol and a corresponding routing algorithm will need to be used so that the packets find paths to the destinations. Figure 7.10 shows an example of such a network.

- *Infrastructure* mode. Here data communication is always between a mobile station (MS) and an access point (AP). The AP is connected to the wired network and provides a service similar to the base station of a cellular network. In this mode, the MSs need to *associate* with an AP using an *association protocol*. An AP and the MSs associated with it form a basic service set (BSS), and a set of BSSs is called an extended service set (ESS). The association, and dissociation, allows the MSs to be mobile within the ESS. Figure 7.11 shows an example deployment of an ESS.

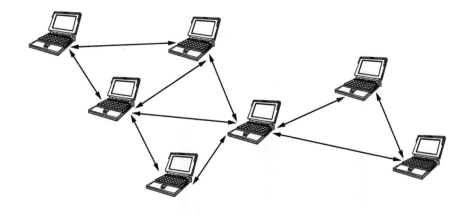

Figure 7.10   An IEEE 802.11 ad hoc network. An arrow between two nodes indicate that they are both in the decode region of each other.

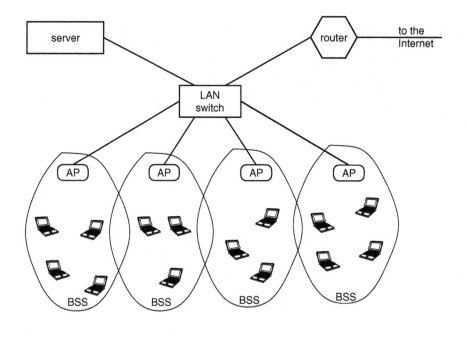

Figure 7.11   A typical IEEE 802.11 ESS architecture.

We first discuss some physical layer (PHY) issues and then describe the medium access control layer (MAC) for these networks. There are many PHY standards corresponding to different frequency bands in which the network operates and also the data rates that can be used by the nodes of the network. The initial 802.11 standard had three PHY standards—(1) infrared, (2) 1 and

2 Mbps over frequency hopping spread spectrum (FHSS) in the 2.4 GHz band, and (3) 1 and 2 Mbps over direct sequence spread spectrum (DSSS) in the 2.4 GHz band. The transmitter and the receiver can choose the data rate to suit the channel conditions.

A second version of the IEEE 802.11 standard defined the following physical layer standards:

- *802.11a* in the 5.0 GHz band using OFDM. Each BSS uses a bandwidth of 20 MHz, which is further divided into 52 OFDM carriers of which 48 are for data. Depending on the channel conditions, the data rates could be any of 6, 9, 12, 18, 24, 36, 48, or 54 Mbps.

- *802.11b* in the 2.4 GHz band using DSSS. Depending on the channel conditions, the data rates could be any of 1, 2, 5.5, or 11 Mbps.

- *802.11g* is an extension of the 802.11b and uses DSSS, OFDM, or both to support data rates in the range of 5.5 to 54 Mbps.

In addition to the data rates that can be changed according to the path loss characteristics, many channels also are defined for the 802.11b and 802.11g. The center frequencies of the channels are separated by 5 MHz and the 30 dB-bandwidth is mandated to be 22 MHz (i.e., the ratio of the peak energy to the energy 22 MHz away from the center frequency is to be more than 30 dB), whereas the 50 dB bandwidth is to be 44 MHz. This means that channels are overlapping but some are functionally nonoverlapping; that is, nodes communicating on these channels can coexist in the same geographical area without causing significant interference to each other. It can be seen that a maximum of three nonoverlapping channels are available—channels 1, 6, and 11. In 802.11a, 12 nonoverlapping channels are defined. Of course, overlapping BSSs need to use nonoverlapping channels. Since they use different frequency bands and different modulation techniques, 802.11a and 802.11b are not interoperable; the latter is more popular and more widely deployed. Table 7.1 summarizes this discussion.

Standard	Band	Data rates	Num of channels
IEEE 802.11 (dated)	2.4 GHz	2 and 1 Mbps	1
IEEE 802.11b	2.4 GHz	11, 5.5, 2 and 1 Mbps	14
IEEE 802.11a	5.0 GHz	54, 48, 36, 24, 18, 12, 9 and 6 Mbps	12
IEEE 802.11g	2.4 GHz	1–54 Mbps	14
IEEE 802.11n		work in progress, data rate up to 540 Mbps	

Table 7.1 Summary of the 802.11 PHY Standards.

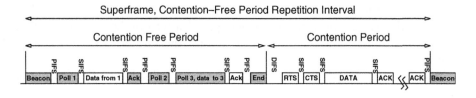

Figure 7.12 Beacon frame, PCF, and DCF periods in an IEEE 802.11 network. The maximum duration for which the contention-free period will last is called the *CFP_Max_Duration*. RTS is transmitted by sender and CTS by the intended receiver of the packet.

We now describe the MAC protocol of an IEEE 802.11 network. Two basic protocols are defined—a polling-based protocol called the point coordination function (PCF) and a random-access protocol called the distributed coordination function (DCF). PCF and DCF can coexist in the same BSS, and we first describe the two protocols assuming that the two coexist.

The BSS has a point coordinator that is typically the AP of the BSS. Time is divided into superframes and each superframe has two parts—(1) the *contention free period* (CFP) and (2) the *contention period* (CP). PCF is used in the CFP and DCF in the CP.

PCF is initiated by the AP by transmitting a beacon frame. The eligible nodes in the BSS then are polled and the data that need to be transmitted to them are transmitted along with the polling message. If the polled node has packets to transmit, it will also transmit them in response to the polling packet. The PCF ends when all the nodes are polled by the AP. The end of the PCF mode of medium access is signaled using the *End* frame. This also marks the end of the contention-free period. This is followed by a contention period using the DCF-based MAC, which continues until the end of the superframe period. Thus we see that in a superframe period, the CFP and the CP alternate. The AP will begin trying to initiate a new CFP, after *target beacon transmission time (TBTT)* has elapsed from the time that the previous one was initiated. Thus TBTT specifies the period of the superframe. Figure 7.12 shows this alternating sequence of CFP and CP and also a sample of the types of packets that are transmitted in the network.

The DCF is derived from the CSMA/CA MACAW protocol that we have described earlier in the section. In addition to the RTS and CTS based handshake mechanism before the transmission of the data packet, the standard specifies the following:

- Minimum silence periods between transmissions; different kinds of packets have to compulsorily wait for different lengths of time after the medium is sensed idle to begin transmissions. The minimum idle sensing time prioritizes different transmissions—shorter minimum waiting time implies higher priority.

- Backoff mechanism to resolve collisions. Like in any backoff mechanism, backoff durations are measured in multiples of a basic slot time, the length of which depends on the version.

Figure 7.13 shows the events during data transfer between four nodes exchanging data during the CP using DCF. A single hop network is considered. At the beginning of the fragment of time shown in the figure, Node 1 is transmitting a MAC data packet, and all the other nodes have deferred to this transmission. At the end of the data transmission, there is a *short inter-frame space (SIFS),* which allows the receiving node to *turn around its radio* and send back a MAC level ACK packet. When this ACK transmission ends, the channel is sensed to be idle by all the nodes, and each one of them starts a *DCF inter-frame space (DIFS)* timer. The DIFS duration is more than SIFS (e.g., in 802.11b, SIFS is 10 μsec and DIFS is 50 μsec). Thus, even though Nodes 2, 3, and 4 did start their DIFS timers

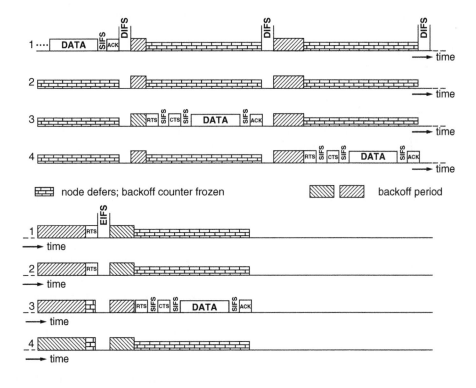

Figure 7.13 Events during data transfer in the IEEE 802.11 DCF MAC protocol with the RTS/CTS mechanism. There are 4 nodes. There is one time line for each node; the time lines start in the top left side of the figure, proceed to the right, and then continue from the left in the bottom part of the figure. Different backoff counts are indicated by a change of the shading.

when Node 1's data packet had completed transmission, since SIFS < DIFS the channel became busy again (with the ACK packet) before the DIFS timers could expire. Thus, we see that the channel essentially is reserved for the receiver and transmitter between which a data packet is being transferred. In the light of our discussions in Section 7.3.1, such reservation is valid only if there are no hidden nodes that are within the interference range of the receiver that cannot decode the CTS, and also cannot sense the carrier from the transmitter. Such nodes will hear the CTS, and hence will defer transmission for a DIFS after the CTS signal ends. After this, they are free to transmit, and hence could nullify the supposed reservation.

When the DIFS timers expire, each node enters a *backoff* phase. Even though the channel is idle, random backoff is used to try to order the transmissions so that a collision does not occur. The node that just completed its data transmission samples a new random backoff value. If a node was already in a backoff when Node 1 started its data transmission, then during Node 1's data transmission, this node's backoff timer is frozen. Upon completion of Node 1's transmission each node that had deferred to Node 1 continues the remainder of its backoff. If a node was idle when Node 1 started its data transmission, and if a packet arrived during Node 1's transmission, then the node with the new arrival defers until the completion of the transmission; it then waits for the DIFS period, and starts a new backoff. The backoff durations are multiples of the basic slot time. When a new backoff is sampled, this multiple is sampled uniformly from the integers $\{0, 1, \ldots, CW_{min} - 1\}$.

When a backoff period of some node expires (for example, Node 3 in Figure 7.13), then this node transmits an RTS packet to its destination node. Upon hearing activity on the medium, all other nodes freeze their backoff timers. The node to which the RTS was directed then sends back a CTS packet (a SIFS elapses in between the two). Node 3 then waits for a SIFS and sends its data packet, after which an ACK is sent by the receiver node. The figure also shows another successful transmission from Node 4.

A collision occurs if two nodes finish their backoffs within one slot of each other. It is assumed that the maximum propagation delay in the network is such that all nodes are able to sense a transmission within one slot time. In this case, both RTS packets collide. A CTS time-out then follows, after which the colliding nodes sample a backoff from a doubled *collision window*; that is, from the window $\{0, 1, 2, \ldots, 2 \cdot CW_{min} - 1\}$. After the collision event, the nodes that were not involved in the collision continue their backoffs with their residual backoff timers. A collision event (e.g., between RTS packets of nodes 1 and 2) is shown in the continued time lines in the bottom of Figure 7.13, where we have taken the additional time after the RTS transmission to be extended inter-frame space (EIFS), before the backoff durations are started. Repeated collisions lead to a doubling of the collision window until it reaches CW_{max}, after which the collision window remains fixed. In the standard, $CW_{min} = 32$, and $CW_{max} = 1024$.

Version	Slot	SIFS	PIFS	DIFS	CW_{min}	CW_{max}
IEEE 802.11a	9 μsec	16 μsec	25 μsec	34 μsec	15	1023
IEEE 802.11b	20 μsec	10 μsec	30 μsec	50 μsec	31	1023
IEEE 802.11g	9 μsec	10 μsec	19 μsec	28 μsec	16	1024

Table 7.2 Interframe spacings and transmission times and the values of the CW_{min} and CW_{max} for the different 802.11a and 802.11b. The handshake packets RTS, CTS and ACK packets are 20, 14, and 14 octets, respectively, and transmitted at the lowest transmission rate. The payload in a packet could be up to 2312 bytes. The MAC header and trailers constitute 34 octets. In addition, there will be PHY headers of 192 bits.

EIFS is longer than DIFS and is used by nodes that cannot decode a transmission. This could either be because of a collision or because the received power is such that the SNR is below the decoding threshold. The duration of EIFS is equal to the sum of the transmission times of SIFS, CTS (and ACK), and DIFS packets and the PHY headers. This prevents collision during reception of CTS and ACK.

The slot lengths and the duration of spacings in the different standards are shown in Table 7.2. Observe that the interframe spacing, the minimum time after a transmission for which a node has to wait before transmitting, is a prioritization mechanism. For example, since the SIFS < DIFS, the CTS, data, and ACK packets have priority over initiation of new transmissions. Also, when the point coordinator wants to initiate the CFP, it waits for PCF interframe spacing (PIFS) after the end of an ongoing transmission before transmitting a beacon or the polling packet. SIFS < PIFS < DIFS implies that the initiation of the CFP has priority over new transmission initiations. We will see later that multivalued DIFS can be used to prioritize among different traffic classes.

7.3.3 HIPERLAN

ETSI has defined a standard that is similar to the IEEE 802.11 and is called the *high performance radio LAN* (HIPERLAN). HIPERLAN is essentially like the IEEE 802.11 at the physical layer but has significant differences with it in the channel access method, which is called the channel access control (CAC). Packets in HIPERLAN are assigned a priority. A brief description of HIPERLAN CAC follows.

The CAC of HIPERLAN uses an elimination-yield, nonpreemptive multiple access (EY-NPMA) mechanism. When a node has a packet to transmit, if the channel is sensed to be idle for a period called the *channel_free_interval,* then the node begins transmission of its packet. If the channel is sensed busy, then a contention resolution mechanism called the *synchronized_channel_access* starts at the end of the current transmission. This access protocol has three phases—prioritization, elimination, and yield phases. The prioritization phase reserves the channel for the highest priority packets that are contending for the channel. There are H slots in this phase and nodes that need to transmit packets of priority p transmit a radio pulse in slot p if the preceding prioritization slots—slots $1, \ldots, (p-1)$—are idle. Of course, Priority 1 is the highest priority. A radio pulse in slot p^* reserves the channel for packets of priority p^* and prevents lower priority packets from competing for access. Let p^* be the highest priority packets that are contending for the channel during an instance of synchronized_channel_access. Only nodes with priority p^* packets contend in the next two phases.

The second phase of the synchronized_channel_access is called the elimination phase, in which nodes transmit for a random duration from the beginning of the phase and those that do not transmit for the longest duration are eliminated. Specifically, each node with priority p^* packets transmits the carrier for a random number of slots, say E_i by node i, and then senses the channel immediately after that. If the channel is idle in the subsequent slot then it has survived the elimination phase and is eligible for the yield phase. The third phase is called the yield phase, in which the nodes that transmit the earliest win that phase. Specifically, each eligible node (i.e., those surviving the elimination phase) listens to the channel for a random number of slots, say Y_i by node i, and then transmits the data packet. E_i and Y_i are chosen according to a truncated geometric distribution by each node. We reiterate that the nodes that choose the largest E_i win in the elimination phase whereas the node(s) that choose the smallest Y_i wins the yield phase. More than one node may transmit simultaneously in the yield phase causing a collision. A collision resolution mechanism is then invoked. Figure 7.14 shows a sample of the channel activities.

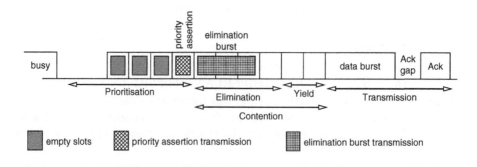

Figure 7.14 The prioritization phase lasts four slots with packets of priority four surviving this phase. Nodes with priority 4 packets transmit in the elimination phase. All of these nodes transmit for three or less slots in this phase. Those nodes that transmitted for three slots in the elimination phase contend in the yield phase and the minimum yield was two slots. Note that the timing of the events is not to scale.

7.4 Saturation Throughput of a Colocated IEEE 802.11-DCF Network

As can be seen from the description of the 802.11-DCF MAC earlier, an exact analytical model of the DCF protocol can be quite complex and possibly intractable. However, by making a few reasonable approximations, a tractable model that can provide insights into the performance of this protocol can be obtained. We discuss one such model that obtains the *saturation throughput*.

Saturation throughput analysis is an important development in understanding the performance of the CSMA/CA protocol in 802.11. Here, we assume that all nodes always have packets to transmit; at a node, a successfully transmitted packet is replaced immediately by another packet that needs to be transmitted. This is also called the *infinite backlog* model. The throughput of the network under this saturation assumption is called the *saturation throughput*. Saturation throughput analysis has been used in systems before, notably in the study of switching systems. It is important to note that in general, the saturation throughput is not the same as the capacity. It is, however, a good indicator of the capacity and for some special systems it has been shown that the queues will be stable if the arrival rate is less than the saturation throughput.

We consider a colocated network of n saturated nodes. This model is applicable in the infrastructure mode where all the nodes are associated with the same access point, or in the ad hoc mode when the geographical spread of the networks is such that each node is within the decode range of all the other nodes. We further assume homogeneous nodes; where the parameters of the backoff process and the state machine that implements it are identical at all the nodes.

To simplify the analysis, we assume that at all the nodes, the backoff times corresponding to both a fresh transmission and after a collision, are sampled from an exponential distribution with mean $\frac{1}{\beta}$. Note that in an implementation of the protocol, in the saturation condition, as the number of nodes increases, the number of collisions will increase and hence the average backoff durations will increase; in a sense, $\frac{1}{\beta}$ captures this average backoff time. We will see a little later how the average backoff time can itself be analyzed. Let δ denote the slot duration specified by the protocol. We will assume that if two nodes transmit within δ of each other, then a collision occurs.

Recall that the nodes freeze their backoff counters when they sense activity in the medium and resume the count after the mandated silence periods. Thus the backoff process is active only when the medium is idle and it is instructive to view the backoff process by considering only the idle times on the channel. This is shown in Figure 7.15 for the transmission sequence of Figure 7.13.

Observe from Figure 7.13 that there are alternating busy and idle periods on the medium. This is abstracted in Figure 7.16. The busy period could correspond to either a successful transmission or to a collision in which two or more nodes transmit an RTS. Now consider the instants of time at which an idle period begins, either after DIFS following a successful transmission, (as at a in Figure 7.16) or

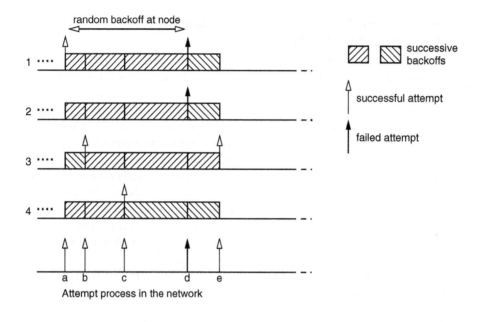

Figure 7.15 Backoff process for the transmissions of Figure 7.13 after removing the channel activity. Interruptions to each countdown by channel activity are shown; e.g., Node 1 was interrupted twice. Aggregate attempt process is shown at the bottom.

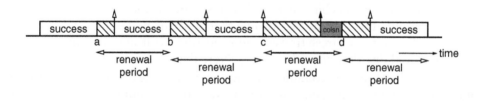

Figure 7.16 The channel alternates between busy and idle periods. The busy periods could correspond to collisions or to successful transmissions.

after a collision period ends (as at d in Figure 7.16). At this time, the nodes involved in the busy period will start a new backoff time (e.g., Node 1 at a and Nodes 1 and 2 at d) and since all nodes have a packet to transmit, the nodes not involved in the busy period resume their backoff timer countdown (e.g., Nodes 2, 3, and 4 at a). From our assumption that the backoff times are all exponentially distributed, the residual backoff times of those that are resuming the backoff and the "fresh" backoff times of those involved in the busy period (the different blocks in Figure 7.15) are all exponentially distributed. Hence, the idle period will last until the completion of the first of these backoffs. This period is the minimum of n i.i.d. exponential random variables of mean $\frac{1}{\beta}$ and is therefore exponentially distributed with mean $\frac{1}{n\beta}$.

At the end of the idle period, a collision will occur if a second backoff completes within δ of the first one. Let T_c denote the average duration of the collision period. If there is no collision, the transmission will be successful. Let T_s denote the length of the average successful period.

Exercise 7.3

Show that the probability that a transmission attempt suffers a collision is given by $\gamma = 1 - e^{-(n-1)\beta\delta}$. Also, show that, for large n, the probability of a collision is approximately given by $\gamma \approx n\beta\delta$.

For the model, we can make the following two observations: (1) idle periods are i.i.d. exponential with mean $\frac{1}{n\beta}$ and (2) the event that an idle period ends in a collision or successful transmission is independent of the event at the end of the previous idle periods.

We will also the assume that the collision durations and the packet lengths (and hence the durations of the successful transmissions) are independent. We can then say that the instants at which idle periods begin are *renewal points*. The mean time between successive renewal points is the sum of the mean idle period and the mean busy period. The mean busy period is given by $\gamma T_c + (1 - \gamma)T_s$. Thus the mean renewal time is

$$\left(\frac{1}{n\beta}\right) + ((1 - \gamma)T_s + \gamma T_c)$$

From the renewal reward theorem (Theorem D.9), the normalized network throughput, $S(\gamma, \beta)$, defined as the fraction of time that the network is involved in transmitting successful packets, is

$$S(\gamma, \beta) = \frac{(1 - \gamma)T_s}{\frac{1}{n\beta} + (1 - \gamma)T_s + \gamma T_c} \tag{7.5}$$

Let the transmission rate of each of the nodes be r bits per second, all packets have L bits, and let T_o be the overhead per packet (the duration of the handshake and the interframe spacings). Then $T_s = L/r + T_o$ and the throughput, in bits per second, is

$$\frac{(1 - \gamma)L}{\frac{1}{n\beta} + (1 - \gamma)\left(\frac{L}{r} + T_o\right) + \gamma T_c} \tag{7.6}$$

From this expression for the normalized throughput, we observe an interesting tradeoff. Notice that the backoff rate parameter, β, appears in two ways: in the mean time until an attempt (i.e., $\frac{1}{n\beta}$), and in the collision probability, γ.

Using a large value of β (i.e., a high attempt rate) reduces the mean time between attempts but increases the probability of collision, and vice versa. Also, observe that as $\beta \to 0$, the normalized throughput goes to zero, since in the limit, there are no attempts; further, as $\beta \to \infty$ the normalized throughput again goes to 0, since the probability of collision goes to 1. It is therefore interesting to seek the value of β that maximizes the normalized throughput.

Exercise 7.4

β, the backoff parameter, as we will see later, is the conditional attempt rate by a node. Show that the normalized throughput is maximized for

$$\beta = \frac{1}{n}\frac{1}{2T_c}\left(\sqrt{1 + \frac{4nT_c}{(n-1)\delta}} - 1\right)$$

Recall a similar exercise for Aloha where we optimized the conditional attempt rate G and also the retransmission probability in (7.1).

This exercise clearly motivates the need for adaptive backoff in the IEEE 802.11 MAC protocol when the number of contending nodes is not known. We observe from this exercise that the optimal value of β is inversely proportional to n, the number of transmitting nodes, similar to $r(n)$ of (7.1). Equivalently, the mean backoff time should be proportional to the number of nodes. According to the standard, the contention window is dropped to CW_{min} after a successful transmission. This is a good approach for the situation when a few nodes are intermittently active. If several continuously active nodes are sharing the network, then it may be better for the nodes to adapt to the optimal value of the contention window and then retain this value.

The analysis when the backoff counter decrements in steps of a slot time, as prescribed in the standard, can be carried out along the same lines. Here we assume that in each slot each node attempts a transmission with probability β independent of the attempts of other nodes. (Note that earlier, β was the reciprocal of the mean backoff period.) Thus the number of slots in an idle period would follow a geometric distribution and have a mean of $\frac{1}{1-(1-\beta)^n}$. Further, the transmission attempt of a node in a slot is successful if none of the other nodes in the network attempt in the same slot. Thus $\gamma = 1 - (1 - \beta)^{n-1}$. The idle period is followed by a successful transmission if exactly one node attempts a transmission in the slot, conditioned on one or more nodes attempting a transmission. The probability for this event, $P_s(n)$, given by

$$P_s(n) = \frac{n\beta(1-\beta)^{n-1}}{1-(1-\beta)^n} \tag{7.7}$$

The normalized throughput is

$$S(\gamma, \beta) = \frac{\frac{n\beta(1-\beta)^{n-1}}{1-(1-\beta)^n} T_s}{\frac{1}{1-(1-\beta)^n} + \frac{n\beta(1-\beta)^{n-1}}{1-(1-\beta)^n} T_s + \left(1 - \frac{n\beta(1-\beta)^{n-1}}{1-(1-\beta)^n}\right) T_c} \tag{7.8}$$

The numerical results that we present later in this section are all obtained using this time slotted model.

We now turn to the problem of obtaining the average backoff duration, $\frac{1}{\beta}$, in the IEEE 802.11 MAC protocol framework. We will first obtain an expression for β in terms of γ, the collision probability. This expression, along with the equation for γ in terms of β, obtained in Exercise 7.3, can be solved to obtain both β and γ and hence to calculate the throughput using (7.5).

Let us now consider a node, called the tagged node, and focus only on those time periods when its backoff timer is counting down during the idle times on the channel. Let t represent this idle time, and let $G(t)$ be the number of attempts by a node up to (idle time) t. Then, β would be the limit

$$\beta = \lim_{t \to \infty} \frac{G(t)}{t}$$

From this, we see that β is the unconditional *attempt rate*. The interpretation of β is the same even in the slotted time model where β is the mean number of attempts per slot.

Most random-access MAC protocols specify an upper bound on the number of transmissions attempts that can be made for each packet. Once this limit is reached the packet is discarded by the node and a new packet is taken up. Now, let $(K + 1)$ be the maximum number of collisions that a packet can experience before it is discarded. Let b_k be the mean backoff duration of a node after the k-th collision, $k = 0, 1, 2, \ldots, K$. As an example, if $K = 1$, then each packet is attempted at most twice. In the first attempt the mean backoff period is b_0; if a collision occurs on this attempt, then one more attempt is made after a random backoff period that has mean b_1. Failure of this second attempt leads to the packet being discarded. In practice, b_k is increased with k.

We will now obtain β when the number of transmission attempts is upper bounded. A sample attempt process, in the idle time t defined earlier, at the tagged node is illustrated in Figure 7.17. Let A_j be the number of transmission attempts and B_j be the total backoff duration (in countdown slots) for the j-th packet from the tagged node. A_j has a truncated (at $(K+1)$) geometric distribution; that is, the probability that the j-th packet makes k transmission attempts is $(\gamma^{k-1}(1 - \gamma))$ for $k = 1, \ldots, K$ and γ^K for $k = K + 1$. B_j is the sum of A_j random backoff periods.

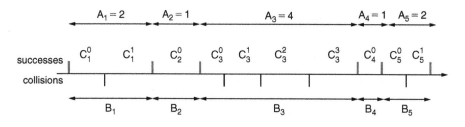

Figure 7.17 The attempts process at a node during the idle time. The busy periods that have been "removed" are also indicated as successes and collisions. Each attempted packet starts a new backoff cycle indicated as C_j^i for the i-th backoff of the j-th packet. A_j and B_j are also indicated.

Therefore, we see that

$$\mathsf{E}(A_j) = 1 + \gamma + \gamma^2 + \cdots + \gamma^K$$

$$\mathsf{E}(B_j) = b_0 + \gamma(b_1 + \gamma(b_2 + \gamma(\cdots(\cdots(\gamma b_K)))))$$

$$= b_0 + \gamma b_1 + \gamma^2 b_2 + \cdots + \gamma^k b_k + \cdots + \gamma^K b_K \qquad (7.9)$$

There will always be one attempt and hence one backoff. There will be a collision on the first attempt with probability γ, resulting in two or more attempts and hence two more more backoffs. Given that there is a second attempt, there will be three or more if there is a collision of the second attempt. And so on.

We can say that for packet j, in an interval B_j, A_j attempts were made. From the assumptions of our model, both $\{A_j\}$ and $\{B_j\}$ are sequences of i.i.d. random variables. Hence we can use the renewal reward theorem (Theorem D.9) again to say that A_j is the reward in the renewal period B_j and obtain

$$\beta = \frac{\mathsf{E}(A_j)}{\mathsf{E}(B_j)}$$

As desired, we have expressed β in terms of quantities specified by the standard, and the (assumed) collision probability γ. Let us define the two functions

$$G(\gamma) := \beta = \frac{1 + \gamma + \gamma^2 \cdots + \gamma^K}{b_0 + \gamma b_1 + \gamma^2 b_2 + \cdots + \gamma^k b_k + \cdots + \gamma^K b_K}$$

$$\Gamma(\beta) := \gamma = 1 - (1 - \beta)^{n-1} \qquad (7.10)$$

In writing the second equation we have essentially made the *decoupling approximation*—the attempt process of the tagged node is independent of the aggregate attempt process of all the other nodes in the network. This means that the success of each transmission attempt from the tagged node is independent of all other attempts and the probability of a success is a constant. (Our initial assumption that the backoff periods are exponential is a further simplification of this decoupling approximation.)

Solving for γ from the two equations in (7.10) is equivalent to asking for a solution to the following *fixed point* equation:

$$\gamma = \Gamma(G(\gamma)) \tag{7.11}$$

Since $\Gamma(G(\cdot))$ is a continuous function that maps the interval $[0, 1]$ into itself, Brouwer's fixed point theorem guarantees that there is a solution to this fixed point equation (see Appendix B, Section B.1).

Exercise 7.5

Let $K = \infty$; a packet is attempted until it succeeds and is never discarded. Further, assume that there is an $m \geq 1$ such that $b_k = \left(\frac{2^k CW_{\min}-1}{2}\right)\delta$, for $0 \leq k \leq m - 1$, and $b_k = \left(\frac{2^m CW_{\min}-1}{2}\right)\delta$, for $k \geq m$. Show that

$$G(\gamma) = \frac{2(1-2\gamma)}{(1-2\gamma)(CW_{\min}-1)+\gamma CW_{\min}(1-(2\gamma)^m)}\frac{1}{\delta}$$

Show also that $G(\gamma)$ is a decreasing function of γ.

Obviously, $\Gamma(\cdot)$ is increasing in its argument. For the case in the preceding exercise, we have seen that $G(\cdot)$ is decreasing in its argument. It follows that $\Gamma(G(\gamma))$ is decreasing in γ, and hence that there is a unique fixed point. Let this fixed point be γ_0. The attempt rate can then be calculated as $G(\gamma_0)$, and the network throughput is calculated as $S(\gamma_0, G(\gamma_0))$.

As must be evident, in the model that we just analyzed, we have made many simplifying assumptions, some of which may seem a bit far fetched. We would first like to check the effect of these simplifications. To do that, we first construct an exact simulation model that captures all the details of the protocol specification and obtain the performance parameters of interest. We then compare these exact results with those obtained from the analytical model. We first study γ as a function of n in Figure 7.18. We have two observations. First, it is heartening to note the close match between the analytical result and that from the simulation over a large range of n. More importantly, the collision probability increases fairly rapidly with n. Our goal, though, was to obtain the saturation throughput. This is shown in Figure 7.19 as a function of n and for different protocol parameters. An interesting observation is that the saturation throughput does not vary much with the number of nodes, thus suggesting that the backoff mechanism correctly adapts the attempt probabilities to keep the throughput roughly constant. As n increases,

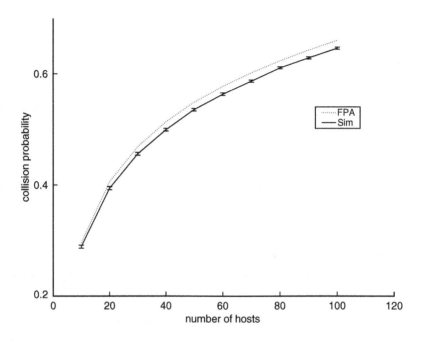

Figure 7.18 The collision probability γ as a function of n from the fixed point analysis and from an exact simulation model.

the high throughput is maintained even as the collision probability increases. This is achieved by decreasing the average idle time.

We find that simple (almost "back-of-the-envelope") analyses can be highly effective in capturing the behavior of complex systems. The decoupling assumption simplifies analysis and can easily be extended to obtain additional insights, some of which are discussed next.

Discussion

1. Although the decoupling approximation is a strong one, with some simplifications, it can be shown to be asymptotically exact; that is, it has been shown that as $n \to \infty$, under saturation, the backoff timers of all the nodes evolve independently.

2. In the slotted time model, under an asymptotic regime, as $n \to \infty$, $n\beta$ (β is the attempt probability of a node in a slot) converges to a positive value. From the relation between the binomial and Poisson distributions, the number of nodes (other than the tagged node) attempting transmission in a slot can be shown to have a Poisson distribution with mean $(n-1)\beta$. Thus in the asymptotic regime, we can write

$$\Gamma(\beta) = 1 - e^{-(n-1)\beta}$$

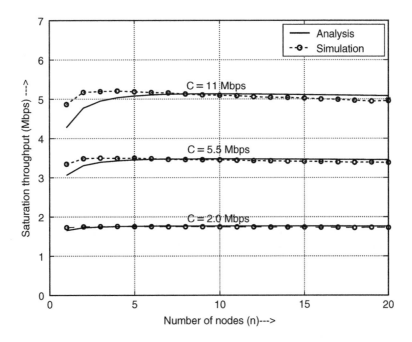

Figure 7.19 Saturation throughput as a function of the number of nodes in the network, for various PHY rates and packet size of 1500 bytes.

3. Note that in the computation of β we did not use the distributions of the backoff times, but only the averages (i.e., the b_k, $k = 0, 1, \ldots, K$); this suggests that the performance may depend on the backoff distributions only through their means. Of course, we have not proved this.

4. In general, it is possible that a fixed point equation has a nonunique solution. If this were the case for (7.11), it could imply that the system has multiple equilibrium points, each one corresponding to a different solution. This could affect the throughput of the protocol. However, it can be shown that (7.11) has a unique fixed point if b_k is a nondecreasing sequence for $k \geq 0$.

5. An important insight from the analysis is that the success probability of an attempt does not depend on the PHY parameters like the transmission rate. It depends only on the backoff parameters. Further, since the backoff process and the parameters are the same at all the nodes, the probability of success is the same for each node.

6. As we mentioned earlier, the approach to obtaining the throughput expression in (7.5) or (7.8) can be used to derive other expressions that yield valuable insights. Recall that many PHY rates are possible in each version of the IEEE 802.11 standard; the standard provides a mechanism for nodes to autonomously vary their rates as they perceive poor or improved

channel performance. If the nodes are mobile, the channel quality between transmitters and receivers will vary, and nodes would need to adapt their transmission rates to suit the channel conditions. Thus, instead of the common rate r in (7.5), a node dependent rate r_i would have to be used. From the previous remark, this means that the node with the lowest transmission rate will govern the network throughput. This is explored further in Problem 7.13.

7. The nodes could also use different backoff parameters. In this case, rather than assume that all the nodes have the same γ, we look for a fixed point solution with γ_i for node i, $1 \le i \le n$. We discuss this in the next section.

8. Most throughput analyses of random access protocols are based on Markov renewal processes. The key idea is the same as here—channel activity alternates between busy periods (collisions or successful transmissions) and idle periods. The channel activity model will identify the following:

   - The renewal points, which are typically the beginnings of the busy periods.

   - The expected length of the renewal period. The model is chosen to make the busy and idle periods i.i.d. and also independent of each other. Further, the events that a busy period is a collision or a successful transmission will also be independent. The expected length of the renewal period is then the sum of the expected lengths of the busy and idle periods.

Let T_i, T_s, and T_c be the expected durations of an idle, successful transmission, and a collision, respectively. Let P_s be the probability that a busy period is a successful transmission. The normalized throughput, S, is then

$$S = \frac{P_s T_s}{P_s T_s + (1 - P_s)T_c + T_i} \tag{7.12}$$

P_s and T_i depend on the transmission attempt model parameters and T_s and T_c depend on the network specifications.

7.5 Service Differentiation and IEEE 802.11e WLANs

QoS at the MAC layer can be provided by either of two mechanisms: (1) *Per-flow time reservation* with admission control, and (2) *Service differentiation* by dividing traffic (or nodes) into different classes and guaranteeing a service quality to each aggregate. In per-flow reservation methods, MAC-level flows are defined and each flow is guaranteed a certain fraction of time during which the node can transmit. The actual rate of transmission will depend on the characteristics of the medium between the transmitter and the receiver. In service differentiation, traffic of the same class compete with one another and receive best-effort-within-class service, and the different classes receive different grades of service in the aggregate.

Absolute guarantees of QoS parameters like delay and loss are not provided. Thus, this is also called "better than best effort" service and is suitable for elastic traffic. In this section, we will consider only service differentiation.

A simple method to provide service differentiation would be to assign absolute priorities to the classes and provide a nonpreemptive priority service. (Recall that this is provided by the prioritization phase of HIPERLAN.) If the packet arrival stream to the higher priorities is a random process, this can cause starvation of the lower priorities and usually is avoided in LAN environments.

A second method to provide service differentiation is to reserve capacity for the different classes. Packets of the same class compete among themselves for channel access. In the IEEE 802.11 WLAN standard, this can be accomplished in two ways, (1) a polling mechanism and (2) an enhanced version of the DCF. We describe these here.

Recall that polled access is provided during CFP. This can be used to serve different classes of traffic in different ratios by varying the polling rate and the duration for which a node is allowed to transmit each time it is polled. This is the method used in the *HCF (hybrid coordination function) controlled channel access (HCCA)*. The AP, which in this case will also be called a hybrid coordination controller (HCC), starts a CFP during which nodes are polled in a predetermined manner. For each node, a *service interval (SI)* is calculated and the node is polled by the HCF with a period equal to SI. Every time it is polled, a node is allowed to transmit for a maximum duration specified by the parameter *transmission opportunity (TxOP)*. Thus, each node is allocated $\frac{TxOP}{SI}$ fraction of time on the channel. This system can be analyzed like a polling system.

The second method, the *enhanced DCF (EDCF)*, is an extension to the IEEE 802.11 DCF. It classifies and prioritizes medium access among the traffic classes, which are called access classes (AC). Recall that the duration of the interframe spacings can be used to provide priority to different types of packets. Different DIFS durations can be used to give different ACs priorities in transmitting the RTS. Further, different backoff windows and backoff window growths are defined for the different classes. This provides service differentiation by changing the probability of obtaining channel access. The following parameters are defined for each AC.

- *Arbitration interframe space* (AIFS) that specifies the minimum number of slots for which the AC should sense the channel to be free before attempting a transmission. Higher priority nodes will start backoff countdown earlier than lower priority nodes and hence, will have a higher success probability.

- Different minimum and maximum contention windows, CW_{min} and CW_{max}, respectively, are specified for each class. Clearly, having a lower CW_{min} will increase the probability of success at the first and subsequent early attempts. Similarly, a lower CW_{max} will increase the success probability if the packet experiences many collisions.

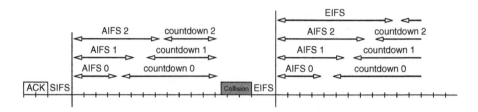

Figure 7.20 Three ACs (0, 1, and 2) are shown. The AIFS for the three classes and the countdown period are shown. The collision could be from either two nodes of the same AC or nodes from different ACs having their backoff counters reach 0 in the same slot.

- The transmission opportunity (TxOP) limit specifies the maximum time for which a node can transmit after acquiring the channel. Allowing high priority nodes a larger TxOP implies that their contention cost per bit (or packet) can be reduced.

Each node maintains a separate queue for each AC. At each node, an internal contention mechanism chooses the AC that will be transmitting a packet. The backoff counter will begin countdown after the channel has been idle for the period specified by the AIFS of the AC. It will be transmitted when the backoff counter has counted down to zero. This allows the higher priority nodes to get a chance to transmit earlier than the lower priority nodes. Figure 7.20 illustrates this.

To analyze the EDCF access mechanism in the same framework as that of the single class network from the previous section, we can just extend the fixed point analysis described to multiple classes. Recall that the parameter in the saturation throughput model is the attempt rate. In the analysis of the EDCF, we assume that the effect of the different MAC parameters described earlier essentially translates to an attempt rate, β_i for class i and hence a different conditional collision probability γ_i. Thus we can generalize (7.9) and (7.10) as follows.

$$E(A_i) = 1 + \gamma_i + \gamma_i^2 + \cdots + \gamma_i^{K_i}$$

$$E(B_i) = b_{i,0} + b_{i,1}\gamma_i + b_{i,2}\gamma_i^2 + \cdots + b_{K_i}\gamma_i^{K_i}$$

$$G_i(\gamma_i) := \beta_i = \frac{1 + \gamma_i + \gamma_i^2 + \cdots + \gamma_i^{K_i}}{b_{i,0} + b_{i,1}\gamma_i + b_{i,2}\gamma_i^2 + \cdots + b_{i,K_i}\gamma_i^{K_i}}$$

$$\gamma_i = \Gamma(\beta_1, \cdots, \beta_n) = 1 - \prod_{j=1, j \neq i}^{n} (1 - \beta_j)$$

$$= \Gamma(G_1(\gamma_1), \cdots, G_n(\gamma_n))$$

Here, $b_{i,j}$ is the average backoff duration for AC i for the $(j + 1)$-th attempt and $(K_i + 1)$ is the maximum number of attempts that a packet of AC i will make. This generalization accounts for the different $CW_{\min}$, $CW_{\max}$, and the multiplier for increasing the value of CW after a collision, but does not take into account the different AIFS and TxOP for the different traffic classes. Using the decoupling approximation that we have made in the single class analysis, we get the expression for γ_i. The last equation can be written compactly as γ

$$\gamma = \Gamma\left(\mathbf{G}\left(\gamma\right)\right) \tag{7.13}$$

This is a vector fixed point equation and it can be shown that the fixed point exists. Depending on the $\{b_{i,j}\}$ and $\{K_i\}$, there will be many interesting properties for the fixed point and we enumerate some of these later. First, we make the following definition. If a solution to (7.13) is such that $\gamma_i = \gamma_j$ for all $1 \leq i, j \leq n$, then we say that such a fixed point is balanced; else we will say that it is unbalanced. A unique balanced fixed point essentially means that the model indicates that all the classes receive the same throughput. We make the following remarks on the solution of (7.13).

- For the homogeneous case (all the nodes have the same backoff parameters) if an unbalanced fixed point exists, then the solution is not unique because any permutation of a solution is also a solution of (7.13).

- If all backoff and attempt parameters are the same for all the ACs, then by symmetry we can look for a balanced fixed point with $\gamma_i = \gamma$ for $1 \leq i \leq n$. If such a balanced fixed point exists, then it will be the unique balanced fixed point.

- However, it is possible that even when all the MAC parameters are the same, there exist unbalanced fixed points. This will correspond to multistability and short-term unfairness in the system, where some nodes get to use the channel for extended durations while locking out the other nodes.

7.6   Data and Voice Sessions over 802.11

Applications on a networked node are typically of one of the following types. A TCP-based session involving elastic data transfer using a closed loop control, for example, web browsing using HTTP over TCP. Another class of application could be a streaming session, for example, a packet voice application, also called Voice over IP (VoIP). Having analyzed the throughput behavior of the 802.11 protocol, we now use the insights developed from that analysis to understand what happens to applications when running on 802.11 based WLANs. We will consider a single hop WLAN in which a number of STAs associate with an AP, which in turn provides Internet access to the STAs via an uplink. This is by far the most widely used configuration for 802.11 based access.

7.6.1   Data over WLAN

A typical data transfer environment will involve a laptop equipped with an 802.11 interface that is downloading files, like the inbox in a mail application, or HTTP documents in a web browsing session. The laptop, or the STA, will be associated with an AP. For both these kinds of downloads, the TCP transport protocol will be used. Of course, there may be many other STAs that may be associated with the same AP. This situation is typical in an airport or a railway station lounge, a hotel lobby, or a cafe that provides Wi-Fi connectivity. This is shown in the top part of Figure 7.21.

For this scenario, we make the following observations. Recall that TCP is an acknowledgment (ACK) based protocol and every data packet that is received has to be acknowledged. The TCP-data packets are transmitted from the AP to the STAs, each of which transmits a TCP-ACK packet for every TCP-data packet that it receives. Thus, there is asymmetry in the traffic pattern with the AP accounting for approximately half of the packets transmitted in the network. Furthermore, the TCP-ACK packets are much smaller than the TCP-data packets. In most deployments, the AP and the STAs use the same backoff parameters; hence, if at a random instant, n of the nodes (AP and STAs) are contending for access to the channel, each of them is likely to succeed with equal probability. However, note

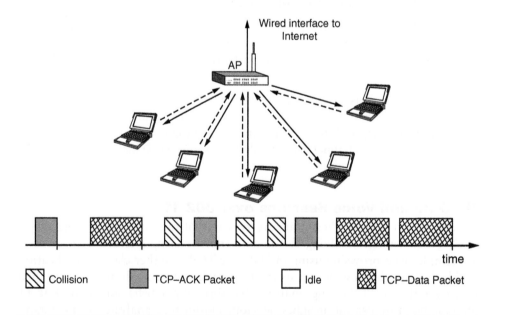

Figure 7.21   The top figure shows five STAs associated with an AP and performing data downloads. Solid lines indicate direction of TCP-data packets and dashed lines that of TCP-ACK packets. The bottom figure shows a sample of channel activity. Renewal instants are marked by vertical bars.

that an STA can transmit an ACK only if it has received a TCP-data packet. Thus, the number of STAs competing for channel access is time varying and it depends on the number of STAs that want to transmit a TCP-ACK packet.

We now make some simplifying assumptions.

- The STA will send a TCP-ACK packet for every TCP-data packet that it receives. (TCP allows delayed ACKs in which the receiver can send one ACK for more than one data packet. We will ignore this in our model here.)

- A TCP-ACK packet is generated at an STA as soon as a TCP-data packet is received. All TCP-ACK packets are immediately queued at the MAC for transmission.

- There is no packet loss due to either channel errors or due to buffer overflows.

- The TCP protocol at the sender does not time out due to late ACKs and the WLAN hop on the path from the sender to the STA is the bottleneck and the AP always has a TCP-data packet to transmit.

Let $S(t)$ denote the number of nodes competing to transmit on the channel at time t. From the model, $S(t)$ depends on the number of outstanding ACKs and this changes after each successful transmission. Let t_k be the instant at which the k-th successful transmission in the network was completed. We can thus define the evolution equation of $S(t_k)$ as follows.

$$S(t_k) = \begin{cases} S(t_{k-1}) + 1 & \text{if AP is successful} \\ S(t_{k-1}) - 1 & \text{if any STA is successful} \end{cases}$$

When $S(t_{k-1}) = 1$, the only transition possible is the transition to $S(t_k) = 2$. Further, because of backoff parameter symmetry between the AP and the STAs, the first transition will occur with probability $1/S(t_{k-1})$, and the second will happen with probability $1 - 1/S(t_{k-1})$. Also, since the AP always has a packet to transmit, $S(t) \geq 1$ for all t. Now, let $T_k := t_{k+1} - t_k$. We will also assume that when n nodes are active, the attempt probability is β_n and the collision probability is γ_n. We will assume that β_n and γ_n are the same as the attempt rate and and collision probability, respectively, in an n-node saturated network. We can then use the saturation throughput model developed in Section 7.4 to obtain these parameters. Further, we will assume that the transmission attempts by the active nodes are independent of all previous transmission attempts. From these assumptions, we see that $S(t_k)$ depends only on $S(t_{k-1})$ and, from our assumptions on the backoff process, T_k depends only on $S(t_k)$. Hence, we can embed a Markov chain at $\{t_k\}$ (see Figure 7.22) and $\{S(t_k), T_k\}$ will be a Markov renewal process. (See Appendix D.3.3.)

Let π_n be the stationary probability of there being n contending nodes, that is, of the event $\{S(t_k) = n\}$. We make yet another approximation—$S(t_k) \in [1, \infty]$.

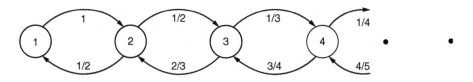

Figure 7.22 The embedded Markov chain S(t_k).

This is not a bad approximation because we see that the probabilities are such that $S(t_k)$ will be near 1 most of the time. We are now ready to solve the Markov chain $\{S_{t_k}\}$. We do this by writing the balance equations $\pi_n/n = n\pi_{n+1}/(n+1)$. From this, we can obtain the recursion

$$\pi_{n+1} = \frac{n+1}{n^2}\pi_n$$

Using the normalizing condition $\sum_{i=1}^{\infty} \pi_i = 1$ gives us

$$\pi_n = \frac{n}{(n-1)!(2e)} \tag{7.14}$$

Exercise 7.6

From (7.14), verify that the AP transmits half the time, which is as it should be.

Now consider the k-th renewal period (t_k, t_{k+1}). Let n nodes including the AP, be contending for use of the channel in this period. Further, let R be the number of collisions in this period. From the assumptions made about the attempt process in state n, T_k would be the sum of $(R+1)$ idle periods each of mean duration $\frac{1}{1-(1-\beta_n)^n}$, R collision periods each of duration T_c and one successful transmission period. The packet would have length T_{DATA} if the AP transmitted, which happens with probability $1/n$, or T_{ACK} if an STA transmitted, which happens with probability $(n-1)/n$. n takes values $1, 2, \ldots$ and R takes values $0, 1, \ldots$. Further, R is geometrically distributed with parameter γ_n and n is distributed as π_n derived earlier.

Now, let $H(t)$ be the number of times that the AP transmits in the interval $(0, t)$. Our interest is in the rate at which TCP packets can be downloaded, i.e., $\lim_{t\to\infty} \frac{H(t)}{t}$. As in the saturation throughput analysis we can use the renewal reward argument to obtain

$$\lim_{t\to\infty} \frac{H(t)}{t} = \frac{\sum_{i=1}^{\infty} \pi_n \frac{1}{n+1}}{E(T_k)}$$

From this discussion we can obtain

$$E(T_k) = \sum_{n=1}^{\infty} \pi_n \left(\left[\frac{1}{1-\gamma_n} \frac{1}{1-(1-\beta_n)^n} \right] + \left[\frac{\gamma_n}{1-\gamma_n} T_c \right] + \left[\frac{1}{n} T_{DATA} + \frac{n-1}{n} T_{ACK} \right] \right)$$

Here the term in the first square brackets corresponds to the mean idle period, the term in the second square brackets corresponds to the mean collision duration, and the term in the third square brackets corresponds to the mean duration of a successful transmission in T_k. We have assumed that β_n and γ_n have the same values as that in an n-node saturated network. We can read β_n from the plot of Figure 7.18.

As with the saturation throughput analysis, there were many simplifying assumptions and we would like to validate the model. Like before, we take recourse to comparing results from a simulation model that captures most details of the protocol specification. The canonical TCP application is a file transfer (FTP) session. The aggregate TCP throughput through the AP is plotted as a function of the number of FTP connections in Figure 7.23. Of course, the analysis does not model the number of connections. Observe that the model is accurate for up to 20 STAs for all three transmission rates. Further, observe the significant reduction in TCP throughput, as compared to the saturation throughput shown in Figure 7.19.

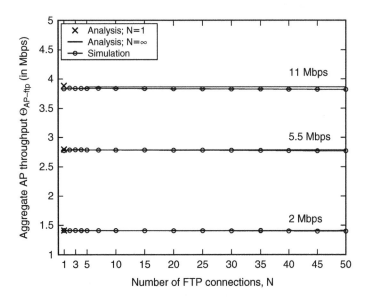

Figure 7.23 Aggregate TCP throughput through the AP as a function of the number of FTP sessions with one session per STA. Results from analysis and simulation are plotted for different transmission rates.

This is because of the transmission of the TCP-ACKs and the time lost from their contentions.

7.6.2 Voice over WLAN

We now consider a VoIP application running over WLAN. An example architecture is shown in Figure 7.24. The analog voice is digitized and may be compressed by the voice encoder using any of the standard codecs. Voice codecs are standardized in the G.7xx series by the ITU with the G.711 (PCM encoding at 64 Kbps), G.721 (ADPCM encoding at 32 Kbps), and G.729 (CS-ACELP encoding at 8 Kbps) being the more widely used codecs. Most codecs have a frame period; that is, they output a speech frame every T_f seconds. Typically, T_f is 10 or 20 ms and can even be 30 ms.

Recall the following from Chapter 3. For packet voice, a packetization interval, T_P, is defined. T_P will typically be an integer multiple of T_f. The encoded speech from every T_P seconds forms the payload of one voice packet. Thus, a voice packet is generated at a constant rate of $\frac{1}{T_P}$ packets per second. This is constant bit rate (CBR) speech. The codec may have voice activity detection (VAD) in which case packets are not generated when there is no speech activity in a packetization interval. This results in variable bit rate (VBR) speech. An important aspect of

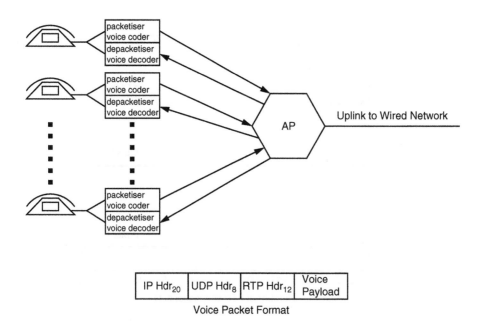

Figure 7.24 **Voice over the WLAN. Arrows indicate the flow of packets between the AP and the STAs. The voice packet format is shown at the bottom.**

transporting voice is that the mouth-to-ear (M to E) delay be less than a specified limit, say, D. This limit is usually in the range of 100–200ms. Packetization, transmission, and propagation delays are three components of this total delay budget. In addition, depending on the nature of the access in the network, there could be queueing (or access) delays. If the end-to-end path for the packet is fixed, the transmission and propagation delays are fixed. Most of the VoIP applications also fix the packetization delay. Thus the only design parameter is the link access delay.

If the access delays on the links are variable, the delay constraint is specified as $\Pr(\text{delay} > D_l) < L_P$. For example, $D_l = 20$ ms and $L_P = 0.01$ imply that at least 99 percent of the packets should experience an access delay of less than 20 ms. D_l is called the MAC-delay budget. Packets whose access delay exceeds D_l are assumed lost because they have arrived at the receiver past their playout time. Receivers can use loss concealment techniques to mitigate the effect of packet loss. Note that packets that arrive late cannot be played but may be useful in concealing future packet losses. Of course, a packet that is lost cannot be of help. However, using the excessively delayed packet requires more intelligent voice decoding and may not always be available. We will therefore assume that excessively delayed packets are also lost. L_P is determined by the codec—the higher the level of compression, the lower the allowable packet loss rate. L_P also is determined by the amount of speech data contained in each packet, on the packetization interval.

A voice service will also define an acceptable quality of speech at the receivers. The quality of speech is measured in terms of the perceived difference between the transmitted speech and that reconstructed at the receiver. The mean opinion score (MOS) metric on a scale of 1–5, with 5 indicating a perfect reconstruction, is typically used. Table 7.3, obtained from extensive experiments, shows the allowable packet loss rates for G.711 and G.729 codecs for different T_P to achieve a MOS of 3.6 and 4.0. Notice the significantly lower allowable packet loss rate for the G.729 (8 Kbps) than for the G.711 (64 Kbps). Also observe the significant reduction in the allowable packet loss rate when T_P is increased.

Codec (T_P)	Minimum MOS	
	4.0	3.6
G.711 (10 ms)	1.0	4.9
G.711 (20 ms)	1.0	3.0
G.729 (10 ms)	Cannot	0.33
G.729 (20 ms)	Cannot	0.19

Table 7.3 Allowable Packet Loss Rates (%) for Different Voice Quality and Packetization Intervals.

The voice payload per *voice over IP* (VoIP) packet is rT_P bits where r is the encoding rate of the codec. In addition, there are other protocols that need to be used and each will add its overhead to form the VoIP packet that is transmitted on the 802.11 network. The real time transport protocol (RTP) is used to convey timing information to the receiver. It introduces a 12 byte overhead. The transport protocol is usually the user datagram protocol (UDP) and it brings an 8 byte overhead. The IP protocol at the network layer introduces another 20 bytes of overhead. This voice-over-RTP-over-UDP-over-IP packet forms the payload for the MAC protocol. Of course, the MAC and the PHY have their own headers and trailers. To amortize the overhead, we can use a larger T_P and pack more voice payload into a packet. However, this will reduce D_l. Also, as can be seen from Table 7.3, this will reduce the acceptable packet loss rate, L_P.

Exercise 7.7

For the 802.11b MAC protocol operating at 11Mbps, obtain the packet transmission times for the G.711 and G.729 codecs with $T_P = 10$ ms. The MAC overhead is 28 bytes and the PHY overhead is 192 μs. Recall that corresponding to every transmission, there will be an ACK packet from the receiver and the SIFS and DIFS between transmissions. Taking all these into account, obtain the total transmission time for a voice packet. Repeat this for $T_P = 20$ ms.

Let us now consider dimensioning of an 802.11 network for voice calls. It is reasonable to assume that the two nodes in conversation are not associated with the same AP. This means that corresponding to each call, there will be packet flows in both directions—from the AP to the STA and vice versa.

Since voice requires delay guarantees and 802.11 supports contention free access, the simplest mechanism to support voice is to use the PCF to transfer voice. This emulates circuit multiplexing. Here, the important parameters are the superframe duration, which we will denote by T_F, and the duration of the contention free period within the superframe, *CFP_max_duration*. The number of calls that can be supported is essentially the number of voice packets that can be accommodated in the *CFP_max_duration*. This calculation and the bounds on the access delay are obtained as follows.

Recall that when the point coordinator (PC) wants to start the CFP, it has to wait for the ongoing transmission to finish. Then, since PIFS is smaller than DIFS, its priority is higher than that of a data packet and a CFP can begin soon after the current transmission ends. Thus, the start of the CFP can be delayed at most by PIFS and the maximum transmission time of a packet, say T_{pkt}. We will assume that all voice nodes in the network have the same packetization interval. Since a packet is generated every T_P seconds, it is reasonable to have $T_F = T_P$. Thus, the maximum delay for a voice packet is $(T_P + T_{pkt} + PIFS)$. From Figure 7.12 observe that in the CFP a poll packet from the PC precedes the voice packet. This poll packet will contain the voice packet from the AP to the STA. Let T_{voice}, T_{poll}, and T_{ACK} denote the transmission time of the voice, poll, and the ACK packets. From Figure 7.12, we can see that the maximum number of simultaneous voice calls that can be supported is the integer part of

$$\frac{CFP\_Max\_Duration - T_{beacon} - T_{end}}{T_{PIFS} + T_{poll} + 2T_{SIFS} + T_{voice} + T_{ACK}}$$

Here T_{beacon} is the transmission time of the beacon frame. The numerator is the maximum time available in a frame for voice packet transmissions. The denominator is the time consumed by each node per poll.

Let us now consider carrying VoIP calls in a network with only contention access. Further, we will assume that the STAs and the APs have the same priority and use the same backoff parameters. Observe that on an average, half the packets will be from the AP to the STAs. Our first interest is in finding N_{max}, the maximum number of active calls that can be supported.

A simple calculation can be done as follows. At high loads, when there are nearly N_{max} active calls, like in saturation analysis, the channel activity will alternate between backoffs of average duration $T_{backoff}$, by the STA and the AP, and transmission periods consisting of SIFS, voice packet transmission, ACK transmission, and DIFS. For every call, on an average, there will be two packets per T_P, one in each direction. Therefore, N_{max} is given by the integer part of

$$\frac{T_P}{2(T_{voice} + T_{SIFS} + T_{ACK} + T_{DIFS}) + T_{backoff}} \qquad (7.15)$$

We can assume that the packets experience very few collisions and that the average duration of the backoff period, $T_{backoff}$, can be approximated by $(\delta \times CW_{min})$, where $\delta \geq 1$ is a constant that depends on the network load.

Note that the N_{max} obtained here does not take into consideration the delays experienced by the packets. We could build more sophisticated models to model the voice packet arrival process. However, numerical results from such models and empirical results suggest that this expression is a very good approximation to evaluate N_{max}. The results obtained by simulation and from (7.15) are tabulated in Table 7.4 for the G.711 and G.729 codecs.

	T_F			
Codec	10ms	20ms	30ms	50ms
G.711	6 (6)	12 (12)	17 (18)	25 (26)
G.729	7 (7)	14 (14)	21 (21)	34 (35)

Table 7.4 Voice Capacity with DCF from Simulation and by Using (7.15) (the latter is shown in parentheses).

7.7 Association in IEEE 802.11 WLANs

Recall that in the infrastructure mode of the 802.11 network, the wireless nodes (STAs) need to connect through an Access Point (AP); it should *bind* to an AP. As we have mentioned earlier, such services may be located in an airport or a railway station lounge, a hotel lobby, or a cafe that provides Wi-Fi connectivity. This is shown in Figure 7.25. For better service, both in terms of the number of simultaneous users that can be supported and geographic coverage, there may be more than one AP. There may even be more than one hot-spot-provider servicing the area. Recall that an 802.11 network can operate over multiple channels, some of which are nonoverlapping. This means that from the same physical location, an STA may have a choice of APs to which it can bind. For example, in Figure 7.25, STAs in the region *AB* can associate with either AP-*A* or AP-*B*. Similarly, those in the region *BC* can associate with either AP-*B* or AP-*C*. Since the radio path between the STA and the APs with which it could associate could be different, it is possible that the transmission rates at which these associations can be made are also different. For example, in Figure 7.25, STA-5 may be able to associate with AP-*A* at 11 Mbps and with AP-*B* at 5.5 Mbps. Which association is better? To answer that, we need to be able to define "better" more precisely and then solve a problem of association of the STAs to APs.

Two kinds of association rules can be used. In the online method, an arriving STA and the AP makes a local decision based on the network condition at the time of the arrival of the STA. In the offline method, each STA submits its bandwidth (or throughput) requirement and also the transmission rate at which it can associate

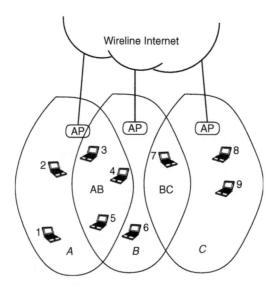

Figure 7.25 *A, B,* and *C* are the three APs covering a geographic area.

with each of the APs. An optimization algorithm is solved offline to determine the association and throughputs that will be allocated to each of the STAs. We will not consider offline algorithms.

We will discuss two simple rules that can be used by a new STA to select the AP to which it can bind. A new STA in the service area will be able to receive the beacons from each of the different APs. Of course, these will be transmitted by the APs on different channels. The strength of the received signal from an AP is an indicator of the SINR and hence, an indicator of the data rate at which the association can be made. Further, this is also an indicator of the proximity of the AP to the STA. The higher the received signal strength indicator (RSSI), the higher the data rate of association. Thus, an obvious rule would be for an STA to associate with the AP whose RSSI is the highest. This is the most commonly employed method.

Note that this association rule does not take into account the number of other STAs associated with the AP and their activity profile. It could be that there are many STAs associated with the AP and, although the transmission rate of the association is good, the actual throughput obtained is low. We now describe a simple rule based on the channel activity that can be used to estimate the throughput available from an AP.

The arriving STA listens to the beacon transmissions from the APs. For each AP, it knows the target beacon transmission time (TBTT) from the information contained in the beacon packet. Assume that the beacon frame does not have priority over other transmissions in the network. Hence, when the AP is ready to transmit the beacon, there will be an access delay that will be statistically identical to the access delay that will be experienced by all the other nodes. The arriving STA can measure the access delays experienced by the beacons over a window and calculate the average access delay. Let T_{access} denote this average access delay. T_{access} will also be the expected access delay (estimate) that will be experienced by a data packet from the arriving STA. After the access delay, there will be a successful transmission consisting of the RTS/CTS handshake, the data and the ACK and also the interframe spacings. Using this, we can calculate the expected time between successful transmissions of packets on the network, T, to be

$$T = T_{\text{access}} + T_{\text{RTS}} + (T_{\text{SIFS}} + T_{\text{CTS}}) + (T_{\text{SIFS}} + T_{\text{data}}) + (T_{\text{SIFS}} + T_{\text{ACK}})$$

Here, T_{data} is the transmission time of the packet at the rate at which it can associate with the AP. The potential throughput from the association will be $1/T$ packets per second. The arriving STA can calculate the potential throughput for all the APs and join the one that offers the highest potential throughput.

7.8 Notes on the Literature

The Aloha protocol was first described by Abramson in [1]. The CSMA protocol was first analyzed by Kleinrock and Tobagi [79] and by Lam [91]. The instability of the slotted Aloha with nonadaptive feedback has been shown by many authors, inlcuding Kleinrock [78], Fayolle et al. [33], and Kelly [73]. Space-time models for local networks were first presented by Molle, Sohraby, and Venetsanopoulos [103]. Many collision resolution algorithms for use in random access networks have been proposed. Bertsekas and Gallager [10] summarize and analyze many of the well-known ones. Rom and Sidi [117] discuss the performance analyses of a large class of random access protocols.

The MACAW protocol was proposed by Bhargavan et al. [11], which in turn was an adaptation of the MACA protocol first described by Karn [68]. The authoritative source for the IEEE 802.11 MAC protocol is the standards document published by IEEE and downloadable from the site of this working group. Mangold et al. [97] have provided a concise description of the basic 802.11 MAC protocol and the 802.11e extension for QoS differentiation.

There is a vast literature on analytical performance modeling of the 802.11 protocol family and even more simulation based studies. The detailed performance analysis that accounts for uniform sampling of the backoff multiplier was provided by Bianchi [12]. The analysis presented in the text is a simplification and generalization of this analysis. Numerical results from a Markovian model without the decoupling assumption have been obtained by Kumar et al. [86]. They also develop the fixed point analysis which is extended to 802.11 networks with service differentiation by Ramaiyan, Kumar, and Altman [114]. Chhaya and Gupta [23] provide an analysis that accounts for the hidden terminals and capture of a signal by a receiver in the event of a collision. Tay and Chua provide a capacity analysis of the basic handshake protocol [129]. Cali, Conti, and Gregori [19] and Carvalho and Garcia-Luna-Aceves [21] develop approximate models for the throughput of the protocol. The idea of adapting the backoff to optimize the throughput was studied by Cali et al. [20]. More recently, Sharma, Ganesh, and Key [121] have developed a Markovian model of the backoff process without the decoupling assumption. Here the system state at time t is described by $\mathbf{X}(t) = [X_i(t)]$ where $X_i(t)$ is the number of nodes in backoff stage i and $\sum_i X_i(t) = N$. It is shown that for a large number of nodes in the network, the system state stays close to a typical state. Renewal reward arguments are used to obtain the throughput in this typical state. Analysis of multiclass networks without the decoupling assumption is also presented.

There is also a considerable amount of literature that goes beyond the saturation throughput analysis. Tickoo and Sikdar [130] present a delay analysis. The many parameters in the protocol, such as superframe duration, ratio of CFP to CP, and backoff intervals, can be adapted to the traffic conditions to improve the performance of the protocol. An example of such an adaptation is described by Dong et al. [28], where past throughputs under the CFP and CP access methods are used as a feedback to adapt the PCF and DCF frame sizes to maximize the network throughput.

In [93], Li and Battiti analyze a network with service differentiation by a suitable adaptation of the model of [12]. Adaptive algorithms also have been proposed to provide service differentiation. In [118], Romdhani, Ni, and Turletti describe an adaptive service differentiation scheme in which after a successful transmission the nodes update their CW adaptively by taking into account the estimated collision rate.

TCP over the 802.11 network also has been extensively studied. Miorandi, Kherani, and Altman [100] describe a queueing model. Harsha, Kumar, and Sharma [53] analyze TCP and VoIP over an IEEE 802.11e networks. The TCP over WLAN analysis that we presented is adapted from the work of Kuriacose et al. [90]. VoIP over WLAN is also of significant interest. Medepalli et al. [99] obtain analytical and simulation results. The analysis that we describe is based on the work of Hole and Tobagi [56]. Tables 7.3 and 7.4 are obtained from [56].

Online association strategies in the 802.11 network have been described and analyzed by Kasbekar, Kuri, and Nuggehalli [69], [70]. Offline association strategies have been described by Tan and Guttag [125], Bejerano, Han, and Li [7], and by Kumar and Kumar [88]. Mussachio and Walrand [105] describe a game theoretic analysis of the association problem.

With the 802.11 hardware becoming a commodity item with widespread availability, there has been a significant interest in using this technology for innovative applications. The Digital Gangetic Plain (DGP) project in Uttar Pradesh and the Ashwini project in Andhra Pradesh, both in India, provide Internet connectivity to large rural areas that do not have any significant communication infrastructure. A hierarchical architecture is used in which each village has an access point and the nodes in the village connect to the access point. A mesh of point-to-point links interconnects the access points. The point-to-point links use the 802.11 protocol to exploit the cost advantage of its hardware. Of course, powerful, specially designed antennas are needed to create these point-to-point links. The design and deployment experience for this project is described by Raman and Chebrolu [115].

Problems

7.1 For a time-slotted network, where the slots are small compared with the packet lengths, consider the following variation of slotted Aloha. When a node has a packet to transmit, it begins transmission at the beginning of a slot. If there was no collision in the first slot, then it has captured the next $(X - 1)$ slots, where X is the packet transmission time and all other stations will defer. If there was a collision in the first slot, then the node makes a randomized retransmission attempt (as in slotted Aloha) and continues to do so until it succeeds. All nodes will know of the end of transmission of this packet when they sense the channel idle again. If slotted Aloha with an adaptive protocol were to yield a throughput of η when the packet length is equal to the slot length, what would be the throughput of this network? Note the similarities with the CSMA/CD protocol.

7.2 Consider a slotted Aloha network where the attempt arrival process is Poisson with rate G. Under the condition of maximum throughput, what is the fraction of empty, successful, and collision slots? If it is observed that the network is not operating under maximum throughput conditions and that the fraction of idle slots is 0.1, what is the throughput of the network? Is this network overloaded or underloaded? Explain.

7.3 Consider the Aloha protocol for multiple access networks. Let X be the number of nodes that are backlogged at the beginning of a slot. Assume X has a Poisson distribution with mean $\hat{x}$. Now assume that each backlogged node transmits in the slot with probability $1/\hat{x}$ independent of the others.

a. Obtain the joint probability of k nodes being backlogged and r transmitting. Also obtain the unconditional probability that the slot is idle.

b. If the slot was observed to be idle, what is the *a posteriori* probability that k nodes were backlogged at the beginning of the slot?

c. Similarly, find the a posteriori probability that given that there was a successful transmission in the slot, there were k backlogged nodes at the beginning of the slot.

d. Using these results suggests a method to continuously estimate $\hat{x}$ based on the event in a slot—success collision, or idle. Suggest an estimation method for $\hat{x}$ when a collision is observed in a slot.

7.4 Consider a pure Aloha network with an infinite number of nodes on a straight line of length a. Each packet transmission is of unit length, and transmissions are attempted according to a Poisson process of rate G. Assume that in this network there are only broadcasts, and that a transmission should be received at all the nodes. A transmission is a success only if it is received correctly at all the nodes.

a. Find the conditional probability that a transmission starting at x, $0 \le x \le a$, is successful.

b. Find the broadcast throughput of the network.

7.5 In an Aloha network, which uses the FCFS collision resolution algorithm a collision occurs in Slot k. Let $T_k = 0$ and $\alpha_k = 2.0$. Consider the arrivals at 0.2, 1.2, 1.6, 1.8, and 2.5. Which of these packets are resolved in the collision resolution period starting in slot k. Apply the FCFS algorithm and obtain the number of slots required to resolve all the collisions in Slot k. Count the number of collisions and idle slots. Repeat for the arrival sequence 0.6, 0.7, 1.5, 1.6, and and 1.8.

7.6 Consider a multiple access channel in which the following situation arises. Two nodes A and B are ready to send a packet at the same time.

This typically happens immediately following a successful transmission. In the k-th round after $(k-1)$ collisions have occurred, the nodes wait for a random period of $w \in [0, 1, \cdots, 2^{k-1} - 1]$ slots of time, with each of the 2^{k-1} choices being equally likely. Let c_k be the probability of a collision in round k given that the previous $(k-1)$ rounds had a collision.

a. Find c_k as a function of k for all k.
b. What is the probability that round k lasts n slots? What is its mean? Assume that a collision has occurred in the round.
c. Find the probability that round k lasts n slots. Do not assume that collision occurred. Also find the mean number of slots in round k.
d. Find p_k, the probability that exactly k rounds are needed to resolve a collision involving only two nodes and no new nodes transmitting until the collision is resolved.
e. Assume that the collision is resolved in favor of A in the third round. In this case A will reset its collision counter. Assume that the packet being transmitted by A is longer than the backoff time chosen by B. Because of 1-persistence, B will transmit soon after A's transmission. Now if A has another packet to transmit, there will be a collision immediately following A's successful transmission and both A and B will increase their collision counters. What is the probability that this collision is resolved in favor of A in the second round?

7.7 In IEEE 802.11 CSMA/CA there is also the Basic Access mode in which after completion of backoff a node just transmits its packet fully and then waits for a MAC ACK. Assume that all nodes use the same packet length L bits and bit rate C. Assume the single cell situation with all nodes saturated.

a. Argue that the analysis of the backoff process proceeds in exactly the same way (as for the RTS/CTS mode) and the same fixed point equations arise, yielding the probability β.
b. Using the notation δ (slot time), T_{SIFS}, T_{ACK}, T_{DIFS}, and T_{EIFS} (and L and C defined earlier) write down an expression for the total network saturation throughput.

7.8 Consider an IEEE 802.11b WLAN with n saturated nodes, all operating at r bps. Let $C(t)$ denote the number of collisions experienced by a node and $A(t)$ the number of attempts of the node in $[0, t]$.

a. State the formal mathematical definitions of node collision rate c, the collision probability γ, and the node attempt rate, a.
b. Given that the attempt probability of each node is β per backoff slot, and taking the decoupling approximation, write down an expression for c. Take the slot length to be δ, the success time to be T_s, and the collision time to be T_c.

c. Similarly write down an expression for a.

d. Verify the well-known simple expression for γ in terms of β from parts (b) and (c).

7.9 For small collision probabilities, let us approximate $1 - \gamma \approx 1$. Obtain the simplified formula for the normalized throughput and the optimal β for this approximation. Compare with the optimal β obtained in Exercise 7.4

7.10 Consider a slotted Aloha system with n nodes. Each packet fits one slot. The queues are saturated. After a packet is transmitted, the next packet is immediately attempted. If there is a collision, the next attempt is made after a number of slots uniformly distributed over 1 to $n-1$ slots. Packets are attempted until they succeed. Let β be the average attempt rate of a node.

a. Using a decoupling approximation, write down a fixed point equation for the attempt probability β of a node in a slot. (Hint: Let γ be the collision probability of a node's attempts.)

b. Argue that this equation has a unique fixed point.

c. Show how the aggregate saturation throughput, $\Theta(n)$, can be obtained from this analysis.

7.11 For the IEEE 802.11 protocol description given in the text, assume that backoff times are real numbers, and the channel propagation delay is zero, and hence that no collisions occur. Ignore the fixed time overheads (such as SIFS and DIFS). Assume that all nodes have the same packet length. Consider an n node network. Node k has an exponentially distributed backoff with mean $\frac{1}{\beta_k}$ and uses the data rate r_k. Under these simplifications show that the normalized throughput for node k is given by $\frac{\Theta_k}{r_k} = \frac{\eta_k L}{1 + \sum_{j=1}^{n} \eta_j L}$, where $\eta_k = \frac{\beta_k}{r_k}$. Θ_k is the throughput in bits per second. Hence argue that when the transmission rates are different (for example, because of different distances of the nodes from the AP), to achieve fair normalized throughputs the mean backoff times of nodes should be adapted to be inversely proportional to their transmission rates.

7.12 Repeat the derivation in Problem 7.11 but without assuming that there are no collisions, and also accounting for the overhead durations T_o and T_c.

7.13 Consider an IEEE 802.11 WLAN operating in the infrastructure mode. n_1 nodes are associated at transmission rate r_1 and n_2 nodes are associated at rate r_2. Assume that all nodes are saturated. Like in our analysis model, assume that all backoff durations are exponentially distributed with mean $1/\beta$. Let γ be the probability that a transmission attempt is successful. Obtain the throughput per node in terms of $\beta, \gamma, n_1, n_2, r_1$ and r_2. Assume all packets are B bits in length.

7.14 In an n node network, under saturation and exponential backoff, notice that if there is a successful packet it could be from any of the n nodes in the network. Assume that node i transmits at rate r_i, all packets are of L bits, and that the overhead per packet is T_o seconds. As before assume that the collision duration is T_c and that the probability that an attempt ends in a collision is γ. Show that the mean renewal period is given by $\frac{1}{n\beta} + \frac{1-\gamma}{n} \sum_{i=1}^{n} \left(\frac{L}{r_i} + T_o \right) + \gamma T_c$. We wish to obtain the throughput of node i. Identify the renewal reward and obtain the throughput of node i.

7.15 Consider a 20-node WLAN. From Figure 7.18 obtain the collision probability γ and hence the exponential backoff rate β. Assume that all the STAs are associated at 11 Mbps. Using the analysis of Section 7.6, obtain the per node TCP throughput.

CHAPTER 8

Mesh Networks: Optimal Routing and Scheduling

I n the previous four chapters we considered access networks. In this and the next two chapters, we consider wireless mesh networks (WMNs) or wireless multihop networks. In this chapter we will study the supporting of point-to-point flows in the mesh networks. We consider the optimal routing of these flows and scheduling of the transmissions on the wireless links.

Overview

In Section 8.1 we first describe the communication graph of a wireless network deployed in a given geographical area. Constraints on the simultaneous transmissions based on SINR, protocol-model, and the network graph are then described. In Section 8.2, for a given set of allowable link activation vectors, we obtain the network stability region, the set of end-to-end packet arrival rates for which the queues in all the network nodes will be stable. In Section 8.3, we consider the joint optimal routing of a set of end-to-end open loop packet flows and the corresponding link scheduling. A static link schedule using graph coloring techniques is derived. In Section 8.4 we develop the important dynamic, queue-length based, backpressure algorithm for joint routing and transmission scheduling on the links. This algorithm is optimal in the sense that it can stabilize any stabilizable end-to-end arrival rate vector. The algorithm is a maximum weight scheduling algorithm and the stability proof makes use of stochastic Lyapunov functions. In Section 8.5 we consider end-to-end elastic traffic and, for a given set of users, we obtain the jointly optimal routing of the packet flows and the transmission schedule on the links. In this case, a utility function on the allocated rate is defined for each user and the sum of the total utilities of all the users is maximized. Using convex programming and Lagrangian duality we obtain optimal joint packet flow rate allocation, routing, and link scheduling policies. In this section, we also consider optimal scheduling of one-hop packet flows in a slotted Aloha network. In the optimal algorithm the nodes update their transmission probabilities using local information to maximize the sum of link utility functions.

8.1   Network Topology and Link Activation Constraints

In the networks that we considered in Chapters 4 through 7, the wireless nodes (mobile hosts or STAs) had an explicit association with a base station or an access point. In this chapter we begin considering networks in which there is no such association between the wireless stations and any fixed infrastructure. Such networks were introduced as wireless mesh networks (WMNs) in Chapter 1.

In WMNs, information transport services are built over a set of arbitrarily located nodes, which are possibly mobile. Every node behaves both like a mobile host and as a wireless router. There are many obvious applications for such networks, such as providing communication services in emergency situations like in areas affected by storms, floods, and earthquakes. A WMN can also provide connectivity to fleets of vehicles operating in areas with no networking infrastructure. Of course, there are also many military applications. In all these applications, we can identify a set of point-to-point packet flows between the nodes in the network with each packet flow having its own QoS requirement, for example, a minimum throughput requirement and, possibly, an average end-to-end packet delay requirement. In this chapter, we analyze the ability of a given network to support a set of throughput requirements and the mechanisms to support them.

Consider a wireless network of N nodes deployed in a two-dimensional area. Let x_i be the coordinate vector of the location of Node i. A wireless link (i, j), $i, j \in \{1, 2, \ldots, N\}$ exists in the network if, in the absence of any other transmission in the network, the transmission from Node i can be decoded by the receiver of j; that is, the SNR for the signal from Node j is above the threshold, say β. From Chapter 7, this also means that x_i is in the decode region of the transmitter at x_j.

A wireless network formed by the N nodes can be represented by a directed graph $G = (\mathcal{V}, \mathcal{E})$ with the vertex set $\mathcal{V}$ representing the N nodes and the edge set $\mathcal{E}$ representing the set of E wireless links in the network. In general, G is not a fully connected graph and the network is a multihop wireless network; packets of end-to-end flows may need to pass through one or more intermediate nodes.

Denote the transmitter and receiver of edge $e \in \mathcal{E}$ by T_e and R_e, respectively. A simple model that often is used in obtaining $\mathcal{E}$ is to assume that the decode region around transmitter i is a circle of radius r_i; $(i, j) \in \mathcal{E}$ if $d_{i,j} := \|x_i - x_j\| < r_i$. Here r_i is a function of the transmission power. A further simplification that is often made is that $r_i = r$ for all i.

8.1.1   Link Activation Constraints

A multihop mesh network exploits spatial reuse; transmissions can occur simultaneously on links that are sufficiently separated in space. We now examine the various models that are used in specifying the set of links that can have simultaneous transmissions. We assume time-slotted networks; all nodes are synchronized in time and time is divided into slots. New transmissions occur at the beginning of a slot and all transmissions are completed at the end of the slot.

The transmission rate on the links is assumed to be such that all packets fit into a slot. All scheduling decisions are taken at the beginning of the slot and, if a transmission is scheduled on a link in a slot, exactly one packet is transmitted in the slot.

As we have just mentioned, the edges in G can be grouped into subsets such that the edges in a subset can be active in the same slot; the receiver of each active edge can decode the transmission from the transmitter of the edge. When such a set, say S, is activated, one packet can be sent across each edge in S. These sets must respect any radio operation constraints and interference constraints. We will use the term *link activation set* to refer to such a set. Let us now consider some models that usually are used to obtain the link activation sets of a network.

Recall from Chapter 2 that the receiver has a minimum SINR requirement to decode the received signal. Let $S \subset \mathcal{E}$ denote a set of links along which transmissions can occur in the same slot. For $e \in S$, let P_{T_e} be the transmit power used by T_e. Let $L(x, y)$ be the path loss function between a transmitter at x and a receiver at y, N_0 the thermal noise spectral density at the receivers, and W the bandwidth allocated to the network. For all $e \in S$, at the receiver R_e, the following minimum SINR requirement should be satisfied.

$$\frac{P_{T_e} L(x_{T_e}, x_{R_e})}{WN_0 + \gamma \sum_{\substack{e_1 \in S \\ e_1 \neq e}} P_{T_{e_1}} L(x_{T_{e_1}}, x_{R_e})} \geq \beta \tag{8.1}$$

γ is called the orthogonality factor and it satisfies $0 \leq \gamma \leq 1$. If the signals are all perfectly orthogonal $\gamma = 0$ and there is no interference. $\gamma = 1$ corresponds to the *physical model*. The simplest model for $L(x, y)$ is the far field attenuation where we assume that the attenuation is inversely proportional to a power of the distance; that is, $L(x, y) = \frac{1}{\|x - y\|^\alpha}$, where α is called the path loss exponent. See Chapter 2 for a more detailed discussion on radio propagation models.

We can also specify the link activation sets using geometric constraints. A simple criterion is to specify that for each $e \in S$, R_e should be further from all the other transmitters in S than it is from T_e. Intuitively, this is to ensure that the interference is lower than the received signal power. For a given $\Delta > 0$, this can be specified as follows. For all $e, e_1 \in S$ and $e_1 \neq e$

$$\|x_{T_{e_1}} - x_{R_e}\| \geq (1 + \Delta)\|x_{T_e} - x_{R_e}\|$$

Here Δ specifies a guard region that should not contain another transmitter. This is called the *protocol model* and is illustrated in Figure 8.2.

A third type of constraint could be those derived from the graph G. Once again, there are many possibilities. The simplest constraint is to ensure that for all $e \in S$, R_e should not be receiving from another node and T_e should not be transmitting to another node. This is called the *primary conflict* constraint. For example, in Figure 8.1, this means that when A is transmitting to B, A should

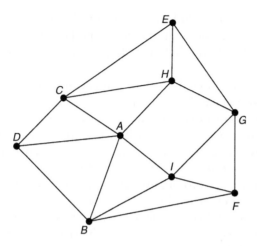

Figure 8.1   Example of a network graph. Vertices represent the nodes in the network. An undirected edge corresponds to directed edges in both directions. For example, undirected edge *AB* implies directed edges *AB* and *BA*; edges represent the half duplex wireless links.

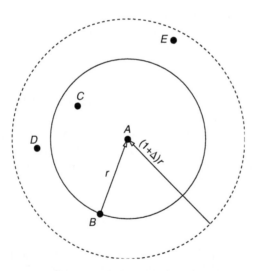

Figure 8.2   Illustrating the guard zone of the protocol model. When *B* is transmitting to *A*, Nodes *C*, *D* and *E*, should not be transmitting simultaneously. *C* can decode the transmission from *B* but *D* cannot decode when no other node is transmitting.

not be transmitting to any other node and B should not be receiving from any other node; when there is communication along directed edge AB, there should be no communication along directed edges AD, AC, AH, AI, DB, IB, and FB. An alternative constraint is as follows. For all $e \in S$, no other neighbor (in-neighbor if G is a directed graph; if (i, j) is a directed edge, then i is the called in-neighbor of j) of R_e should be transmitting. For example, in Figure 8.1, when A is transmitting to B, nodes D, I, and F should not be transmitting at the same time. This is also called the *receiver conflict* constraint. A third type of constraint is the *transmitter-receiver conflict* constraint. Here, in addition to the receiver conflict constraint, we also add the constraint that the neighbors of T_e should not be simultaneously transmitting. As an example, under this constraint, for the network of Figure 8.1, when A is transmitting to B, in addition to D, I and F, C and H should also not transmit. This is useful in IEEE 802.11 like protocols (see Chapter 7) where the transmitter expects a link layer acknowledgment from the receiver after the packet has been transmitted.

In general, a link activation set S can be represented by an E-dimensional vector of nonnegative rates, with r_e specifying the link layer data rate for the edge e when the set S is active. The modulation scheme can be adapted to suit the interference (from the active links) and the noise. In much of this chapter we will assume only one transmission rate on all the links. This means that for a desired bit error rate and a given modulation scheme (this fixes the link transmission rate), the SINR requirement is given by (8.1). In this case, the link activation set only specifies whether a link is allowed to transmit or not allowed to transmit and it suffices to describe S using an E-dimensional 0/1 column vector $[\mu_1, \mu_2, \ldots, \mu_E]^T$. We will work with this model in the rest this chapter. $\mathcal{S}$ will denote the set of all possible link activation vectors, including the all-zero vector, and will be called the *link activation constraint set* of the network.

If the network topology is changing with time, $\mathcal{S}$ could also be a function of time. In this chapter we will consider only static wireless networks; the geographical location of the nodes is assumed fixed.

8.2 Link Scheduling and Schedulable Region

Let S_t denote the link activation vector for slot t. Of course, $S_t \in \mathcal{S}$ for $t = 0, 1, 2, \ldots$. A sequence $\{S_t\}_{t \geq 0}$ is called a *schedule*. The schedule in a network could be static (predetermined) or it could be dynamic. Static scheduling is usually a periodic schedule and is analogous to a time division multiplexed (TDM) link. A set of T slots form a frame and S_t is defined for $0 \leq t \leq (T - 1)$. The sequence is repeated in every frame. In dynamic scheduling a schedule is decided for each slot. The schedule could be computed in a centralized manner in which a central entity decides the transmission schedule and distributes it to all the nodes in the network. The schedule could also be computed in a distributed manner in which each node executes a distributed algorithm that will determine the schedule. A centralized dynamic scheduling algorithm would typically use the complete network topology information and the queue length information at every node at the beginning of every slot. The scheduling decision for a slot could use the current state of the network and, possibly, the recent history.

Consider an arbitrary schedule Π. Let $\phi_S(\Pi)$, $S \in \mathcal{S}$, represent the fraction of time that the link activation set S is used in the schedule Π; that is,

$$\phi_S(\Pi) := \lim_{\tau \to \infty} \frac{1}{\tau} \sum_{t=0}^{\tau-1} I_{\{S_t=S\}} \tag{8.2}$$

where $I_{\{X\}}$ is the indicator function for the event X. Define $\Phi(\Pi)$ to be the $|\mathcal{S}|$-dimensional vector with elements $\phi_S(\Pi)$. Schedules for which the limit in (8.2) exists for all $S \in \mathcal{S}$ will be called *ergodic schedules* and we will only consider such schedules. In this case, $0 \le \phi_S(\Pi) \le 1$ and $\sum_{S \in \mathcal{S}} \phi_S(\Pi) = 1$.

We can also construct a randomized schedule; S_t could be chosen according to a probability distribution. In this case, the schedule can be thought of in terms of the probabilities $\phi_S, S \in \mathcal{S}$. In fact, given a probability vector $\tilde{\Phi}$ on $\mathcal{S}$, it is easy to see that we can obtain a schedule by simply choosing activation set S with probability ϕ_S in a slot independently of the activation set chosen in all the other slots. This is a static randomized schedule. Note that in this case $\Phi = \tilde{\Phi}$.

The *packet-flow capacity* of an edge $e \in \mathcal{E}$ is the maximum rate at which packets can flow along the edge. For an ergodic schedule, this capacity can be obtained as follows. Consider a schedule Π and for this schedule define $C_e(\Pi)$ as follows.

$$C_e(\Pi) := \sum_{\{S:e \in S\}} \phi_S(\Pi)$$

$C_e(\Pi)$ is the long term fraction of time that transmissions are scheduled on edge e. Thus $C_e(\Pi)$ is the maximum rate at which packets can flow along edge e under the ergodic schedule Π; that is, it is the packet-flow capacity of edge e in the ergodic schedule Π; $\mathbf{C}(\Pi) := [C_e(\Pi)]_{\{e \in \mathcal{E}\}}$ is the vector of edge capacities for schedule Π. This means that to transport packet flows in the network using schedule Π, the flow rate allocated to edge e can be no more than C_e. Thus, for a given schedule, we can think of the wireless network as a *capacitated network,* which we represent by $G(\mathcal{V}, \mathcal{E}, \mathbf{C}(\Pi))$. Thus, unlike wireline networks where the link capacities are fixed and given, in wireless networks they depend on the schedule. For a given network graph $G(\mathcal{V}, \mathcal{E})$ and the set of possible link activation vectors $\mathcal{S}$, the set of possible schedules Π defines the set of possible link capacities. The schedule is therefore an important variable in the optimization of a wireless network.

Let us now characterize the set of edge capacity vectors that can arise from all possible ergodic schedules. Let $\mathbf{C} := [C_e]_{\{e \in \mathcal{E}\}}$ be a vector of link capacities realized by an ergodic schedule. Consider a two-link network shown in the top part of Figure 8.3. In a slot only one of the two links can be activated; $S_1 = [1, 0]$ and $S_2 = [0, 1]$. Of course $[0, 0]$ is also possible. Consider a requirement that capacities

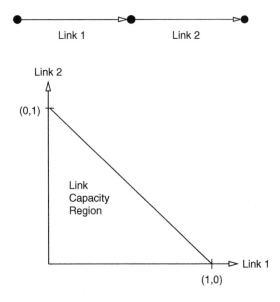

Figure 8.3 Illustrating the link capacity region for the two link network shown in the top part of the figure.

on the two links must be C_1 and C_2 packets per slot, respectively. Assume C_1 and C_2 are rational numbers. We can then express $C_1 = m_1/m$ and $C_2 = m_2/m$ for integers m_1, m_2, and m, and construct a periodic schedule with a frame of m slots in which S_1 is activated in m_1 slots and S_2 is activated in m_2 slots. Notice that if $(m_1 + m_2) > m$ then the link capacity requirement cannot be satisfied. The requirement that C_1 and C_2 be rational was for convenience of illustration. In Problem 8.1 we devise a mechanism to allocate capacity to a link that is an irrational number.

In the previous example, observe that an ergodic schedule can be constructed to achieve any link capacity that satisfies

$$C_1 \geq 0, \quad C_2 \geq 0 \quad \text{and} \quad C_1 + C_2 \leq 1 \tag{8.3}$$

This means that any capacity vector that is less than or equal to the convex combination of $[1, 0]$ and $[0, 1]$ can be achieved by an ergodic schedule. Thus we can say that the region in R^2 specified by (8.3) is the *link capacity region* for the network. This is illustrated in Figure 8.3.

Generalizing this notion, for a given $\mathcal{S}$, and a schedule Π,

$$\mathbf{C}(\Pi) = \sum_{S \in \mathcal{S}} \phi_S S$$

Thus $\mathbf{C}(\Pi)$ is a convex combination of the elements of $\mathcal{S}$. If we now consider the set of all possible ergodic schedules, we get the set of all possible link capacities.

This in turn is the set of all convex combinations of $S \in \mathcal{S}$. This is the *convex hull* of $\mathcal{S}$ and is denoted by $\mathsf{Co}(\mathcal{S})$. We can thus state the following result.

Lemma 8.1

For a wireless network with link activation constraint $\mathcal{S}$, the set of link capacities C that can be achieved by an ergodic schedule is the same as $\mathsf{Co}(\mathcal{S})$. ∎

We reiterate that any link-capacity vector that is in $\mathsf{Co}(\mathcal{S})$ can be achieved by a stationary randomized schedule.

We have just described the service capacity of the links in the wireless network that can be achieved by a schedule. We next describe the arrival rates of packets that can be stabilized by these service rates.

8.2.1 Stability of Queues

Consider a discrete time packet queueing system. Assume that the server capacity in each slot is a random sequence. Let $A(t)$ be the number of packets that arrive in slot t, $\mu(t)$ the server capacity (the number of packets that the server could have served) in slot t, and $Q(t)$ the number of packets in the queue at the beginning of slot t, just after the arrival instant. The number of departures in slot t would be less than or equal to $\mu(t)$. This queue is unlike that in a traditional queueing system analysis where $\mu(t)$ is assumed constant for all t. In fact, $\mu(t)$ could even depend on arrivals and queue occupancies up to time t. Similarly, $A(t)$ could also depend on the queue occupancies up to time t. We need to consider this generalization because, in a wireless mesh network, the arrival and service processes of packets at the nodes depends on the schedule and would have such dependencies.

Let $A(t) \leq A_{\max}$ for all $t \geq 0$ and $\lim_{T \to \infty} \frac{1}{T} \sum_{t=0}^{T-1} \mathsf{E}(A(t)) = \lambda$. Similarly, let $\mu(t) \leq \mu_{\max}$ for all $t \geq 0$ and $\lim_{T \to \infty} \frac{1}{T} \sum_{t=0}^{T-1} \mathsf{E}(\mu(t)) = \mu$. λ and μ correspond to the arrival rate and service rate, respectively, for the queue. Let $H(t)$ represent the history of the queue up to and including time t; that is, the sequence of arrivals into, and service from, the queue up to time t. Knowing $A(t)$ and $\mu(t)$ implies knowing $Q(t)$. We also impose the following restriction on the arrival and service processes. For any $\epsilon_1 > 0$, there exists an interval of T slots such that for every t_0 the following property is satisfied.

$$\mathsf{E}\left(\frac{1}{T} \sum_{t=t_0}^{t_0+T-1} A(t) \mid H(t_0) \right) \leq \lambda + \epsilon_1$$

$$\mathsf{E}\left(\frac{1}{T} \sum_{t=t_0}^{t_0+T-1} \mu(t) \mid H(t_0) \right) \geq \mu - \epsilon_1 \tag{8.4}$$

This is not a very restrictive property and is satisfied, for example, if $A(t)$ (and $\mu(t)$) is an i.i.d. sequence. Informally, this means that the conditional time average of the arrival and service rates have the same limit as the unconditional rates. We are imposing a restriction on how this convergence to the limit occurs.

Let us assume that arrivals occur at the beginning of the slot, departures at the end of the slot, and the queue is observed at the beginning of the slot, just after the arrival instant. This is shown in Figure 8.4. We can see that $Q(t)$ evolves as follows.

$$Q(t + 1) = \max(Q(t) - \mu(t), 0) + A(t + 1)$$

For the queue to provide useful service, the queue should be *stable*. Informally, stability of $Q(t)$ means that it does not grow to infinity with time. There are many ways in which the notion of stability can be formalized. For example, if $Q(t)$ evolves as a Markov chain, then $Q(t)$ is stable if it is positive recurrent (see Appendix D). In this chapter, we will use the following notion. $Q(t)$ is said to be *strongly stable* if

$$\limsup_{T \to \infty} \frac{1}{T} \sum_{t=0}^{T-1} E(Q(t)) < \infty \tag{8.5}$$

Let us first obtain a sufficient condition for $Q(t)$ to be strongly stable. Recall our analysis of the stability of the s-Aloha protocol in Chapter 7 using the *drift* from a state. There the drift from a state was defined as the conditional expectation of the one-step change in the state. We generalize that notion and define a *Lyapunov drift* (see Appendix D). Define a nonnegative function $L(\cdot)$ from the integers (the values that $Q(t)$ can take) to the reals that are increasing in its argument; $L(Q(t))$: $\mathcal{Z}_+ \to R_+$. $\mathcal{Z}_+$ is the set of nonnegative integers and R_+ is the set of nonnegative reals. Let $0 < B < \infty$ and $\epsilon_2 > 0$ be constants such that

$$E(L(Q(t + 1)) - L(Q(t)) \mid Q(t)) \leq B - \epsilon_2 Q(t) \tag{8.6}$$

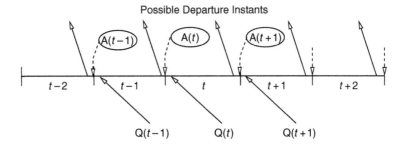

Figure 8.4 Observing the queue. Arrivals occur at the beginning of the slot, departures at the end of the slot, and the queue is observed after the departures at the end of the slot.

for all timeslots t. Taking expectations, we obtain the following.

$$E(L(Q(t+1))) - E(L(Q(t))) \leq B - \epsilon_2 E(Q(t))$$

Summing over slots $t = 0, \ldots, T-1$ we get

$$E(L(Q(T))) - E(L(Q(0))) \leq BT - \epsilon_2 \sum_{t=0}^{T-1} E(Q(t))$$

Dividing by T, rearranging the terms, and noting that $L(Q(T)) \geq 0$ we get

$$\frac{1}{T} \sum_{t=0}^{T} E(Q(t)) \leq \frac{B}{\epsilon_2} + \frac{1}{T\epsilon_2} E(L(Q(0)))$$

Taking lim sup as $T \to \infty$ (see Appendix B for definition of lim sup), we get

$$\limsup_{T \to \infty} \frac{1}{T} \sum_{t=0}^{T} E(Q(t)) \leq \frac{B}{\epsilon_2} < \infty$$

This means that $Q(t)$ is strongly stable if (8.6) is satisfied; that is, $Q(t)$ is strongly stable if we can find a scalar function of the state space for which the expected drift in a slot is strictly negative for all large queue lengths and is finite for the remaining queue lengths. The negativity requirement essentially means that queue lengths are being pushed back towards the lower values when it becomes too large. The finiteness ensures that on the average, the process does not jump to very large values in one step when the queue lengths are small.

Now consider the queue at timeslot t_0. Along the lines of the one-slot Lyapunov drift defined earlier, we can define a T-slot Lyapunov drift as follows.

$$E(L(Q(t_0 + T)) - L(Q(t_0)) \mid Q(t_0))$$

The arguments leading to (8.6) being a sufficient condition for stability of $Q(t)$ can be extended to obtain the following T-slot stability condition. If there exists a T such that $E(Q(t)) < \infty$ for $t = 0, \ldots T-1$, and there exist constants $B > 0$ and $\epsilon_3 > 0$ such that

$$E(L(Q(t_0 + T)) - L(Q(t_0)) \mid Q(t_0)) \leq B - \epsilon_3 Q(t_0) \tag{8.7}$$

then $Q(t)$ is strongly stable. (8.7) is essentially generalizing (8.6) to allow a negative drift over T slots rather than over one slot.

We will need the following identity in our analyses. Let W, X, Y, and Z be four nonnegative numbers.

$$\text{If} \qquad W \le \max\{X - Y, 0\} + Z \qquad\qquad (8.8)$$

$$\text{then} \qquad W^2 \le X^2 + Y^2 + Z^2 - 2X(Y - Z) \qquad (8.9)$$

To show this, we need to consider two cases. First, consider the case of $(X - Y) > 0$. For this case,

$$W^2 \le (X - Y + Z)^2$$

$$= X^2 + Y^2 + Z^2 - 2X(Y - Z) - 2YZ$$

Since Y and Z are nonnegative $2YZ > 0$ and (8.9) follows. Now consider the second case of $(X - Y) \le 0$.

$$W^2 \le Z^2$$

$$\le Z^2 + ((X - Y)^2 + 2XZ)$$

$$= X^2 + Y^2 + Z^2 - 2X(Y - Z)$$

The second inequality follows because $((X - Y)^2 + 2XZ)$ is the sum of two nonnegative quantities.

Let us now get back to analyzing $Q(t)$. Our objective is to obtain a condition on $A(t)$ and $\mu(t)$ that will make $Q(t)$ stable. Consider the evolution of $Q(t)$ over T slots. An upper bound on $Q(t_0 + T)$ can be obtained as follows. Assume that in slots $t_0, \ldots, t_0 + T - 1$, we serve only packets that were present at t_0. In a slot at most $\mu(t)$ packets depart. Therefore $Q(t_0)$ can decrement by at most $\left(\sum_{t_0}^{t_0+T-1} \mu(t)\right)$ packets in T slots. All packets that have arrived in slots $t_0 + 1, \ldots t_0 + T$ will be in the queue at the beginning of timeslot $(t_0 + T)$. We thus have

$$Q(t_0 + T) \le \max \left(Q(t_0) - \sum_{t=t_0}^{t_0+T-1} \mu(t), \ 0 \right) + \sum_{t=t_0+1}^{t_0+T} A(t)$$

We will use the Lyapunov function $L(Q(t)) = Q^2(t)$. Using (8.8) and (8.9), and simplifying, we get

$$Q^2(t_0 + T) \le Q^2(t_0) + T^2 \mu_{\max}^2 + T^2 A_{\max}^2$$

$$- 2TQ(t_0) \left(\frac{1}{T} \sum_{t=t_0}^{t_0+T-1} \mu(t) - \frac{1}{T} \sum_{t=t_0+1}^{t_0+T} A(t) \right)$$

Taking conditional expectation on $Q(t_0)$ we get

$$\mathsf{E}(Q^2(t+T) - Q^2(t_0) \mid Q(t_0)) \leq T^2 \mu_{\max}^2 + T^2 A_{\max}^2$$

$$- 2TQ(t_0) \left(\mathsf{E}\left(\frac{1}{T} \sum_{t=t_0}^{t_0+T-1} \mu(t) \mid Q(t_0) \right) - \mathsf{E}\left(\frac{1}{T} \sum_{t=t_0+1}^{t_0+T} A(t) \mid Q(t_0) \right) \right)$$

$$(8.10)$$

Let $\lambda < \mu$, (i.e., $\mu = \lambda + \epsilon_4$ where $\epsilon_4 > 0$). For $\epsilon_3 > 0$, if we choose $\epsilon_1 = \frac{\epsilon_3}{4T}$ in (8.4) and $\epsilon_4 = \frac{\epsilon_3}{T}$ we see that there exists a T such that

$$\mathsf{E}\left(\frac{1}{T} \sum_{t=t_0}^{t_0+T-1} \mu(t) \mid Q(t_0) \right) - \mathsf{E}\left(\frac{1}{T} \sum_{t=t_0+1}^{t_0+T} A(t) \mid Q(t_0) \right) \geq \frac{\epsilon_3}{2T}$$

From this and writing $B = T^2 \mu_{\max}^2 + T^2 A_{\max}^2$ in (8.10), we see that (8.7) is satisfied. Thus $\lambda < \mu$ ensures that $Q(t)$ is stable.

It can also be shown that if $\lambda > \mu$ then the queue is unstable. An informal argument is that the rate at which packets exit the queue is lower than the rate at which they arrive and eventually the queue will build up.

With this background on the stability of a single queue, we now consider the stability of the network of queues in the wireless network.

8.2.2 Link Flows and Link Stability Region

Consider a WMN in which there are J users indexed by j, $1 \leq j \leq J$, each with their end-to-end packet flows that need to be transported by the network. (In the rest of this chapter, the terms user, flow, and session will be used synonymously.) User j has source node s_j and destination node d_j. We will assume that the packet flows are open loop flows; that is, they have an intrinsic arrival rate (see Chapter 3). Let $A_j(t)$ denote the number of new packets of User j arriving at node s_j in slot t. We assume that $A_j(t)$ are i.i.d. for each j, $\mathsf{E}(A_j(1)) = \lambda_j$, and $\mathsf{E}((A_j(1))^2) < \infty$, for $1 \leq j \leq J$. λ_j is the packet arrival rate of User j in packets per slot. Let $\lambda = [\lambda_1, \lambda_2, \ldots, \lambda_J]^T$ be the column vector of the packet arrival rates of the J users. Let us assume that each packet can be transported to its destination over any route in the network that begins at the source and ends at the destination.

A *routing policy* determines the sequence of links to be traversed by the packet. We will be interested in *ergodic routing policies,* which route the packets such that we can define an arrival rate for the packets that arrive at the transmitting node of a link. Thus, for a given λ, the routing policy determines the rate at which packets arrive at the transmitter of a link. Now consider a routing policy $\mathcal{R}$ that routes the user flows with arrival rates λ such that $f_{e,j}(\mathcal{R})$ is the rate at which packets of User j are to be transmitted on edge e. Define

$$\mathbf{f}(\mathcal{R}, j) := [f_{1,j}(\mathcal{R}), f_{2,j}(\mathcal{R}), \ldots f_{E,j}(\mathcal{R})]^T$$

to be the E-dimensional column vector indicating the rate of User j on link e, $1 \le e \le E$. $f_{e,j}(\mathcal{R})$ also depends on λ but we do not express the dependence explicitly. Let

$$\hat{\mathbf{f}}(\mathcal{R}) := \sum_{j=1}^{J} \mathbf{f}(\mathcal{R}, j) = [\hat{f}_1(\mathcal{R}), \hat{f}_2(\mathcal{R}), \ldots, \hat{f}_E(\mathcal{R})]^T$$

be the column vector of total flows on the links; $\hat{f}_e(\mathcal{R})$ is the total flow on link e under routing policy $\mathcal{R}$. Also define

$$\mathbf{f}(\mathcal{R}) := \left[\mathbf{f}(\mathcal{R}, 1)^T, \mathbf{f}(\mathcal{R}, 2)^T, \ldots, \mathbf{f}(\mathcal{R}, J^T) \right]^T$$

This is simply a column vector whose first E elements specify the link-flow vector for User 1, the next E elements specify the link-flow vector for User 2,..., and the last E elements specify the link-flow vector for user j. Thus, it is a column vector of dimension $JE \times 1$. To simplify the notation, many a time, we will drop the reference to $\mathcal{R}$ when referring to the flow vectors; the dependence will be implicit.

Since we are considering a multihop network, and also since packet arrivals are random, nodes will need to store packets of different users before forwarding them on the next hop toward the destination. Let $Q_{i,j}(t)$ denote the number of packets of User j that are in the queue at node i at the beginning of slot t. Let $\mathbf{Q}(t) := [Q_{1,1}(t), \ldots, Q_{1,J}(t), \ldots, Q_{N,J}(t)]^T$. We say that $\mathbf{Q}(t)$ is strongly stable if $Q_{i,j}(t)$ for $1 \le i \le N$ and $1 \le j \le J$ are strongly stable.

The arrival rate of packets of User j to be transmitted on link e will be $f_{e,j}(\mathcal{R})$. The total arrival rate of packets at node T_e to be transmitted on link e is $\hat{f}_e(\mathcal{R}) = \sum_{j=1}^{J} f_{e,j}(\mathcal{R})$. Let $C_{e,j}(\Pi)$ be the capacity allocated to packets of User j on link e. Let Λ_1 denote the set of all λ for which there exists a routing $\mathcal{R}$ and a schedule Π such that

$$f_{e,j}(\mathcal{R}) < C_{e,j}(\Pi) \tag{8.11}$$

Similarly, let $\bar{\Lambda}_1$ denote the set of all λ for which there exists a routing $\mathcal{R}$ and a schedule Π such that

$$f_{e,j}(\mathcal{R}) \le C_{e,j}(\Pi) \tag{8.12}$$

It can be shown that $\bar{\Lambda}_1$ is the closure of the set Λ_1.

Since the set of feasible link capacities are in $\mathsf{Co}(\mathcal{S})$, a feasible routing policy will result in link flows that are in $\mathsf{Co}(\mathcal{S})$.

Let P denote a *routing and scheduling policy*, a set of rules that determines the routing and scheduling of the packets in each slot for a realization of a random process of packet arrivals into the network. Let Λ denote the set of all λ for which there exists a P such that $Q(t)$ is stable; Λ is called the *schedulable region* for the network.

It can be shown that if $\lambda \notin \bar{\Lambda}_1$, then $Q(t)$ is unstable, i.e., $\Lambda \subset \bar{\Lambda}_1$. Informally, this is because if $\lambda \notin \bar{\Lambda}_1$, then for every P, there will be at least one link, say e_1, for which the arrival rate of User j packets is strictly greater than the capacity allocated to User j on the link (i.e., $f_{e_1,j}(P) > C_{e_1,j}(P)$ for all P). Hence, at Node T_{e_1}, the queue of User j packets that are to be transmitted on link e_1 will build up leading to instability of $Q_{T_e,j}$, and hence of $Q(t)$. In Section 8.4 we will develop the maximum weight routing and scheduling algorithm, for which the network will be stable if $\lambda \in \Lambda_1$. This implies $\Lambda_1 \subset \Lambda$. We can thus say the following.

Lemma 8.2

If $\lambda \in \Lambda_1$, then $Q(t)$ is schedulable from the maximum weight schedule and hence $\Lambda_1 \subset \Lambda$. If $\lambda \notin \bar{\Lambda}_1$, then $Q(t)$ is unstable and hence $\Lambda \subset \bar{\Lambda}_1$. Thus,

$$\Lambda_1 \subset \Lambda \subset \bar{\Lambda}_1$$

■

Exercise 8.1
Show that Λ is a convex set, and if $\lambda \in \Lambda$ then for any λ' such that $\lambda'_j < \lambda_j$, for all j, $\lambda' \in \Lambda$.

Given λ and a schedule Π (and hence $C(\Pi)$), the problem of determining the optimal link flows $f_{e,j}$ is well known and is called the *multicommodity flow* problem. (The flow from each user is treated as a distinct commodity; hence the name.) There are many candidates for optimality: maximize the minimum spare capacity on the links, minimize the maximum flow allocated to a link, minimize the average delay, and so on. Alternatively, given λ and $\mathcal{R}$ for which a schedule exists, it may be possible to obtain the schedule from graph coloring techniques. We will discuss one such technique in the next section. The most general case, of course, is to find both simultaneously. In the rest of this chapter we will consider three basic kinds of problems in which we need to obtain $\mathcal{R}$ and Π simultaneously.

- First, we consider the case of a given λ for which we obtain the routing and a static scheduling scheme.

- Second, we consider the case when λ is unknown but we assume that $\lambda \in \Lambda_1$. For this we devise a dynamic routing and scheduling algorithm that is guaranteed to stabilize the network.

- Finally we consider elastic flows in which User j will define a utility function on λ_j. We will choose $\lambda \in \bar{\Lambda}_1$ to maximize the sum of the utility functions of all the users, over all possible $\mathcal{R}$ and Π and also derive the corresponding optimal $\mathcal{R}$ and Π.

8.3 Routing and Scheduling a Given Flow Vector

In this section we consider the scheduling and routing of a given flow vector λ in a network with the *primary conflict constraint*. The packets of User j arriving at Node s_j are routed over multiple paths toward the destination d_j. The routing problem will determine the fraction of User j packets that will be transported on each of the edges. The sum of such fractions from all the flows on an edge is the flow assignment for the edge. Recall that the edge flows should be in $\mathsf{Co}(\mathcal{S})$. The cardinality of $\mathcal{S}$ is usually very large and it is hard to check for $\hat{\mathbf{f}} \in \mathsf{Co}(\mathcal{S})$. We therefore obtain a sufficient condition that ensures $\hat{\mathbf{f}} \in \mathsf{Co}(\mathcal{S})$ and use this sufficient condition as a constraint in a multicommodity routing problem formulation.

Let us first see how to find the elements of $\mathcal{S}$. Since determining $\mathcal{S}$ is hard, we will just obtain a subset of $\mathcal{S}$. Recall that under the primary conflict constraint, in a slot, a node can be either receiving from one transmitter or transmitting to one receiver. Edges that have a node in common are called *adjacent* edges—if two edges e_1 and e_2 are adjacent then one or more of the following are satisfied:

$$T_{e_1} = T_{e_2} \quad T_{e_1} = R_{e_2}$$

$$R_{e_1} = R_{e_2} \quad R_{e_1} = T_{e_2}$$

Color the edges of G using a minimum number of colors $1, 2, \ldots$, such that no two adjacent edges have the same color. Two edges that have the same color satisfy the primary conflict constraint and can transmit simultaneously. Thus, a link activation vector can be obtained by doing the following for each color c. If edge e is colored c, then set $\mu_e = 1$, otherwise set $\mu_e = 0$. Each color gives us one activation vector. Using fewer colors corresponds to higher spatial reuse. Let $\chi(G)$ be the number of colors used. The minimum $\chi(G)$ is called the *edge-chromatic number* of G. A flow of at least $\frac{1}{\chi(G)}$ is possible on each link. (See Problem 8.3 to see how there could be some links with capacity more than $\frac{1}{\chi(G)}$.) Thus, a total flow allocation of less than $\frac{1}{\chi(G)}$ on all the links is a sufficient condition for the schedulability of a flow assignment. We can thus formulate an optimal routing problem with this capacity constraint. In this scheme no link gets a capacity more than $\frac{1}{\chi(G)}$. There will surely be some $\lambda \in \Lambda_1$ that may require that the flow on some of the edges be greater than $\frac{1}{\chi(G)}$. Thus this capacity constraint on the links will not be able to route and schedule all $\lambda \in \Lambda_1$. Further, it is very restrictive in its allocation of link capacity. We therefore look for alternate sufficient conditions for achievable link flows. Also note that there is no unique edge coloring. Thus another coloring of the graph would give us another set of link activation sets. To obtain S we need to find all possible colorings of G. This is, in general, a hard problem.

For now assume that we know the edge flows. Let $\hat{f}_e$ be the required total flow on edge e; the vector of link flows will be denoted by $\hat{\mathbf{f}}$. We will first consider the scheduling of the links into a static schedule to satisfy a specified link flow vector $\hat{\mathbf{f}}$.

Let $\mathcal{N}_{in}(v)$ and $\mathcal{N}_{out}(v)$, respectively, denote the set of in-neighbors and set of out-neighbors (if there is an edge (i,j) in G, i is the in-neighbor of j and j is the out-neighbor of i) of node v, and $\mathcal{N}(v)$ the set of neighbors of v:

$$\mathcal{N}_{in}(v) = \{v_1 : T_e = v_1 \text{ and } R_e = v\}$$

$$\mathcal{N}_{out}(v) = \{v_1 : R_e = v_1 \text{ and } T_e = v\}$$

$$\mathcal{N}(v) = \mathcal{N}_{in}(v) \cup \mathcal{N}_{out}(v)$$

From the primary conflict assumption, in a slot, a node can be transmitting on at most one edge or receiving on at most one edge but not both. Hence, it is necessary that for all $e \in \mathcal{E}$, $\hat{f}_e$ satisfies the following inequality.

$$\sum_{e:T_e=v} \hat{f}_e + \sum_{e:R_e=v} \hat{f}_e \leq 1 \qquad (8.13)$$

This is identical to the clique constraint for resource allocation that we discussed in Chapter 4. The resource here is a timeslot; in every clique of the graph G, in any timeslot, at most one node can be active in a slot. Note that (8.13) is not a sufficient condition.

To achieve a specified $\hat{\mathbf{f}}$, we will assume that all $\hat{f}_e$ are rational and find an integer τ such that we can express all $\hat{f}_e$ as

$$\hat{f}_e = \frac{w_e}{\tau}$$

where w_e is also an integer. This essentially says that we can achieve a link flow vector $\hat{\mathbf{f}}$ with a periodic schedule using a frame of τ slots, and, in each frame, edge e is activated at least w_e times. Let us now see how to construct such a schedule. In the process we will be able to obtain sufficient conditions for $\hat{\mathbf{f}}$ to be schedulable that will be easy to use as a constraint in an optimization problem.

We will once again use graph coloring techniques to obtain the schedule. First convert the network graph G into a *scheduling multigraph* $G_1(\mathcal{V}_1, \mathcal{E}_1)$ as follows. A multigraph is a graph in which there can be multiple edges between two nodes. The vertex set of G and G_1 will be the same; $\mathcal{V}_1 = \mathcal{V}$. Corresponding to every edge $e \in \mathcal{E}$, $\mathcal{E}_1$ will have w_e edges (T_e, R_e).

Let D denote the maximum degree, D_{out} the maximum out-degree and D_{in} the maximum in-degree of G_1;

$$D = \max_v \sum_{e \in \mathcal{N}(v)} w_e$$

$$D_{\text{out}} = \max_v \sum_{e \in \mathcal{N}_{\text{out}}(v)} w_e$$

$$D_{\text{in}} = \max_v \sum_{e \in \mathcal{N}_{\text{in}}(v)} w_e$$

Now color the edges of G_1 using a minimum number of colors $1, 2, \ldots,$ using the adjacency constraint that we used earlier. Let $\chi(G_1)$ denote the chromatic number of the multigraph G_1.

Let us now interpret the colored graph. Like in the example of the two-link network of the previous section, consider a frame of $\chi(G_1)$ slots. For $1 \le t \le \chi(G_1)$, let the edges that are colored t transmit in slot t. Consider any node v in the network. From the coloring constraint, we see that none of the incoming or outgoing edges of v have the same color. Hence, in a slot, node v will either be receiving from a node, or transmitting to a node, or neither. Thus the primary conflict constraint is satisfied by the schedule. From this we can conclude that a link flow vector $\hat{\mathbf{f}}$ is schedulable if $\chi(G_1) \le \tau$. This is because the activation of every color over a frame of length $\chi(G_1)$ achieves $\hat{f}_e \ge \frac{w_e}{\tau}$.

For a given G_1, let us now characterize $\chi(G_1)$. Since the edges that have a vertex in common cannot have the same color, $\chi(G_1)$ is lower bounded by the maximum degree of G_1; $\chi(G_1) \ge D$. It can also be shown that $\chi(G_1)$ is upper bounded by $\frac{3D}{2}$. We thus have

$$D \le \chi(G_1) \le \frac{3D}{2} \tag{8.14}$$

We can now use this discussion to determine a sufficient condition for $\hat{\mathbf{f}}$ to be schedulable on G. $\hat{\mathbf{f}}$ is schedulable if $\tau \ge \chi(G_1)$. This is the same as saying that the following be satisfied for all $v \in V$.

$$\frac{3}{2} \left(\sum_{e:T_e=v} w_e + \sum_{e:R_e=v} w_e \right) \le \tau$$

Divide both sides of this inequality by τ and observe that $\frac{w(e)}{\tau}$ is the flow rate on edge e in packets per slot. We can therefore say that if for all $v \in V$,

$$\left(\sum_{e:T_e=v} \hat{f}_e + \sum_{e:R_e=v} \hat{f}_e \right) < \frac{2}{3}$$

then $\hat{\mathbf{f}}$ is schedulable. This condition implies that a flow assignment in which each node is active (receiving or transmitting) for at most two-thirds of the slots is achievable. We thus have a sufficient condition for a flow assignment $\hat{\mathbf{f}}$ to be schedulable and also a mechanism to obtain this schedule.

Comparing with (8.13), we have a gap between the necessary and sufficient conditions for the schedulability of $\hat{\mathbf{f}}$. In the rest of the discussion we will use the sufficient condition to determine the link flow allocation. We will comment on reducing the gap between necessary and sufficient conditions in practice later in the section.

We are now ready to formulate the optimal route assignment problem. We assume that the packets from each flow may be split arbitrarily across all possible paths between the source and the destination. This assumption allows a simple formulation. An alternative is to define a set of paths for User j and then split the flow across these paths, but we will not pursue that.

We need some more notation. The network graph G can be summarized using its *node-link incidence matrix* $\mathbf{A}$. $\mathbf{A}$ is an $N \times E$ matrix with a row for each node and a column for each edge. Let $A_{i,e}$ represent the (i, e)-th element of $\mathbf{A}$. Let $\mathbf{A}_{i,\cdot}$ represent the i-th row of $\mathbf{A}$ and $\mathbf{A}_{\cdot,e}$ represent the e-th column of $\mathbf{A}$. Then, the column corresponding to edge e has the following entries:

$$A_{i,e} = \begin{cases} +1 & \text{if } i = T_e \\ -1 & \text{if } i = R_e \\ 0 & \text{otherwise} \end{cases}$$

Consider the product $\mathbf{Af}(j)$. The product is a column vector with N elements. The i-th element of this vector is the product of the row $\mathbf{A}_{i,\cdot}$ and the vector $\mathbf{f}(j)$. It can be seen that $\mathbf{A}_{i,\cdot}\mathbf{f}(j)$ is the *net outgoing traffic* of User j at Node i.

At any node i other than s_j and d_j, the net outgoing traffic corresponding to User j will be zero because such a node neither sources nor sinks packets of User j. The same argument shows that when $i = s_j$, the product $\mathbf{A}_{i,\cdot}\mathbf{f}(j)$ should be λ_j, and when $i = d_j$, the product $\mathbf{A}_{i,\cdot}\mathbf{f}(j)$ should be $-\lambda_j$. These *flow conservation equations* are

$$\mathbf{A}_{i,\cdot}\mathbf{f}(j) = \begin{cases} \lambda_j & \text{if } i = s_j \\ -\lambda_j & \text{if } i = d_j \\ 0 & \text{otherwise} \end{cases}$$

If we now consider all the rows of $\mathbf{A}$ together, then we have the following compact equation:

$$\mathbf{Af}(j) = \mathbf{v}(j) \tag{8.15}$$

where $\mathbf{v}(j)$ is an $N \times 1$ vector with the following entries:

$$v_i(j) = \begin{cases} \lambda_j & \text{if } i = s_j \\ -\lambda_j & \text{if } i = d_j \\ 0 & \text{otherwise} \end{cases}$$

From what we just saw, $\mathbf{v}(j)$ is a vector specifying the amount of net User j traffic from each node in the network.

Equation (8.15) holds for all j, $1 \leq j \leq J$. Thus, there are J equations of the form of (8.15), one for each $j = 1, 2, \ldots, J$. We can obtain a single compact equation that expresses the J equalities together. Consider the matrix

$$\mathbb{A} = \begin{bmatrix} \mathbf{A} & 0 & 0 & \cdots & 0 \\ 0 & \mathbf{A} & 0 & \cdots & 0 \\ \vdots & \vdots & \vdots & \ddots & \vdots \\ 0 & 0 & \cdots & 0 & \mathbf{A} \end{bmatrix}$$

There are J block-elements in each row and J block-elements in each column. $\mathbf{A}$ is the node-link incidence matrix defined earlier and is of dimension $N \times E$. 0 is also a matrix of dimension $N \times E$. Hence $\mathbb{A}$ is a matrix of dimension $JN \times JE$.

With these definitions, consider the equation

$$\begin{bmatrix} \mathbf{A} & 0 & 0 & \cdots & 0 \\ 0 & \mathbf{A} & 0 & \cdots & 0 \\ \vdots & \vdots & \vdots & \ddots & \vdots \\ 0 & 0 & \cdots & 0 & \mathbf{A} \end{bmatrix} \begin{bmatrix} \mathbf{f}(1) \\ \mathbf{f}(2) \\ \vdots \\ \mathbf{f}(J) \end{bmatrix} = \begin{bmatrix} \mathbf{v}(1) \\ \mathbf{v}(2) \\ \vdots \\ \mathbf{v}(J) \end{bmatrix} \tag{8.16}$$

Since $\mathbf{v}(j)$ is a vector of dimension $N \times 1$ for each $j \in \{1, 2, \ldots, J\}$, the vector on the right-hand side is of dimension $JN \times 1$. This is what we expect when a $JN \times JE$ matrix is multiplied with a $JE \times 1$ vector. Equation (8.16) is the compact flow conservation equation we were looking for. Clearly, it is nothing but J equations of the form $\mathbf{Af}(j) = \mathbf{v}(j)$, with $1 \leq j \leq J$.

Now consider any node in the network. Recall from our earlier discussion that for a feasible routing, if the sum of all flows into and out of a node is less than 2/3 packets per slot, then the link-flow assignment is schedulable. Let ρ denote an N element column vector with every element being equal to 2/3. ρ is analogous to the capacity vector of a multicommodity flow problem or the capacity vector of a routing problem of wireline networks. However, note that unlike in traditional routing problems, the capacity is defined in terms of the nodes and not in terms of the links. Let

$$\psi(j) = |\mathbf{A}| \, \mathbf{f}(j) = [\psi_1(j), \psi_2(j), \ldots \psi_N(j)]^T$$

where $|\mathbf{A}|$ is obtained by taking the magnitudes of the corresponding elements of $\mathbf{A}$. Notice that $\psi(j)$ is a $N \times 1$ column vector and $\psi_i(j)$ denotes the sum of the incoming and outgoing rates of User j at Node i. Define

$$\mathbf{\Phi} := \sum_{j=1}^{J} \psi(j) = \sum_{j=1}^{J} |\mathbf{A}| \, \mathbf{f}(j) = [\phi_1, \phi_2, \ldots \phi_N]^T$$

We see that ϕ_i is the sum of the incoming and outgoing flows at node i.

The vector of *spare node-capacities*, denoted by $\mathbf{z}$, is given by

$$\mathbf{z} = \rho - \boldsymbol{\Phi}$$

Let $z := \min_{1 \le i \le N} z_i$ be the *smallest* spare node-capacity corresponding to a given feasible routing. Then, the following inequality holds:

$$\boldsymbol{\Phi} \le \rho - z\mathbf{1}$$

where $\mathbf{1}$ is a column vector of N elements, all of which are 1.

For a given network and a set of end-to-end flow rate vectors of the J users, there may be many feasible routings. To choose one routing from this set, we need to define an *objective function* and then choose the routing that optimizes this objective function. There are many objective functions possible. Let us define the objective function as the quantity z earlier; the objective function is the smallest spare capacity at a node resulting from a routing. Then, an optimal routing would be that which maximizes the smallest spare capacity. Defining an optimal routing in this way is reasonable because *any* node in the network has a spare capacity of at least z. This increases the chance that a future demand between any pair of nodes in the network would find sufficient free capacity. In other words, we avoid routings that lead to a bottleneck node with very little spare capacity. Further, this objective promotes a balanced utilization of capacity and does not create hot spots.

Putting together all these elements, we have the following optimization problem:

$$\max z$$

subject to

$$\begin{bmatrix} \mathbf{A} & 0 & 0 & \cdots & 0 \\ 0 & \mathbf{A} & 0 & \cdots & 0 \\ \vdots & \vdots & \vdots & \ddots & \vdots \\ 0 & 0 & \cdots & 0 & \mathbf{A} \end{bmatrix} \begin{bmatrix} \mathbf{f}(1) \\ \mathbf{f}(2) \\ \vdots \\ \mathbf{f}(J) \end{bmatrix} = \begin{bmatrix} \mathbf{v}(1) \\ \mathbf{v}(2) \\ \vdots \\ \mathbf{v}(J) \end{bmatrix} \qquad (8.17)$$

$$\sum_{j=1}^{J} |\mathbf{A}| \, \mathbf{f}(j) + z \cdot \mathbf{1} \le \rho \qquad (8.18)$$

$$\mathbf{f}(j) \ge 0, \quad 1 \le j \le J, \quad z \ge 0$$

We can see that this is a linear program, with the variables being $\mathbf{f}(j)$, $1 \le j \le J$, and z. The objective is a linear function of the variables, with z being the sole variable determining its value. This is the final form of the optimization problem that defines the optimal routing. Since this is a linear program, efficient algorithms for computing its solution are available and one can actually obtain the optimal routing. We will not discuss the solution technique.

Discussion

1. If the linear program results in a solution in which the flow allocation to some of the nodes is not rational, we will have to round it to the next rational number. Further, this method will not be able to route and schedule all $\lambda \in \Lambda_1$.

2. Recall that there is a significant gap between the sufficient and necessary conditions for a set of node flows to be feasible. Also, from Problem 8.3 we know that link flows greater than that derived from the sufficient condition can be supported. Thus it has been advocated that one could use $\rho = [1, \ldots, 1]$ rather than $\rho = [2/3, \ldots, 2/3]$. In this case, it is possible that the clique constraint is not satisfied. Thus, this is only a heuristic and does not guarantee a flow allocation for which a schedule can be obtained.

3. The optimal routing problem that we formulated earlier is just one example of the many alternative formulations possible. One other popular formulation is based on the cost of the utilization of a link. We describe this briefly.

    Let $D_e(x)$ be a cost function associated with Edge e when the rate of packet flow on the edge is x. Let P_j be the set of paths on which the User j packets can be routed. Each path $p \in P_j$ is a sequence of connected edges starting at the source s_j and ending at the destination d_j of User j. Let $\mathcal{P} := \bigcup_{j=1}^{J} P_j$ denote the set of all paths defined in the network. The total flow λ_j is to be split among the paths in P_j with $x_j(p)$ allocated to path p subject to the link flow constraints. Let y_e be total traffic rate on link e. y_e is just the sum of rates $(x_j(p))$ allocated to the paths that use link e. For a given path allocation $\mathbf{x} := \{x_j(p) : j = 1, \ldots, J, \ p \in \mathcal{P}\}$, the sum of the inflow into and outflow from node v in the network, denoted by $\phi_v(\mathbf{x})$, will be given by

$$\phi_v(\mathbf{x}) = \sum_{\{e:T_e=v\}} y_e + \sum_{\{e:R_e=v\}} y_e$$

The first term is the inflow into the node and the second term is the outflow from the node. The optimal routing problem is thus,

$$\text{Minimize} \quad \sum_{e \in \mathcal{E}} D_e(y_e)$$

subject to

$$x_j(p) \geq 0 \qquad \text{for } p \in P_j, \ j = 1, \ldots J$$

$$\sum_{p \in P_j} x_j(p) = \lambda_j \quad \text{for } 1 \leq j \leq J,$$

$$\phi_v(\mathbf{x}) \leq \frac{2}{3} \qquad \text{for } 1 \leq v \leq N$$

This is a well-known optimal routing problem except for additional linear constraints on the variables from $\phi_v(\mathbf{x})$. These constraints just restrict the state-space. We can use well-known methods (e.g., flow deviation method) to solve obtain the optimal link-flow assignment vector.

8.4 Maximum Weight Scheduling

In the previous section we considered a static routing and scheduling algorithm for scheduling user flow rates $\lambda \in \Lambda_1$. We assumed λ was known. In this section we assume that λ is unknown but is in the open set Λ_1. We will derive a dynamic routing and scheduling algorithm for which the network will be strongly stable. This provides a constructive proof to show that $\Lambda_1 \subset \Lambda$.

It would seem that if the end-to-end flow vectors are known to be in the schedulable region, finding the optimal routes for the flows and a corresponding static link activation schedule would be possible. We saw in the previous section that even for a simple constraint, this is not easy. Although we could convert the network constraint into a set of necessary and sufficient conditions for the link flows, there is still a significant gap between the two.

In this section, we will study a dynamic routing and scheduling algorithm to route a $\lambda \in \Lambda_1$. We will show that an *optimal centralized, dynamic scheduling and routing algorithm* exists that will stabilize the network for any $\lambda \in \Lambda_1$. Interestingly, we do not even need to know λ. Before we develop this important algorithm, we will need some assumptions and notation.

We begin with some notation. Let $A_{i,j}(t)$ denote the number of packets of User j arriving into the network at Node i in slot t. $A_{i,j}(t)$ are bounded i.i.d. random variables; $A_{i,j}(t) \le A_{max} < \infty$ for all $t > 0$ and $1 \le j \le J$ and $\mathsf{E}(A_{i,j}(1)) = \lambda_{i,j}$. Note that we are allowing new packets of User j to arrive into more than one node in the network. This is a generalization from the previous (and the next) section where each user has only one source node. Packets of User j have destination d_j.

Each packet can be transported to its destination over any route in the network that begins at the source and ends at the destination. In each slot, either exactly one packet or no packet is transmitted on a link. Further, in a slot, each node receives at most one packet or transmits at most one packet. In the rest of the section it is convenient to represent a directed edge from node i to node k by (i, k) and we will follow this notation. Let $\mu_{(i,k),j}(t)$ be the indicator variable for the transmission of packet of User j on link (i, k) being scheduled in slot t; $\mu_{(i,k),j}(t) \in \{0, 1\}$ and $\sum_{j=1}^{J} \mu_{(i,k),j}(t) = \mu_{(i,k)}(t)$ where $\mu_{(i,k)}(t)$ is the indicator variable for a packet transmission on link (i, k) in slot t. The sum will have at most one nonzero term because, in a slot, we serve one full packet from at most one queue. Thus $\mu_{(i,k)}(t) \in \{0, 1\}$. It will be implicit that $\mu_{(i,k),j}(t) = 0$ if $(i, k) \notin \mathcal{E}$.

The sequence $\{\mu_{(i,k),j}(t)\}_{t \ge 0}$ represents the routing and scheduling of packets of User j on link (i, k) and is governed by the routing and scheduling policy P. If $\mu_{(i,k),j}(t) = 1$, then one packet of User j is transmitted from Node i and received at Node k at the end of slot t. If $k = d_j$ then the packet is removed from the network.

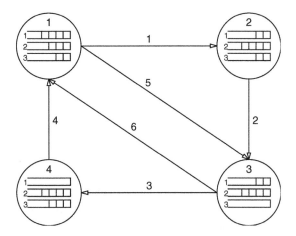

Figure 8.5 Illustrating the variables for dynamic scheduling in a WMN with six links and 3 flows of packets. The link number is marked against the links. The queue occupancies at the beginning of a slot are shown. Some example values for the variables are $Q_{1,1} = 4$, $Q_{2,1} = 2$, $Q_{1,2} = 3$, $Q_{4,3} = 3$, $w_{1,1} = 4 - 2 = 2$, $w_{1,2} = 3 - 4 = -1$, $w_{1,3} = 2 - 2 = 0$, $\bar{w}_1 = 2$.

Otherwise, the packet is queued at Node k. In this section, we will develop an optimum policy to choose $\mu_{(i,k),j}(t)$.

Each node maintains a separate queue for packets of User j; that is, there are NJ queues in the network with $Q_{i,j}(t)$ denoting the number of packets of User j queued at Node i and at the beginning of slot t. Collect the $Q_{i,j}(t)$ into the vector $Q(t) := [Q_{1,1}(t), \ldots, Q_{1,J}(t), \ldots, Q_{N,J}(t)]$. Figure 8.5 illustrates the queues and links in a network with four nodes and six links.

Multicommodity Flow Criteria

From our discussion in Section 8.3, the packet flow rates on each of the links should satisfy the following flow conservation equation for $1 \le j \le J$ and $1 \le i \le N, i \ne d_j$.

$$\sum_{k \in \mathcal{N}_{\text{out}}(i)} f_{(i,k),j} - \sum_{k \in \mathcal{N}_{\text{in}}(i)} f_{(k,i),j} = \lambda_{i,j} \qquad (8.19)$$

In addition, we will require that

- Flows $f_{(k,i),j}$ are all nonnegative and positive flows are not assigned to nonexistent links.

- Packets that have reached the destination are not injected back into the network.

We can show the following lemma.

Lemma 8.3

If $\lambda \in \Lambda_1$ then there exists a stationary randomized algorithm to choose $\mu_{(i,k),j}(t)$ independent of the history of the arrivals and departures up to timeslot t, such that

$$\mathsf{E}\left(\sum_{k \in \mathcal{N}_{\mathrm{out}}} \mu_{(i,k),j}(t) - \sum_{k \in \mathcal{N}_{\mathrm{in}}} \mu_{(k,i),j}(t) \right) > \lambda_{i,j} \qquad (8.20)$$

for all $1 \le i \le N$, and $1 \le j \le J$.   ∎

The following is an informal proof. Recall that any link capacity vector in $\mathsf{Co}(\mathcal{S})$ can be achieved by a stationary randomized schedule that chooses the link activation vector in each slot independently of all previous selections and the current state of the network. From the definition, Λ_1 is the set of λ for which the link flow rate vectors are in the interior of $\mathsf{Co}(\mathcal{S})$. If there exist flows $f_{(i,k),j}$ that satisfy (8.19) and the corresponding $\hat{f}$ is in the interior of $\mathsf{Co}(\mathcal{S})$, then there is a link capacity vector $\mathbf{C}$ that is equal to these flows that can be achieved by a stationary randomized schedule. Since Λ_1 is an open set, there exists another capacity vector $\mathbf{C}_1$ in which every edge has capacity strictly larger than the capacity of the same edge in $\mathbf{C}$. The $\mu_{(i,k),j}(t)$ in (8.20) can be thought of as coming from the stationary randomized schedule for $\mathbf{C}_1$.

Lyapunov Stability of Network of Queues

Our interest is in ensuring the stability of the vector of queue lengths, $\mathbf{Q}(t)$. We will use the Lyapunov technique introduced in Section 8.2.1 to derive the conditions for the strong stability of $\mathbf{Q}(t)$. The stability conditions will yield a routing and scheduling policy.

Like in Section 8.2.1 for single queues, we can define a nonnegative Lyapunov function on the vector $\mathbf{Q}(t)$, $L(\cdot) : \mathcal{Z}_+^{NJ} \to R_+$. $L(\cdot)$ defines a scalar value for every value that $\{\mathbf{Q}(t)\}$ can take. Along the same lines as the derivation of (8.6), we can make the following claim. If there exist constants $B > 0$ and $\epsilon > 0$ such that

$$\mathsf{E}(L(\mathbf{Q}(t+1)) - L(\mathbf{Q}(t)) \mid \mathbf{Q}(t)) \le B - \epsilon \sum_{i=1}^{N} \sum_{j=1}^{J} Q_{i,j}(t) \qquad (8.21)$$

then, $\mathbf{Q}(t)$ is stable.

Exercise 8.2

Show that if (8.21) is satisfied, then the queues $Q_{i,j}(t)$, for $1 \le i \le N$ and $1 \le j \le J$, are all stable.

In our analysis, we will be using the following quadratic Lyapunov function.

$$L(\mathbf{Q}(t)) := \sum_{i,j} (Q_{i,j}(t))^2$$

This is just the sum of the squares of the queue occupancies in a slot.

The Algorithm and Its Analysis

Our goal in this section is to develop a centralized dynamic routing and scheduling algorithm that will stabilize all $\lambda \in \Lambda_1$. The routes for the packets of the users and the sequence of activation vectors will not be statically determined. The centralized algorithm will decide both of these—the link activation vector to be used in a slot (scheduling) and the packets to be transmitted on the activated links (routing)—dynamically in every slot. We will consider a policy in which the routing and scheduling decision depends on the queue occupancies at the nodes in the network. Further, we will only consider stationary policies in which the scheduling algorithm is the same in every slot.

At nodes $i \neq d_j$, in each slot, $Q_{i,j}(t)$ is incremented by the packets of User j transmitted to i by its neighbors and also by the new User j packets arriving into the network at Node i. $Q_{i,j}(t)$ is decremented by a transmission of User j packets by Node i to its neighbors in slot t. We can thus write the evolution equation for $Q_{i,j}(t)$ as follows.

$$Q_{i,j}(t+1) = \max\left\{ Q_{i,j}(t) - \sum_{k \in \mathcal{N}_{\text{out}}(i)} \mu_{(i,k),j}(t), 0 \right\}$$

$$+ \left(A_{i,j}(t+1) + \sum_{k \in \mathcal{N}_{\text{in}}(i)} \mu_{(k,i),j}(t) \right) \qquad (8.22)$$

There is no queue for User j at d_j and the preceding equation applies only for $i \neq d_j$. We will now see what it takes to make $\{\mathbf{Q}(t)\}$, and hence the network, strongly stable.

We begin by analyzing the Lyapunov drift for our network. Recognizing the analogy between the variables of (8.8) and those of the queue evolution equation of (8.22) and applying (8.9) we can write

$$(Q_{i,j}(t+1))^2 \leq (Q_{i,j}(t))^2 + \left(\sum_{k \in \mathcal{N}_{\text{out}}(i)} \mu_{(i,k),j}(t) \right)^2 + \left(A_{i,j}(t) + \sum_{k \in \mathcal{N}_{\text{in}}(i)} \mu_{(k,i),j}(t) \right)^2$$

$$- 2 Q_{i,j}(t) \left(\sum_{k \in \mathcal{N}_{\text{out}}(i)} \mu_{(i,k),j}(t) - A_{i,j}(t) - \sum_{k \in \mathcal{N}_{\text{in}}(i)} \mu_{(k,i),j}(t) \right)$$

Summing over $1 \le j \le J$ and $1 \le i \le N$, we get

$$
\sum_{i=1}^{N}\sum_{j=1}^{J}\left((Q_{i,j}(t+1))^2 - (Q_{i,j}(t))^2\right) \le \sum_{i=1}^{N}\sum_{j=1}^{J}\left(\sum_{k\in\mathcal{N}_{\text{out}}(i)}\mu_{(i,k),j}(t)\right)^2
$$

$$
+ \sum_{i=1}^{N}\sum_{j=1}^{J}\left(A_{i,j}(t) + \sum_{k\in\mathcal{N}_{\text{in}}(i)}\mu_{(k,i),j}(t)\right)^2
$$

$$
- \sum_{i=1}^{N}\sum_{j=1}^{J}2Q_{i,j}(t)\left(\sum_{k\in\mathcal{N}_{\text{out}}(i)}\mu_{(i,k),j}(t) - A_{i,j}(t) - \sum_{k\in\mathcal{N}_{\text{in}}(i)}\mu_{(k,i),j}(t)\right)
$$

Recall the following from our discussion earlier on the properties of $\mu_{(i,k),j}(t)$: In a slot, at most one packet can be transmitted by a node, at most one packet can be received by a node, and at most $A_{\max}$ packets of User j will arrive into the network at Node i. We can therefore say

$$
\sum_{j=1}^{J}\left(\sum_{k\in\mathcal{N}_{\text{out}}(i)}\mu_{(i,k),j}(t)\right)^2 \le 1
$$

$$
\sum_{j=1}^{J}\left(A_{i,j}(t) + \sum_{k\in\mathcal{N}_{\text{in}}(i)}\mu_{(k,i),j}(t)\right)^2 \le J(A_{\max}+1)^2
$$

Writing

$$
B := \sum_{i=1}^{N}\left(1 + J(A_{\max}+1)^2\right) = N\left(1 + J(A_{\max}+1)^2\right)
$$

and rearranging the terms, we get

$$
\sum_{i=1}^{N}\sum_{j=1}^{J}\left((Q_{i,j}(t+1))^2 - (Q_{i,j}(t))^2\right) \le B + 2\sum_{i=1}^{N}\sum_{j=1}^{J}\left(Q_{i,j}(t)A_{i,j}(t)\right)
$$

$$
- 2\sum_{i=1}^{N}\sum_{j=1}^{J}\left(Q_{i,j}(t)\left(\sum_{k\in\mathcal{N}_{\text{out}}(i)}\mu_{(i,k),j}(t) - \sum_{k\in\mathcal{N}_{\text{in}}(i)}\mu_{(k,i),j}(t)\right)\right)
$$

$$\tag{8.23}$$

Now consider the last term in this inequality. Since the summation is overall $1 \leq i, k \leq N$, $\mu_{(i,k),j}$ appears twice in the sum—once multiplied by $Q_{i,j}(t)$ and another time multiplied by $(-Q_{k,j}(t))$. Further, all the edges appear in the summation. We can see this more easily by letting k range from 1 to N and changing the order of the summation. From this observation we can rewrite the last term in the preceding inequality as

$$2 \sum_{i=1}^{N} \sum_{j=1}^{J} Q_{i,j}(t) \left(\sum_{k=1}^{N} \mu_{(i,k),j}(t) - \sum_{k=1}^{N} \mu_{(k,i),j}(t) \right)$$

$$= 2 \sum_{i=1}^{N} \sum_{k=1}^{N} \sum_{j=1}^{J} Q_{i,j}(t) \mu_{(i,k),j}(t) - \sum_{i=1}^{N} \sum_{k=1}^{N} \sum_{j=1}^{J} \mu_{(k,i),j}(t) Q_{i,j}(t)$$

$$= 2 \sum_{i=1}^{N} \sum_{k=1}^{N} \sum_{j=1}^{J} \left(\mu_{(i,k),j}(t) \left(Q_{i,j}(t) - Q_{k,j}(t) \right) \right) \tag{8.24}$$

Using (8.24) and taking conditional expectation in (8.23) we get

$$E(L\left(\mathbf{Q}(t+1) \right) - L\left(\mathbf{Q}(t) \right) \mid \mathbf{Q}(t))$$

$$\leq B + 2E \left(\sum_{i=1}^{N} \sum_{j=1}^{J} Q_{i,j}(t) A_{i,j}(t) \mid \mathbf{Q}(t) \right)$$

$$- 2E \left(\sum_{i=1}^{N} \sum_{k=1}^{N} \sum_{j=1}^{J} \mu_{(i,k),j}(t) \left(Q_{i,j}(t) - Q_{k,j}(t) \right) \mid \mathbf{Q}(t) \right)$$

Since $A_{i,j}(t)$ are independent of $Q_{i,j}(t)$, the second term in the preceding inequality is simplified as follows.

$$2E \left(\sum_{i=1}^{N} \sum_{j=1}^{J} Q_{i,j}(t) A_{i,j}(t) \mid \mathbf{Q}(t) \right) = 2 \sum_{i=1}^{N} \sum_{j=1}^{J} Q_{i,j}(t) E(A_{i,j}(t))$$

$$= 2 \sum_{i=1}^{N} \sum_{j=1}^{J} Q_{i,j}(t) \lambda_{i,j}$$

We thus get,

$$E(L\left(\mathbf{Q}(t+1)\right) - L\left(\mathbf{Q}(t)\right) \mid \mathbf{Q}(t)) \leq B + 2\sum_{i=1}^{N}\sum_{j=1}^{J} Q_{i,j}(t)\lambda_{i,j}$$

$$- 2E\left(\sum_{i=1}^{N}\sum_{k=1}^{N}\sum_{j=1}^{J}\left(\mu_{(i,k),j}(t)\left(Q_{i,j}(t) - Q_{k,j}(t)\right)\right) \mid \mathbf{Q}(t)\right) \qquad (8.25)$$

The last term in (8.25) involves the routing and link scheduling algorithm. If we choose an algorithm that makes

$$\sum_{i=1}^{N}\sum_{k=1}^{N}\sum_{j=1}^{J}\left(\mu_{(i,k),j}(t)\left(Q_{i,j}(t) - Q_{k,j}(t)\right)\right)$$

as large as possible in every slot, then the expectation in the last term in (8.25) will also be large, and hence the right-hand side of (8.25) will be made small. This can lead to the queues being stable for larger values of $\lambda_{i,j}$. We now describe a routing and scheduling algorithm that achieves this.

For slot t, define the weight, $w_{(i,k)}(t)$, of each edge $(i, k) \in \mathcal{E}$ as follows. First we define

$$w_{(i,k),j}(t) := \begin{cases} Q_{i,j}(t) - Q_{k,j}(t) & \text{if } k \neq d_j \\ Q_{i,j}(t) & \text{if } k = d_j \end{cases}$$

This is illustrated in Figure 8.5. From this, define the weight of link (i, k) as follows.

$$w_{(i,k)}(t) := \max_{j} w_{(i,k),j}(t)$$

Thus the weight of a link is the maximum of the difference in the queue lengths of User j packets at the transmitter and receiver of the link. For each $S \in \mathcal{S}$ calculate W_S, the weight of S, and use it to choose the link activation vector for slot t, $S^*(t)$. This is done as follows.

$$W_S(t) = \sum_{(i,k)\in\mathcal{E}} w_{(i,k)}(t)\mu_{(i,k)}(S)$$

$$S^*(t) = \arg\max_{S\in\mathcal{S}} W_S(t)$$

Here $\mu_{(i,k)}(S)$ is the value of $\mu_{(i,k)}$ in the link activation vector S. This decides the scheduling of the link transmissions in slot t. To decide the routing, we need to determine which of the J flows are to be transmitted on each of the active links. On each active link $(i,k) \in S^*(t)$, select the j for which $w_{(i,k),j}(t)$ is maximum.

The routing and scheduling algorithm that we just described is called the *maximum weight scheduling* (MWS) algorithm. Since the weights are chosen based on the queue lengths in the NJ queues in each slot, this is a *queue-length-based* scheduling algorithm. Notice that the packets do not move forward toward the destination if the queues ahead have a higher occupancy. Thus we can see that a queue exerts a backpressure toward the source till its backlog is cleared. Hence this is also called a *queue-length-based backpressure* (QLB) algorithm.

Let $\mu^*_{(i,k),j}(t)$ denote the scheduling in slot t in the MWS algorithm. Using this notation in (8.25), we can write

$$\mathsf{E}(L\left(\mathbf{Q}(t+1)\right) - L\left(\mathbf{Q}(t)\right) \mid \mathbf{Q}(t)) \le B + 2\sum_{i=1}^{N}\sum_{j=1}^{J} Q_{i,j}(t)\lambda_{i,j}$$
$$- 2\mathsf{E}\left(\sum_{i=1}^{N}\sum_{k=1}^{N}\sum_{j=1}^{J} \mu^*_{(i,k),j}(t)\left(Q_{i,j}(t) - Q_{k,j}(t)\right) \mid \mathbf{Q}(t)\right) \qquad (8.26)$$

Let $\tilde{\mu}_{(i,k),j}(t)$ be any other routing and scheduling algorithm that selects the activation vector and the packet to transmit on each link according to our assumptions made in the beginning of the section. From our choice of $\mu^*_{(i,k),j}(t)$ we can say the following.

$$\mathsf{E}\left(\sum_{i=1}^{N}\sum_{k=1}^{N}\sum_{j=1}^{J} \mu^*_{(i,k),j}(t)\left(Q_{i,j}(t) - Q_{k,j}(t)\right) \mid \mathbf{Q}(t)\right)$$
$$= \mathsf{E}\left(\sum_{i=1}^{N}\sum_{k=1}^{N} \mu^*_{(i,k)}(t)w_{(i,k)}(t) \mid \mathbf{Q}(t)\right)$$
$$\ge \mathsf{E}\left(\sum_{i=1}^{N}\sum_{k=1}^{N} \tilde{\mu}_{(i,k),j}(t)w_{(i,k)}(t) \mid \mathbf{Q}(t)\right)$$

The first equality is obtained from the definition of $w_{(i,k)}(t)$, and from our assumption that $\tilde{\mu}_{(i,k),j} \in \{0,1\}$ and $\sum_{j=1}^{J} \tilde{\mu}_{(i,k),j} = \tilde{\mu}_{(i,k)}$. Using the same reasoning we also see that the following is true.

$$w_{(i,k)}(t)\tilde{\mu}_{(i,k),j}(t) \ge \sum_{j=1}^{J}\left(Q_{i,j}(t) - Q_{k,j}(t)\right)\tilde{\mu}_{(i,k),j}(t)$$

Using this and continuing with our previous calculations, we get

$$E\left(\sum_{i=1}^{N}\sum_{k=1}^{N}\sum_{j=1}^{J}\mu^*_{(i,k),j}(t)\left(Q_{i,j}(t)-Q_{k,j}(t)\right)\mid\mathbf{Q}(t)\right)$$

$$\geq E\left(\sum_{i=1}^{N}\sum_{k=1}^{N}\sum_{j=1}^{J}\tilde{\mu}_{(i,k),j}(t)\left(Q_{i,j}(t)-Q_{k,j}(t)\right)\mid\mathbf{Q}(t)\right)$$

$$= E\left(\sum_{i=1}^{N}\sum_{j=1}^{J}Q_{i,j}(t)\left(\sum_{k=1}^{N}\tilde{\mu}_{(i,k),j}-\sum_{k=1}^{N}\tilde{\mu}_{(k,i),j}\right)\mid\mathbf{Q}(t)\right)$$

$$= \sum_{i=1}^{N}\sum_{j=1}^{J}Q_{i,j}(t)\left(E\left(\left(\sum_{k=1}^{N}\tilde{\mu}_{(i,k),j}-\sum_{k=1}^{N}\tilde{\mu}_{(k,i),j}\right)\mid\mathbf{Q}(t)\right)\right)$$

The first equality is obtained by switching the order of the summation and rearranging like we did in (8.24).

From Lemma 8.3, we know that if $\lambda\in\Lambda_1$, then there exists a stationary randomized schedule that satisfies (8.20) and is not dependent on $\mathbf{Q}(t)$. Let $\tilde{\mu}_{(i,k),j}(t)$ in the preceding discussion come from such a schedule. Then, from (8.20), we have

$$E\left(\left(\sum_{k=1}^{N}\tilde{\mu}_{(i,k),j}-\sum_{k=1}^{N}\tilde{\mu}_{(k,i),j}\right)\mid\mathbf{Q}(t)\right)=E\left(\sum_{k=1}^{N}\tilde{\mu}_{(i,k),j}-\sum_{k=1}^{N}\tilde{\mu}_{(k,i),j}\right)\geq\lambda_{i,j}+\epsilon$$

for some $\epsilon>0$, $1\leq i\leq N$, and $1\leq j\leq J$. Resuming our earlier calculations, we get,

$$E\left(\sum_{i=1}^{N}\sum_{k=1}^{N}\sum_{j=1}^{J}\mu^*_{(i,k),j}(t)\left(Q_{i,j}(t)-Q_{k,j}(t)\right)\mid\mathbf{Q}(t)\right)\geq\sum_{i=1}^{N}\sum_{j=1}^{J}Q_{i,j}(t)\left(\lambda_{i,j}+\epsilon\right)$$

Using this last relation in (8.26), we get

$$E(L\left(\mathbf{Q}(t+1)\right)-L\left(\mathbf{Q}(t)\right)\mid\mathbf{Q}(t))$$

$$\leq B+2\sum_{i=1}^{N}\sum_{j=1}^{J}Q_{i,j}(t)\lambda_{i,j}-2\sum_{i=1}^{N}\sum_{j=1}^{J}Q_{i,j}(t)(\lambda_{i,j}+\epsilon)$$

$$= B-2\epsilon\sum_{i=1}^{N}\sum_{j=1}^{J}Q_{i,j}(t)$$

Thus we see that (8.21) is satisfied. When $\sum_{i=1}^{N} \sum_{j=1}^{J} Q_{i,j}(t) > \frac{B}{2\epsilon}$ the right-hand becomes negative and the drift is negative pushing the queues towards smaller values. Thus the maximum weight scheduling algorithm stabilizes $\lambda \in \Lambda_1$.

Discussion

1. In deriving the scheduling algorithm, we could also consider the error probability on the link. Of course, if the probability of a packet error on a link is nonzero, then the stability region and also the queue evolution equation would need to be changed. However, the MWS algorithm is only slightly different. This is explored in Problem 8.9.

2. With a suitable choice of edge weights, the MWS routing and scheduling algorithm is applicable in considerably more general scenarios. For example, we could use the same algorithm when the topology is time varying in a manner that a time average probability for a link to exist can be defined.

3. The link activation vectors S could be nonnegative reals. Recall that the transmission bit rate could be a function of the SINR at the receiver. This in turn depends on the transmission power used by the transmitters in the link activation vector. Thus corresponding to a transmission rate vector S, we also need to specify the transmission powers. In such cases, an obvious optimization criterion could be to minimize the energy or power consumption.

4. The MWS algorithm is complex to implement. Further, what we have described is a centralized algorithm that requires complete knowledge of the network state. Hence this is not quite a practical algorithm. Many distributed and randomized algorithms have been proposed in the literature.

5. The MWS algorithm is a significantly general algorithm and can be applied to a large class of problems. The most notable use is in developing maximum throughput scheduling algorithms in input queued switches.

8.5 Routing and Scheduling for Elastic Traffic

In the discussion in the previous two sections our concern had been to support a given end-to-end flow rate requirement λ through appropriate routing and link scheduling. We assumed that a requirement of λ exists due to applications involving stream traffic like interactive voice and streaming video. Much of the traffic in networks is due to client-server based data exchanges like those of `ftp` and `http` applications. See Chapter 3 for a discussion of different types of traffic and their characteristics. Figure 8.6 shows a network with several application sessions between `http` or `ftp` servers and clients. The servers could be directly connected to the WMN or they could be connected via a node that in turn connects to a wireline network. We have discussed issues associated with these applications

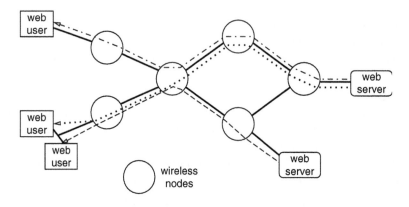

Figure 8.6   Elastic transfers between web servers and clients over a multihop wireless network.

in Chapter 3. These data transfer sessions are elastic sessions and there is no intrinsic rate that the applications demand.

In this section we assume that a number of elastic sessions are sharing the network resources and that *each session is transferring a single file with infinitely large volume of data.* This is the file transfer abstraction in analyzing a network with elastic sessions. Of course, in practice, sessions have finite lifetimes, and sessions arrive and depart. Hence even if the network topology and routing do not change, the session topology is constantly changing. Thus, in a sense, we are considering a situation in which the session topology and the network variations occur over a timescale that is slower than the file transfer time. However, we will assume that the stabilization of the dynamics of the congestion control scheme is on a faster timescale than the file transfer time. This simplifying assumption helps us develop an understanding of some of the basic issues in bandwidth sharing and congestion control in wireless networks.

Fair bandwidth sharing in a network is a complex issue. The complexity is compounded in wireless networks because the link capacity is itself a variable. Further, there are many notions of fairness that can be defined as being desirable, or achievable by specific resource sharing mechanisms. Of course, the network will also need to allocate bandwidth *efficiently* while being fair. Once again many definitions of efficiency are possible.

Fair sharing and efficiency have been studied extensively in wireline networks. As we have mentioned earlier, an important difference between wireline and wireless networks is that in the latter, link capacity is a variable in the bandwidth sharing algorithm. To illustrate this difference, consider the two-link network shown in Figure 8.7. Assume that the physical layer transmission rates on the links are equal, which we think of as unity. If it were a wireline network, assigning equal rates to the sessions could lead to a rate of one-third being assigned to each session, thus having unutilized bandwidth on link 1. However, if it were a time slotted wireless network, with the primary conflict constraint, each session

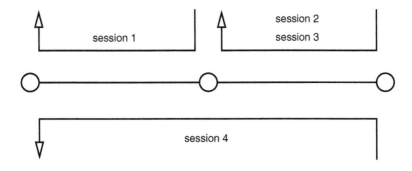

Figure 8.7 Example of a network and flows in which equal rate allocation to all flows is possible in wireless networks but not in wireline networks.

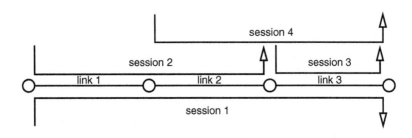

Figure 8.8 Fair sharing is not equal sharing in a network of links and transfers.

could be scheduled once in every five slots, and hence would be allocated a rate of 0.2 packets per slot. In doing so, links 1 and 2 have been allocated capacity 0.4 and 0.6, respectively. Since each clique has unit capacity, this implies that the clique capacity is not wasted.

Allocating equal rates to all the flows will not always be efficient from the network point of view. To illustrate, consider the network and sessions shown in Figure 8.8. Assuming the primary conflict scheduling constraint we have $S_1 = [1, 0, 1]$ and $S_2 = [0, 1, 0]$ as the link activation vectors. Let ϕ_1 and ϕ_2 be fractions of time that, respectively, S_1 and S_2 are activated. Let x_i be the rate allocated to Session i. The x_i needs to satisfy the following inequalities:

$$\phi_1 + \phi_2 \leq 1$$

$$x_1 + x_2 \leq \phi_1$$

$$x_1 + x_2 + x_4 \leq \phi_2$$

$$x_1 + x_3 + x_4 \leq \phi_1$$

With equal rate allocation to all the sessions (i.e., $x_i = x$ for all i), these inequalities reduce to

$$x \le \frac{\phi_1}{2} \quad x \le \frac{\phi_2}{3} \quad x \le \frac{\phi_1}{3}$$

and $x \le \frac{1}{3} \min(\phi_1, \phi_2)$. Since $\phi_1 + \phi_2 \le 1$ we get $x = \frac{1}{6}$ as the largest possible solution such that all x_i are equal. Clearly, allocating equal rates would be inefficient because while Session 4 is using link c, link a could be underutilized. Once again many notions of efficiency are possible. A simple efficiency objective could be that of *Pareto efficiency*, which, somewhat informally, is defined as follows: *An allocation of resources in a system is Pareto efficient if there does not exist another allocation in which some individual is better off while no individual is worse off.*

The preceding examples illustrate another point. In the example of Figure 8.7, notice that with the rate allocation of 0.2 packets per slot, Session 4 requires resources from both links; hence for each unit of Session 4 traffic carried we take away a unit of bandwidth from each of the two links. Yet, we need to be fair in some sense between the various sessions.

As in the previous sections, we will assume J users but each of them now has an infinite backlog of data to send. The network will have to decide the rate to be allocated to User j, say x_j; $\mathbf{x} := [x_1, \ldots, x_J]^T$. We thus have added a third dimension to the problem of routing and scheduling—determining the optimal fair rates for the users that can be routed and scheduled on a network with network graph G and link activation vector set S. We are, of course, constrained that the flow allocation be within the feasible region determined by the graph G and the link activation vector set S. An obvious question that arises now is the definition of fair sharing of the network capacity by the different users.

Max-min fairness (MMF) is a popular fairness notion. In a network represented by the network graph G and the constraint S, a stabilizable rate vector $\mathbf{x}$ is **max-min fair (MMF)** if it is not possible to increase the rate of a user j, *while maintaining the feasibility property*, without reducing the rate of some session j_1 with $x_{j_1} \le x_j$.

An important property of the MMF allocation is the following: Consider a feasible rate vector and look at the smallest rate in this vector. The MMF rate vector has the largest value of this minimum rate. Further, among all feasible rate vectors with this value of the minimum rate, consider the next larger rate. The MMF rate vector has the largest value of the next larger rate as well, and so on.

Another important and more general approach to achieve fairness is to assume that each flow j has a *utility function* $U_j(x_j)$, that defines the utility that User j obtains when it is allocated a flow rate of x_j. Then, if the assigned rate vector is $\mathbf{x} = [x_1, x_2, \ldots, x_J]^T$, the total utility of all the users in the network is $\sum_{j=1}^{J} U_j(x_j)$. For simplicity in the discussion, let us assume that all sources have the same utility function $U(\cdot)$. The following is a popular utility function:

$$U(x) := \log(x)$$

An important property of this utility function is that it is a nondecreasing and concave function of x. Concavity is related to the practical fact that any incremental value of additional rate decreases with increasing rate that a user already has—the law of diminishing returns. In this framework, the optimal bandwidth sharing is provided by the solution of the following utility maximization problem:

$$\max \sum_{j=1}^{J} U(x_j)$$

subject to

$$\mathbf{x} \in \Lambda$$

(8.27)

If the flows and capacity allocations are all deterministic, then we can replace Λ by $\bar{\Lambda}_1$. In this section we will consider the utility function approach to fair sharing of wireless network resources by elastic flows. We will first consider single hop flows in Aloha networks and then consider multihop flows.

8.5.1 Fair Allocation for Single Hop Flows

We begin with a simple network in which all users have end-to-end flows that are just one hop; that is, we have only single-hop sessions. Consider a network with all the edges that have flows. We will assume that the network uses the s-Aloha protocol at the MAC layer.

Recall from Chapter 7 that the s-Aloha is a distributed MAC protocol in which if a node has a packet to transmit, it just transmits it. If another neighboring node of the receiver also transmits in the same slot, then there is a collision and the receiver cannot decode the packet that was transmitted to it. The assumption is that a node will have a packet to transmit in a slot with some probability. Since we are considering elastic flows in which all the sources are infinitely backlogged, a node always has a packet to transmit to all its neighbors. Further, there is a backlog on all the links and a node can transmit on at most one link in a slot. We cannot have the nodes transmitting all the time. We therefore assign a transmission probability to every edge. The probability that the packet was decoded successfully at the receiver without a collision is the flow rate on the link. Thus the scheduling decision essentially consists of choosing the transmission probability for each link in a slot. Since all flows are from single-hop sessions, we do not need to consider routing. Also, only single-hop sessions mean that the flow rate allocated to a link is the rate allocated to the session. Thus, this is a simple example of distributed fair scheduling in wireless networks.

In the following, we will assume that in each slot, every node transmits independently of other nodes and also independently of its own previous transmission attempts. This is similar to the attempt model used in the analysis of s-Aloha in Chapter 7. Further, given that a node transmits, we will also assume that it chooses a receiver from among its neighbors independently. We thus have an attempt probability along each edge. Let $G_{(i,k)}$ be the attempt probability for edge (i,k); $\mathbf{G}$ is the vector of the edge attempt probabilities. From our assumptions, the attempt probability for Node i, $\hat{G}_i$ is then given by

$$\hat{G}_i = \sum_{k \in \mathcal{N}_{\text{out}}(i)} G_{(i,k)}$$

Of course, $0 \le G_{(i,k)} \le 1$ and $0 \le \hat{G}_i \le 1$.

Consider the network shown in Figure 8.1. If node A transmits to B, for B to successfully decode this transmission, neither B nor any of its neighbors B should be transmitting in the slot. Generalizing, for a transmission on edge (i,k) to be successfully decoded, k should not be transmitting and none of the neighbors of k (except i) should be transmitting in the slot. As we have said earlier, the flow rate on edge (i,k), denoted by $x_{i,k}$, is equal to the probability of a successful transmission along edge (i,k). Thus we have

$$x_{(i,k)} = G_{(i,k)} \, (1 - \hat{G}_k) \prod_{\substack{m \in \mathcal{N}_{\text{in}}(k) \\ m \ne k}} (1 - \hat{G}_m)$$

If we use the logarithmic utility function introduced earlier for the links, then the network utility, which is the sum of the edge utilities, is given by

$$\sum_{(i,k) \in \mathcal{E}} \log \left(x_{(i,k)} \right)$$

Let $G^*_{(i,k)}$ be the optimum edge attempt probabilities and let $\mathbf{G}^*$ denote the vector of $G^*_{(i,k)}$. $\mathbf{G}^*$ is obtained as

$$\arg \max_{\substack{0 < G_{(i,k)} < 1 \\ (i,k) \in \mathcal{E}}} \left(\sum_{(i,k) \in \mathcal{E}} \log \left(x_{(i,k)} \right) \right)$$

Exercise 8.3

Note the strict inequality in the range for $G_{(i,k)}$. What are the implications of allowing $G_{(i,k)} = 0$ and $G_{(i,k)} = 1$?

Let us now consider the network utility function $U(\mathbf{G})$.

$$U(\mathbf{G}) = \sum_{(i,k)\in\mathcal{E}} \log\left(G_{(i,k)} \, (1 - \hat{G}_k) \prod_{\substack{m\in\mathcal{N}_{\text{in}}(k) \\ m\neq i}} (1 - \hat{G}_m) \right)$$

$$= \sum_{(i,k)\in\mathcal{E}} \left(\log\left(G_{(i,k)}\right) + \log\left(1 - \hat{G}_k\right) + \sum_{\substack{m\in\mathcal{N}_{\text{in}}(k) \\ m\neq i,}} \log\left(1 - \hat{G}_m\right) \right) \qquad (8.28)$$

Exercise 8.4
Show that $\log\left(G_{(i,k)}\right)$ and $\log\left(1 - \hat{G}_i\right) = \log\left(1 - \sum_{k\in\mathcal{N}_{\text{out}}(i)} G_{(i,k)}\right)$ are strictly concave functions in $G_{(i,k)}$.

From Exercise 8.4 we see that $U(\mathbf{G})$ is a sum of concave functions and hence it also is concave with a unique maximum. Hence, the value of $\mathbf{G}$ which maximizes $U(\cdot)$, $\mathbf{G}^*$, can be obtained by observing that a strict concave function over a compact set has a unique maximum (see Theorem C.2 in Appendix C). Let us now obtain $G^*_{(i,k)}$.

Since $1 - \hat{G}_i = 1 - \sum_{k_1\in\mathcal{N}_{\text{out}}(i)} G_{(i,k_1)}$,

$$\frac{\partial(\log(1 - \hat{G}_i))}{\partial G_{(i,k)}} = -\frac{1}{1 - \hat{G}_i}$$

Let us now see the terms that will have $G_{(i,k)}$ when the summation in (8.28) is expanded. The second term in (8.28) corresponds to the receiver not transmitting. Thus $(1 - \hat{G}_i)$ will appear in terms that correspond to any of the in-neighbors of i transmitting. Hence we will have $|\mathcal{N}_{\text{in}}(i)|$ terms of the form $\frac{-1}{1-\hat{G}_i}$. The third term in (8.28) corresponds to the out-neighbors of the receiver not transmitting. Thus $(1 - \hat{G}_i)$ will appear whenever the neighbors of i have to receive. Hence we have $\sum_{m\in\mathcal{N}_{\text{out}(i)}} |\mathcal{N}_{\text{in}}(m)| - |\mathcal{N}_{\text{out}}(i)|$ terms of the form $\frac{-1}{1-\hat{G}_i}$. We subtract $|\mathcal{N}_{\text{out}}(i)|$ because i should not be counted. Thus

$$\frac{\partial U}{\partial G_{(i,k)}} = \frac{1}{G_{(i,k)}} - \left(\frac{G_{(i,k)}}{1 - \hat{G}_i}\right)\left(|\mathcal{N}_{\text{in}}(i)| + \sum_{m\in\mathcal{N}_{\text{out}(i)}} |\mathcal{N}_{\text{in}}(m)| - |\mathcal{N}_{\text{out}}(i)|\right)$$

$$G^*_{(i,k)} = \frac{1 - \hat{G}^*_i}{|\mathcal{N}_{\text{in}}(i)| + \sum_{m\in\mathcal{N}_{\text{out}(i)}} |\mathcal{N}_{\text{in}}(m)| - |\mathcal{N}_{\text{out}}(i)|}$$

Now observe that $G^*_{(i,k)}$ is independent of k and, for a given i, it is the same for all k; that is, all outgoing edges are activated with equal probability. Therefore $G^*_{(i,k)} = \frac{\hat{G}^*_i}{N_{\text{out}}(i)}$. Substituting in the preceding and simplifying we get

$$G^*_{i,k} = \frac{1}{|\mathcal{N}_{\text{in}}(i)| + \sum_{m \in \mathcal{N}_{\text{out}}(i)} |\mathcal{N}_{\text{in}}(m)|} \tag{8.29}$$

Exercise 8.5
Verify that $G^*_{i,j}$ obtained in (8.29) is a valid probability by verifying that $0 \le G^*_{(i,k)} \le 1$ for $(i,k) \in \mathcal{E}$ and $0 \le \hat{G}^*_i \le 1$ for $1 \le i \le N$.

Observe that the attempt probabilities for each edge can be conveniently obtained by the transmitter of the edge using local information, the in-degree of the node and the in-degrees of the neighbors. This means that even when the topology of the network is changing, the optimum attempt probabilities can be obtained quickly and the network can quickly settle into its optimum operating point. This is an important requirement in networking protocols in general and in wireless networks in particular because of the dynamic nature of the topology. We will also be seeking such capability in the algorithms that we will explore next.

8.5.2   Fair Allocation for Multihop Flows

Let us now consider multihop flows. Our interest will be in determining a routing and a schedule to support this fair allocation. As in Section 8.3, we will assume that the flows can be routed along any possible path in the network and follow the same notation. The optimization that we consider is similar to that in Section 8.3 except that we are maximizing the network utility and not the "minimum spare capacity"; that is, we have

$$\max \ U = \sum_{j=1}^{J} U(x_j)$$
$$\text{subject to} \tag{8.30}$$
$$\mathbf{x} \in \mathbf{\Lambda}$$

Here x_j is the flow rate allocated to User j. Since all users have only elastic traffic and the flow volume allocated to a user is variable, we will use x rather than λ to indicate the flow rate.

The constraint is the schedulability constraint for the user rates. Previously, we have also referred to $\mathbf{\Lambda}$ as the *capacity region*. Note that this is not a linear program. The objective function is a sum of concave functions and is hence concave. The constraints also define a convex region. Thus this is a convex program.

Recall that the capacity region Λ is the set of user rates $\mathbf{x}$ ($\mathbf{x} = (x_1, x_2, \ldots, x_J)^T$) such that for each such vector in the set, there exists a scheduling and routing that ensures that the vector can feasibly be carried in the network. It is also worth recalling that if $\mathbf{x} \notin \Lambda$, then there is *no* scheduling and routing such that the rate vector $\mathbf{x}$ can be transported through the network.

What is the capacity region for our problem? This can be defined in terms of a *capacitated network*; recall from Section 8.2 that given an ergodic schedule Π, each link e in the wireless network has an effective capacity $C_e(\Pi)$ that is determined by the fraction of slots in which e gets activated in the schedule.

Let $C_{e,j} \geq 0$, $e \in \mathcal{E}$, $1 \leq j \leq J$ denote the capacity (as yet unknown) that is earmarked for User j traffic on link e. Then the capacity region Λ is characterized by the following: $\Lambda = \{\mathbf{x} = (x_1, x_2, \ldots, x_J)^T \geq 0\}$ such that there exist capacity vectors $C_{e,j} \geq 0, e \in \mathcal{E}, 1 \leq j \leq J$ satisfying

$$\sum_{\{e:R_e=i\}} C_{e,j} + x_j I_{\{i=s_j\}} \leq \sum_{\{e:T_e=i\}} C_{e,j},$$

$$\text{for } 1 \leq j \leq J \text{ and } 1 \leq i \leq N, i \neq d_j \tag{8.31}$$

$$\left(\sum_{j=1}^{J} C_{e,j} \right)_{\{e \in \mathcal{E}\}} \in \mathsf{Co}(\mathcal{S}) \tag{8.32}$$

The first inequality is written for all users (indexed by j) and all nodes (indexed by i) except the destination node of User j, d_j. The first term on the left-hand side of (8.31) gives the sum of the capacities allocated for User j traffic on all *incoming* edges terminating at Node i; similarly, the term on the right gives the sum of User j capacities allocated on all *outgoing* edges leaving Node i. Therefore, the inequality asserts that the total incoming User j capacity at Node i, plus the rate injected into the network by User j (if Node i is the source node of j, viz., s_j) must be less than the total outgoing User j capacity at Node i.

When $i \neq s_j$, there is no User j traffic inserted into the network at Node i, so the second term on the left in (8.31) disappears. When $i = s_j$, we do not need to allocate any capacity for j on the incoming edges at Node i, so the first term on the left in (8.31) may actually be suppressed. However, as $C_{e,j} \geq 0$, allocating $C_{e,j} = 0$ on the incoming links achieves the same effect. Hence we retain the first term on the left in (8.31) even when $i = s_j$.

We note that in (8.31), we do not insist on strict inequality, as was done earlier in (8.11). In writing (8.11), we considered *open-loop* stochastic traffic. For ensuring stability with open-loop stochastic traffic, it is necessary to allocate, on each link, strictly higher capacity than the average aggregate traffic on that link, so that the queues in the network are stable. In this section, however, we consider *elastic* traffic that is subject to closed-loop control. Elastic traffic sources react to feedback signals from the network. As long as the feedback from the network remains unchanged, traffic is injected at constant rates. Because the input traffic is not varying stochastically, it is possible to allocate outgoing capacity that is not just strictly less than but even *equal to* the aggregate inflow into a node.

The second constraint of (8.32) says that the aggregate link capacity vector (with $|\mathcal{E}|$ elements), obtained after adding all the allocated User j capacities on each link, must belong to the convex hull of all the possible link activation vectors in the set S. This ensures that an ergodic schedule can be found that achieves the desired aggregate link capacities.

Another point worth noting about the equations characterizing Λ is that for every $j \in \{1, 2, \ldots, J\}$, (8.31) is written for all nodes i except d_j. Why do we omit the Node d_j? First, by omitting Node d_j, we are not omitting any links. This is because every link e has an associated transmitting node and a receiving node (T_e and R_e, respectively), and the incoming links at d_j have been considered when we wrote the equations for the corresponding transmitting nodes. Second, we know that the rate of User j traffic leaving the network at d_j cannot be more than the sum of the j capacities of the incoming links at d_j. But this conclusion already follows from the equations written for the other nodes, and hence presents no new information.

To solve the optimization problem (8.30), we will follow the method outlined in Section 6.2. Let $\mathbf{p}(j) := [p_{1,j}, \ldots, p_{N,j}]^T$, with $p_{d_j,j} = 0$. As before in Section 6.2 we define $\mathbf{p} := [\mathbf{p}^T(1), \ldots, \mathbf{p}^T(J)]^T \geq 0$. $p_{i,j}, 1 \leq i \leq N, 1 \leq j \leq J$ are the *Lagrangian* or *dual* variables. Relaxing the $J \times (N-1)$ constraints in (8.31) that define Λ, we get the following Lagrangian function:

$$L(\mathbf{x}, \mathbf{p}) = \sum_{j=1}^{J} U(x_j) - \sum_{j=1}^{J} \sum_{i=1, i \neq d_j, i \neq s_j}^{N} p_{i,j} \left(\sum_{\{e:R_e=i\}} C_{e,j} - \sum_{\{e:T_e=i\}} C_{e,j} \right)$$

$$- \sum_{j=1}^{J} p_{s_j,j} \left(\sum_{\{e:R_e=s_j\}} C_{e,j} + x_j - \sum_{\{e:T_e=s_j\}} C_{e,j} \right)$$

$$= \sum_{j=1}^{J} \left(U(x_j) - p_{s_j,j} x_j \right) - \sum_{j=1}^{J} \sum_{i=1, i \neq d_j}^{N} p_{i,j} \left(\sum_{\{e:R_e=i\}} C_{e,j} - \sum_{\{e:T_e=i\}} C_{e,j} \right)$$

$$= \sum_{j=1}^{J} \left(U(x_j) - p_{s_j,j} x_j \right) + \sum_{j=1}^{J} \sum_{e \in \mathcal{E}} \left(p_{T_e,j} - p_{R_e,j} \right) C_{e,j}$$

In arriving at the last line, we have computed the sum

$$\sum_{i=1, i \neq d_j}^{N} p_{i,j} \left(\sum_{\{e:R_e=i\}} C_{e,j} - \sum_{\{e:T_e=i\}} C_{e,j} \right)$$

by collecting together terms referring to the same edge. As each edge e has a transmitting node T_e and a receiving node R_e, we get the second term in the last line.

Thus, the relaxed problem is

$$\max \quad \sum_{j=1}^{J} \left(U(x_j) - p_{s_j,j} x_j \right) + \sum_{j=1}^{J} \sum_{e \in \mathcal{E}} \left(p_{T_e,j} - p_{R_e,j} \right) C_{e,j}$$

subject to

$$\left(\sum_{j=1}^{J} C_{e,j} \right)_{\{e \in \mathcal{E}\}} \in \mathrm{Co}(\mathcal{S}), \quad \mathbf{x} \geq 0$$

$$(8.33)$$

In this problem, the maximization is carried out over x_j, $1 \leq j \leq J$ and $C_{e,j}$, $e \in \mathcal{E}$, $1 \leq j \leq J$. The quantity to be maximized in this problem is just the Lagrangian that we obtained before; let us denote the maximum value by $D(\mathbf{p})$. This notation is justified because after the maximization, the objective becomes a function of $\mathbf{p}$ only. It may be noted that the constraint has changed from $\mathbf{x} \in \Lambda$ to $(\sum_{j=1}^{J} C_{e,j})_{\{e \in \mathcal{E}\}} \in \mathrm{Co}(\mathcal{S})$ after the relaxation. Recall also that $\mathrm{Co}(\mathcal{S})$ is a convex set.

It is interesting to note that in the relaxed problem, the constraint $\mathbf{x} \geq 0$ can affect the first term (the summation) only; the constraint $(\sum_{j=1}^{J} C_{e,j})_{\{e \in \mathcal{E}\}} \in \mathrm{Co}(\mathcal{S})$ can affect the second term only. Further, the constraint $x_j \geq 0$ can affect the j-th term in the summation only. Thus, there is a nice decomposition of the relaxed problem into several subproblems. For a given vector of dual variables $\mathbf{p}$, we can therefore solve the flow control problem for each User j, $1 \leq j \leq J$, and the scheduling problem independently.

The relaxed problem suggests a simple interpretation of the dual variables $\mathbf{p}$. Consider the j-th flow control problem. The term $p_{s_j,j} x_j$ can be thought of as the total cost that User j has to pay for sending x_j amount of traffic into the network. It is as if upon entry into the network via node s_j, User j pays a *price* of $p_{s_j,j}$ for every unit of traffic that it sends into the network. Thus, given $p_{s_j,j}$, User j's flow control problem is to maximize the *net utility*, where the latter is defined as the difference between the utility $U(x_j)$ and the total cost $p_{s_j,j} x_j$.

Consider now the last subproblem, which is the scheduling subproblem. We can think of $(p_{T_e,j} - p_{R_e,j})$ as the price associated with link e for User j traffic, and the term $(p_{T_e,j} - p_{R_e,j})C_{e,j}$ becomes the *weighted* capacity allocated to j on link e, with the weight being the link price. Thus, given $p_{i,j}$, $1 \leq i \leq N$, $1 \leq j \leq J$, the last subproblem asks us to find $C_{e,j}$, $e \in \mathcal{E}$, $1 \leq j \leq J$, such that $(\sum_{j=1}^{J} C_{e,j})_{e \in \mathcal{E}}$ lies in $\mathrm{Co}(\mathcal{S})$, and the sum of weighted link capacities over all j and over all links in the network is maximized.

From the preceding discussion, given $\mathbf{p}$, the network needs to solve the scheduling problem, with the objective being to maximize the sum of weighted link capacities, while each user needs to solve its individual net utility maximization problem. We note that according to this simple view, the solution of the scheduling problem must be found by a centralized entity that is aware of *all* prices $p_{i,j}$ and *all* possible capacity allocations $C_{e,j}$.

Now if we solve the subproblems independently for an *arbitrary* $\mathbf{p} \geq 0$, are we sure to get a feasible solution to the original problem in (8.30)? The answer is no. What comes to our rescue, however, is the fact that there *is* at least one particular value of the price vector $\mathbf{p}$ such that *a* solution of the relaxed problem in (8.33) is, indeed, feasible for the original problem in (8.30); moreover, that solution is *optimal* for the original problem in (8.30). This conclusion is based on the Strong Duality Theorem (see Appendix C), which is applicable here because

- The objective function in (8.30) is concave, and therefore, the negative of the objective function is convex.

- The first constraint in (8.31) is *linear* in the variables $x_j, 1 \leq j \leq J$, and $C_{e,j}, e \in \mathcal{E}, 1 \leq j \leq J$, and thus trivially convex.

- The vector of unknowns $[x_j, C_{e,j}]_{\{e \in \mathcal{E}, 1 \leq j \leq J\}}$ lies in a convex set.

Exercise 8.6

Given that $[x_j]_{\{1 \leq j \leq J\}}$ lies in a convex set and $[C_{e,j}]_{\{e \in \mathcal{E}, 1 \leq j \leq J\}}$ lies in a convex set, show that the vector $[x_j, C_{e,j}]_{\{e \in \mathcal{E}, 1 \leq j \leq J\}}$ lies in a convex set also.

This motivates us to consider the *Dual Problem*

$$\min D(\mathbf{p})$$
$$\text{subject to} \qquad\qquad\qquad (8.34)$$
$$\mathbf{p} \geq 0$$

The Strong Duality Theorem assures us that there is no *duality gap*, and therefore, the objective function values of the primal and dual problems are equal. Moreover, if we can find an optimal price vector $\mathbf{p}^*$ at which $D(\mathbf{p})$ is minimized, then we just need to solve the relaxed problem (8.33) for *that* $\mathbf{p}^*$, and the optimal solution to the original problem (8.30) will be obtained.

Let us consider the problem in (8.33) again. As we noted before, for a given $\mathbf{p}$, the problem is decomposed into several subproblems that can be solved independently. For a given $\mathbf{p}$, the maximizing $\mathbf{x}$, denoted by $\mathbf{x}^*(\mathbf{p})$, can be obtained without difficulty when each User j solves its individual net utility maximization problem independently. For solving the scheduling problem, let us consider some aggregate capacity vector $(C_e)_{e \in \mathcal{E}} \in \mathbf{Co}(\mathcal{S})$. For each $e \in \mathcal{E}$, the first question is about how C_e should be split into $C_{e,j}, 1 \leq j \leq J$. For this, we need to note the value of j for which $(p_{T_e,j} - p_{R_e,j})$ is largest. Letting

$$j^*(e, \mathbf{p}) = \arg \max_{1 \leq j \leq J} (p_{T_e,j} - p_{R_e,j})$$

the best split is given by

$$C_{e,j} = C_e \quad \text{for } j = j^*(e, \mathbf{p})$$

$$C_{e,j} = 0 \quad \text{for } j \neq j^*(e, \mathbf{p})$$

In other words, the best split is obtained by allocating, on edge e, the entire capacity C_e to *that* flow $j^*(e, \mathbf{p})$ that exhibits the largest price differential between the transmitter and receiver nodes of e. With this observation, the objective function of the scheduling problem becomes

$$\sum_{e \in \mathcal{E}} (p_{T_e, j^*(e, \mathbf{p})} - p_{R_e, j^*(e, \mathbf{p})}) C_e \tag{8.35}$$

Hence, to maximize the weighted sum of link capacities, it is necessary to select a vector $(C_e)_{e \in \mathcal{E}}$ in $\mathsf{Co}(\mathcal{S})$ such that this sum is maximized. We observe that this problem is actually a linear program because we are maximizing a linear function of C_e, $e \in \mathcal{E}$, over the convex hull of $\mathcal{S}$, the set of all link activation vectors. Hence, an optimizing vector can always be found at some extreme point of $\mathsf{Co}(\mathcal{S})$, that is, at some vector in $\mathcal{S}$ itself. Denoting such a vector by $(C_e^*(\mathbf{p}))_{e \in \mathcal{E}}$, the optimal solution to the scheduling problem is seen to be

$$C_{e,j}^*(\mathbf{p}) = C_e^*(\mathbf{p}) \quad \text{for } j = j^*(e, \mathbf{p})$$

$$C_{e,j}^*(\mathbf{p}) = 0 \quad \text{for } j \neq j^*(e, \mathbf{p}) \tag{8.36}$$

As remarked before, if we solve the scheduling and flow control problems for an arbitrary price vector $\mathbf{p}$, there is no guarantee that the solution so obtained will even be feasible for the original problem. The question that arises then is how to get the "right" price vector $\mathbf{p}^*$. Such a price vector $\mathbf{p}^*$ would constitute the *optimal dual variables*. What we need is an algorithm that, starting from some initial price vector $\mathbf{p}(0)$ at slot 0, updates the price vector in each slot such that $\mathbf{p}(t)$ converges to $\mathbf{p}^*$. Such an algorithm is:

$$p_{i,j}(k+1) =$$

$$\left(p_{i,j}(k) - h_k \left(\sum_{\{e:T_e=i\}} C_{e,j}^*(\mathbf{p}(k)) - \sum_{\{e:R_e=i\}} C_{e,j}^*(\mathbf{p}(k)) - I_{\{i=s_j\}} x_j^*(\mathbf{p}(k)) \right) \right)^+ \tag{8.37}$$

where $C_{e,j}^*(\mathbf{p})$ are obtained from (8.36) and h_k, $k = 1, 2, \ldots$ is a sequence of positive step-sizes. The factor multiplying h_k is known as the *subgradient* of the dual objective function $D(\mathbf{p})$ at $\mathbf{p}$. It can be shown that if the sequence h_k satisfies the two conditions

1. $h_k \to 0$ as $k \to \infty$

2. $\sum_k h_k = \infty$

then the iteration in (8.37) converges to the optimal price vector $\mathbf{p}^*$. For example, the sequence $h_k = 1/k$ satisfies the two preceding conditions. Finally, solving the relaxed problem in (8.33) with this $\mathbf{p}^*$ yields the optimal solution to the original problem in (8.30).

Discussion

We provide an overview here of how the solution to the problem of sum-utility maximization for elastic flows is obtained. In slot k, we have the price vector $\mathbf{p}(k)$. Using this, the network solves the scheduling problem. For $\mathbf{p}(k)$, the optimal aggregate capacity vector $(C_e^*(\mathbf{p}(k))_{e \in \mathcal{E}}$ is obtained by solving the linear program whose objective function is given in (8.35), and the optimal split of this among the users j is obtained as in (8.36). The price vector $\mathbf{p}(k)$ is now fed back to the users, and each user now solves its own net utility maximization problem, yielding $\mathbf{x}^*(\mathbf{p}(k))$.

Accordingly, in slot k, users j, $1 \leq j \leq J$ inject the appropriate amounts of traffic into the network. The network activates the links in the vector $(C_e^*(\mathbf{p}(k))_{e \in \mathcal{E}}$ and transfers data from the users $j^*(e, \mathbf{p}(k))$ over the duration of slot k. At the end of slot k, the network evaluates the right side of (8.37) and new price variables for slot $(k+1)$ are obtained.

An interesting conclusion follows from (8.37). Consider a Node i and a User j for which the factor multiplying h_k in (8.37) is negative. This means that the total inflow rate of User j traffic into Node i is more than the corresponding outflow rate; that is, packets of User j are queueing up at Node i. Under these circumstances, $p_{i,j}(k+1)$ is more than $p_{i,j}(k)$. This means that in slot $(k+1)$, the pair (i, j) is likely to be part of the vector that maximizes the weighted sum of link capacities. In that case, the network would schedule this node and user pair in slot $(k+1)$, thereby depleting the queue of User j packets that had started to build up in Node i. Thus, the schedule computed by the network tends to keep queue lengths small.

The observation that queue lengths tend to be small also suggests that the schedule computed by the network leads *implicitly* to a routing in which traffic from the sources does, ultimately, reach the respective destinations. If this were not true, then somewhere in the network, queues would start building up.

We have an iterative process in which the network computes prices and informs these to the users, who react by sending traffic into the network. Next, the network schedules links according to the prevailing prices and users' traffic gets transferred across links. At the end of this, the network computes fresh prices, and the cycle repeats. It can be shown that if this process is allowed to run for many slots, then, as long as the conditions on h_k are satisfied, the price vector $\mathbf{p}$ and the users' rate vector $\mathbf{x}$ both converge to their respective optimal values. After convergence, we would therefore have a vector of user rates that can be transported through the network, and, at the same time, achieve sum-utility maximization, which was our original objective.

Another conclusion from (8.37) is as follows. Consider (8.37) *after convergence*, and suppose $p_{i,j}^*(\infty)$ is positive. This implies that on the right side of (8.37), the factor multiplying h_k must be zero. This says that the total rate of traffic from User j coming into and going out of Node i are equal; the first constraint in (8.31) of the original problem is satisfied with equality. We note that this is exactly the same conclusion that follows from the Complementary Slackness conditions (see Appendix C).

In the case where open-loop traffic was to be transported, the problem for the network was to determine a schedule and routing such that traffic could be carried in the network. The input traffic was stochastically characterized and given. In the case of elastic traffic, we recall that the input traffic to the network was *not given*. We just had an objective stating that the sum of users' utilities was to be maximized, subject to the constraint that the users' injected traffic should be supportable by the network. In contrast to the wired network case where link capacities are given and fixed, in the wireless network, not even the link capacities are known; in fact, they depend on the scheduling strategy followed by the network. It is somewhat remarkable that the method outlined in this section is able to provide a scheduling, routing, and rate control that manages to actually achieve the original objective of sum-utility maximization.

8.6 Notes on the Literature

Much of the recent work on optimization in wireless networks has its roots in the work of Tassiulas and Ephremides [126, 127, 128]. The schedulable region or the stability region of a wireless mesh network is characterized in [127]. A less general version of the schedulable is used in [51]. The discussion on the stability of queues is adapted from [40].

Hajek and Sasaki [51] first addressed the problem of simultaneous routing and scheduling in wireless networks. The discussion of Section 8.3 is adapted from [80]. The optimal routing formulation is adapted from Chapter 14 of [89].

The maximum weight scheduling algorithm of Section 8.4 was first described in [127]. This has been considerably generalized and many new applications found. The analysis in [127] assumes that the arrival of new packets into the network are i.i.d. in every slot. Since the routing and scheduling in a slot in the MWS algorithm depend only on the queue occupancies at the beginning of the slot, positive recurrence of the resulting Markov chain implies stability of the queues. This is shown using a technique similar to what is described here. Neely, Modiano, and Rohrs [107] consider significant generalizations, for example, non i.i.d. arrivals and stationary time varying network topology. Georgiadis, Neely, and Tassiulas [40] provide a comprehensive overview of the recent developments and generalizations. Our discussion of Section 8.2.1 and 8.4 is based on [40]. We have made some simplifying assumptions for pedagogical convenience but the results can be generalized using the framework that we have provided. An important application of the MWS algorithm is in the maximum weight matching algorithm for input queued switches [98].

Ever since the proportional-fairness paradigm for congestion control in wireline networks was introduced by Kelly, Maulloo, and Tan [75], there has been significant interest in extending that to wireless networks. Kar, Sarkar, and Tassiulas [66] apply this congestion control principle to select the transmission probabilities to optimize single-hop flows in Aloha networks. This is an interesting introduction to this problem. This discussion is adapted from there. Extending it to multihop flows has been the focus of, Lin and Shroff [95] and Lo Presti [112] among others. The discussion in Section 8.5.2 is based on [95]. Lin, Shroff, and Srikant [94] provide an excellent tutorial on these techniques.

Since the optimizations involve multiple layers of the network stack, these are also called *cross-layer optimizations*. Kawadia and Kumar [72] provide important insights into the pitfalls that could accompany cross-layer optimisations.

Problems

8.1 Consider a network in which the frames are not equal. Let T_k be the number of slots in the k-th frame and B_k the number of slots allocated to a link in the k-th frame. Find the expression for the bandwidth allocated to the link. Now consider the following sequence the continued fraction expansion of $(1 + \sqrt{2})$ which is given by

$$2 + 1/(2 + 1/(2 + 1/(2 + \cdots)))$$

Identify suitable T_k and B_k so that the link is allocated a bandwidth of $\frac{1}{1+\sqrt{2}}$. Since every irrational number can be expressed as a continued fraction, this gives you a method to allocate irrational capacities to links. Explore irrational allocations using this method.

8.2 Given a network graph G, describe a graph coloring algorithm for each of the three graph-based constraints that determine the schedule. Let $\chi(G)$ be the vertex-chromatic number of G, the minimum number of colors required to color the vertices such that adjacent vertices have different colors. Let $\Delta(G)$ and $\omega(G)$ be the maximum vertex degree and the clique number, respectively. It can be shown that $\omega(G) \leq \chi(G) \leq \Delta(G) + 1$. Derive the corresponding inequalities for each of the three graph-based constraints.

8.3 Devise a greedy algorithm for edge-coloring of a graph. For the network of Figure 8.1, perform a greedy coloring and use the coloring to devise a scheme to provide all edges a capacity greater than $\frac{1}{C_1}$ where C_1 is the number of colors used. Observe that the coloring is not unique. Perform a second distinct coloring of this network. If C_2 is the number of colors required for the second coloring, number the colors $C_1+1,\ldots,$ $C_1 + C_2$. Comment on the change in $\mathsf{Co}(S)$.

8.4 Consider a routing and scheduling policy, say P, in a slotted WMN. Let Λ_P denote the set of arrival rate vectors λ that are stabilized by this policy. Let P_0 denote the queue-length-based centralized scheduling policy discussed in Section 8.4. Argue that $\Lambda_{P_0} = \cup_P \Lambda_P$. This means that if there exists any policy that stabilizes the queues under the arrival rate vector λ, then the queues will be stable for this arrival rate vector under policy P_0.

8.5 Construct an example to illustrate that (8.13) is not a sufficient condition.

8.6 Consider a two-link network. Let $S_1 = [1,0]$, $S_2 = [0,1]$, $S_3 = [0.25, 0.75]$, and $S_4 = [0.75, 0.25]$ be the four possible schedules in the network. Draw the link layer capacity region for this network.

8.7 Consider a two-link network. The links operate in a fading environment and in each slot, link i is either available with probability p_i or not available with probability $(1 - p_i)$, independently of the other link and of its availability in other slots. Characterize the link-capacity region for this system.

8.8 Consider the network shown in Figure 8.1. Choose ten arbitrary source-destination pairs and designate them as users $1,\ldots 10$. Assume $A_{i,j}(t)$ to be i.i.d. Bernoulli with probability of arrival λ in every slot. Write a program to simulate MWS algorithm on this network. Let $h_j(\lambda)$ be the average hop length for user j. Compare $h_j(\lambda)$ with the minimum hop distance.

8.9 Consider the MWS algorithm in a network in which there are link errors. Let $p_{(i,k)}$ be the packet error probability on link (i,k). Find the link capacity region for this network. Adapt the MWS algorithm for this case and show that it will stabilize the network for all $\lambda \in \Lambda_1$.

8.10 Consider a network in which all flows are single hop flows. All flows are to be routed over the single-hop path from the source to the destination. Given a network graph G obtain a schedule that maximizes the total utility.

8.11 Consider the single-hop s-Aloha network with N nodes that we studied in Chapter 7. In each slot, all nodes transmit independently with probability p. Derive the proportionally fair p when all nodes have the same utility function. Generalize to the case when Node i has utility function $a_i \log (x_i)$.

CHAPTER 9

Mesh Networks: Fundamental Limits

W e continue with our analysis of ad hoc internets. In the previous chapter we were concerned with routing and transmission scheduling to support point-to-point flows. We had assumed that the network graph was given and that it was connected. We had also assumed that the set of allowable link activation vectors was given. Since these can be derived from the node locations, we essentially assumed that the latter were given. In this chapter we will explore the limits of two fundamental properties of wireless networks—connectivity and capacity—without assuming that the node locations are known. For a random deployment, we will first derive the requirement to make the network connected. In the connectivity regime (i.e., when the network is connected), we explore the ability of the random network to transport data; we obtain the so-called transport capacity of the network. We will also consider the transport capacity of arbitrary networks. Like in the previous chapter, we will assume that time on the network is slotted and in each slot, one packet can be transmitted.

Overview

In Section 9.1 we provide an overview of connectivity and capacity issues in wireless mesh networks. We also provide an overview of node distribution models and the random graph models—random geometric graphs (RGGs) based on a Boolean model, and signal-to-interference ratio graphs (STIRGs). In Section 9.2 we consider the random geometric graph model. We first derive the exact probability for the connectivity of a one-dimensional network with a finite number of nodes and then carry out an asymptotic analysis of the connectivity of a network in two dimensions. In this asymptotic analysis, we first obtain a simple sufficient condition to make isolated nodes disappear. Ignoring edge effects, necessary and sufficient conditions for asymptotic connectivity are obtained. The derivation of the necessary condition is elementary (not simple!), but involved, and is very instructive of the proof techniques. In contrast, the sufficient condition would seem fairly simple. In Section 9.3, connectivity in STIRGs is discussed using percolation models.

The capacity analysis begins with a discussion on the connection between the different spatial models that were used in Chapter 8. In Section 9.5, we consider arbitrary networks where we do not make any assumptions on the node locations. For the protocol model, the transport capacity of an arbitrary network is obtained using simple geometric arguments. Section 9.6 first obtains the capacity of a randomly deployed network under the protocol model. The analysis is based on a routing problem in parallel computers known as $k \times k$ *permutation on an $n \times n$ mesh*. This is followed by a detailed discussion on the issues with the protocol model. We then discuss transport capacity in STIRG models using percolation. Finally, informal arguments are presented that describe the effective exploitation of mobility.

9.1   Preliminaries

An important feature of wireless networks is that the node locations in a network are random, because of either mobility or deployment constraints. Further, most wireless nodes will be battery operated and energy conservation will be an important objective of network operation. It is known that a battery recovers some energy when it is not in use. Hence, wireless nodes may frequently turn off their transceivers to conserve, and also to recover, energy. Even when a wireless node is switched on, depending upon its communication needs and battery level, the transmission power may vary with time. Thus, in exploring the fundamental performance limits of wireless networks, it is reasonable to model the node locations as random variables; the node locations can be assumed to be a realization of a *point process* with the points corresponding to the node locations. The distribution of the node locations could correspond to the ensemble of randomly deployed nodes that are static, or to the stationary distribution of networks with time varying locations.

Two models for the node locations are widely used. In one model a unit area (usually a square or a circle) is designated as the operational area of the wireless network and n nodes are distributed uniformly inside this area. This corresponds to a node density of n nodes per unit area. In a second model, the wireless nodes are assumed to be distributed in $\mathbb{R}^2$ and the node locations form a spatial Poisson process of intensity λ.

Each wireless network is allocated a frequency band for operation and the transmission energy from all the nodes should be restricted to this band. We will assume that all transmissions occupy the entire bandwidth available to the network.

The network graph captures the communication capabilities among the nodes in the network. Thus connectivity is an important network graph property. For a wireless network, the network graph also captures the communication constraints; for example, it can be used to specify the nodes that can transmit simultaneously. Thus it determines the rate at which nodes in the network can exchange information. Since the nodes in the network are randomly located, the network graph will be a *random graph*. We will now develop some random graph models for wireless networks.

9.1.1 Random Graph Models for Wireless Networks

In most wireless mesh networks, the nodes use omnidirectional antennas and multiple access protocols to access the channel. Hence, the channel becomes a broadcast channel. Transmission energy from a node reaches all nodes in the network. The received power at these other nodes will depend on the distance and the characteristics of the radio channel between the transmitter and the receiver. If node i is located at x_i and is transmitting with power P_i, then the SINR at a receiver at location x_j is given by

$$\text{SINR}_{i,j} := \frac{P_i \, L(x_i, x_j)}{N_0 W + \gamma \sum_{\substack{k \neq i \\ k \text{ transmitting}}} P_k \, L(x_k, x_j)} \tag{9.1}$$

where $L(x, y)$ is the path loss function when the transmitter is at x and the receiver is at y, N_0 is the thermal noise spectral density at the receiver, W is the channel bandwidth, and γ, $0 \leq \gamma \leq 1$, is the orthogonality factor between the transmissions. If the transmissions are all perfectly orthogonal, or if no other node is transmitting when Node i is transmitting, then $\gamma = 0$. Typically, $L(x, y)$ takes the form $L(d)$ where d is the distance between the points x and y. The standard model for path loss is the far field model where $L(x, y) = L(\|x - y\|)^{-\alpha}$, $\|x - y\|$ is the Euclidean distance between x and y, and $\alpha > 0$ is called the path loss exponent. See Section 2.1.4 for a more detailed discussion on the signal attenuation and delay phenomena on the path from the transmitter to the receiver.

To see how to construct the network graph from the path loss model, assume that all nodes transmit with the same power, say P, and that the minimum SNR required at a receiver is β. Define

$$r_1 := \sup_{d > 0} \left\{ \frac{P L(d)}{N_0 W} > \beta \right\}$$

When no other node in the network is transmitting, all nodes at a distance less than r_1 from the transmitter, say Node 0, can decode the transmission with acceptable error probability; the decode region is a circle of radius r_1 centered at the location of Node 0. r_1 is called the *SNR-cutoff*.

Exercise 9.1

Obtain the SNR-cutoff for $L(d) = d^{-\alpha}$.

Now consider the situation when Node 0 is transmitting. When there are other nodes transmitting in the same slot as Node 0, the extent of the decode region of Node 0 depends on the location of these other transmitters. This is because the signal from them will be the interference at the receivers of Node 0, and this reduces their SINR. This is evident from (9.1). We will first simplify the effect of this interference and assume that a transmission from node i, located at x_i, can be decoded at node j, located at x_j, if

$$\|x_i - x_j\| < r \tag{9.2}$$

$\| \cdot \|$ denotes the Euclidean distance. r is called the *transmission range* or the *cutoff*, and essentially captures the effect of interferences. The network graph $G = (V, E)$ is constructed as follows. The vertex set corresponds to the n nodes in the network. The edge set in the network graph, representing the links in the network, is given by

$$E := \{(i, j) : \|x_i - x_j\| < r\}$$

Figure 9.1 illustrates the edges obtained using (9.2) at two sample nodes in a network. Note that (9.2) only fixes the links in the network. The set of links that can simultaneously be active can now be derived from graph-based constraints like those described in Chapter 7.

When the node locations are random, let X_i denote the random location of node i and let $\mathbf{X}_n := (X_i, i \in \{1, \dots, n\})$ denote the set of node locations in the n-node network. Since $\mathbf{X}_n$ is random, the network graph is a *random graph*. Let $G_n(r_n)$

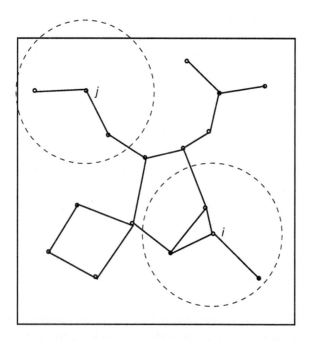

Figure 9.1 A sample realization of a random geometric graph. The cutoff region around nodes *i* and *j* is shown as dashed circles.

denote a realization of a network graph obtained when there are n nodes in the network and the cutoff is r_n. $G_n(r_n)$ is a graph-valued random variable.

Random graphs have been studied for more than 50 years, primarily as Erdös-Renyi random graphs. In this classical random graph model, in an n-node graph, edge (i,j), $1 \le i < j \le n$, occurs with probability p_n, $0 < p_n < 1$, independently of all the other edges. Observe that in the graph $G_n(r_n)$, the edges are not independent. To see this, observe that if (i,j) and (j,k) are edges in G, then we have more information on the relative locations of node pair (i,k). Therefore, we cannot say that the existence of edge (i,k) is independent of the existence of edges (i,j) and (j,k). Random graphs of the type $G_n(r_n)$ are called *random geometric graphs*. From the method used in determining the edges in the network, this is also called the *Boolean model*.

A network graph that captures the effect of interferences in more detail can also be constructed. Assume that signals transmitted by the different nodes are orthogonal, albeit with some imperfections, like in CDMA networks. Let each node randomly choose a color represented by an integer in $[1, T]$. Divide time into frames of T slots each. In slot t of each frame, $1 \le t \le T$, nodes that chose color t will transmit. Let S_t denote the set of nodes that transmit in slot t. Thus each node gets a chance to transmit once in every T slot. We will assume that even if a node, say Node j, is transmitting in a slot, it can decode a signal transmitted by another node, say Node i, if the SINR threshold is met. We will also assume that there is sufficient orthogonality between the transmissions so that Node j can decode all the simultaneous transmissions for which the SINR is above the threshold. Now consider Node i, transmitting with power P_i and a possible receiver Node j. Let $\text{SINR}_{i,j}$ be the SINR at Node j for the transmission from Node i. To obtain $\text{SINR}_{i,j}$, observe that the received power from Node i is the signal and the received power from all other transmitters in the slot is the interference. Thus

$$\text{SINR}_{i,j} = \frac{P_i L(d_{i,j})}{N_0 W + \gamma \sum_{\substack{k \in S_t \\ k \neq i}} P_k L(d_{k,j})}$$

Let E_t denote the set of directed edges for slot t, those along which packets could be exchanged in slot t. E_t is obtained as follows:

$$E_t := \{(i,j) : \text{SINR}_{i,j} > \beta\}$$

Let $E'_T := \cup_{t=1}^{T} E_t$, where E'_T is the set of all directed edges along which communication could take place in at least one of the T slots of a frame. Let E_T be the set of bidirectional edges in E'_T,

$$E_T := \{(i,j) : (i,j) \in E'_T \text{ and } (j,i) \in E'_T\}$$

E_T is the set of undirected edges and the network graph, $G_T = (N, E_T)$, is called a *signal-to-interference-ratio graph (STIRG)*. It is easier to study STIRGs by assuming that the nodes are distributed according to a Poisson spatial process in $\mathbb{R}^2$. There are many parameters that define the STIRG—the node density λ, the transmission power P, the receiver noise power WN_0, the SINR threshold β, the orthogonality factor γ, and the path loss function $L(\cdot)$. We will study the properties of STIRGs as a function of λ and γ.

The properties of random graphs that are of interest to us will essentially be events in a probability space. Many a time, we will be analyzing the asymptotic behavior of the properties of the network graph. In this we will be using the order notation (see Section A.3 in Appendix A) with some modifications.

Consider a sequence of random experiments indexed by n. Let X be an event of interest and let $p_n(X)$ be the probability that X occurs in the n-th experiment of the sequence. If $p_n(X) \to 1$ as $n \to \infty$, we say that X occurs *with high probability* (w.h.p.). This means that for any $\epsilon > 0$, we can find an $n^*(\epsilon)$ such that $\Pr(X_n) > 1 - \epsilon$ for all $n \geq n^*(\epsilon)$. We now apply this notion to the order notation. Let $X(n)$ be a random sequence. For example, $X(n)$ could be the random variable representing the execution time of a protocol in a random network; the protocol itself could be a randomized protocol. If

$$\Pr\big(X(n) \leq a\, g(n)\big) \to 1 \text{ as } n \to \infty$$

for some positive constant a, then $X(n) = O(g(n))$ w.h.p. and we write $X(n) = \tilde{O}(g(n))$. Similarly, we say $X(n) = \tilde{\Omega}(g(n))$ if $X(n) = \Omega(g(n))$ w.h.p. And $X(n) = \tilde{o}(g(n))$ implies that $X(n) = o(g(n))$ w.h.p.

9.1.2   Spatial Reuse, Network Capacity, and Connectivity

We have seen in earlier chapters that using the same part of the spectrum simultaneously in many parts of the network is an important mechanism to increase the capacity of a wireless network. The transmission power determines the *spatial reuse*, the number of simultaneous transmissions that are possible in the network. In cellular networks (see Chapters 4 and 5), the effect of spatial reuse on the capacity is obvious. However, for wireless mesh networks, we first need to formalize the notion of capacity.

Consider a mesh network of n nodes. Let each node have a flow to an arbitrary destination giving us a set of n end-to-end flows in the network. Our first interest is in the *per node capacity*, the maximum rate at which the n node pairs can exchange information. Since all flows will reach the destinations over multiple hops in the network, it is important to capture the spatial distance covered by the flows in the definition of capacity. We define the *transport capacity* of the mesh network as the sum of the distances toward the destination traveled by every bit per unit time. Thus the unit of transport capacity is bit-meters per second. For example, let d_i be the destination for the flow from node i. If $b_i(\tau)$ bits reach from i to d_i in τ seconds, then $\sum_{i=1}^{n} b_i(\tau)\|X_i - X_{d_i}\|$ bit-meters are transported in τ seconds. The transport capacity of the network, C_T, is given by

$$C_T := \lim_{\tau \to \infty} \frac{1}{\tau} \sum_{i=1}^{n} b_i(\tau)\|X_i - X_{d_i}\|$$

bit-meters per second. It is important to note that C_T is a random variable for each protocol because the node locations, and hence the network graph, are random.

To maximize spatial reuse, the transmission power, and hence the transmission range, should be as low as possible. This increases the number of hops for each flow. One might argue that the multiple hops can actually reduce capacity because a packet has to be transmitted multiple times. Informally, let D be the average source-destination distance, R the average distance toward the destination covered by a hop, and S the number of nodes that are transmitting simultaneously in a slot. R is proportional to r, the transmission range. If each packet is l bits, then $C_T = RlS$ bit-meters per slot. Now observe that increasing the transmission range increases R, and hence decreases the average number of hops, D/R, linearly. If the network graph is a random geometric graph and the receiver conflict constraint for spatial reuse (see Section 8.1.1) is used, in a circle of radius r around a receiver, only one node can transmit. Thus increasing r decreases S in proportion to r^2. This means that the transport capacity increases as the transmission range decreases. Hence, as low a transmission range as possible should be used. However, the transmission range cannot be made too low because the network will then become disconnected; that is, there may not be a path between every pair of nodes in the network. Thus, keeping the network connected becomes an important issue. This in turn means that capacity and connectivity need to be studied together and we will do just that in this chapter.

We begin by studying the connectivity of a randomly deployed network using the random geometric graph or the Boolean Model in the next section.

9.2 Connectivity in the Random Geometric Graph Model

In this section we will analyze the connectivity property of a wireless network with nodes randomly deployed in a finite area. We will use the random geometric graph model for the network graph. For a given cutoff, r, obtaining exact expressions for the probability that the network is connected is hard except for some simple cases. We will first develop one such simple case in which the nodes are distributed on a finite line segment. Numerical analysis of the results points to an interesting threshold behavior of the connectivity as a function of the cutoff. For a wireless network in two dimensions, we will analyze this behavior and obtain the cutoff required for asymptotic connectivity.

9.2.1 Finite Networks in One Dimension

Consider a two-node, one-dimensional network with the location of each node uniformly distributed in $[0, z]$ and chosen independently of each other. Let the transmission range of both nodes be r. We now obtain the probability that the two nodes are connected. Without loss of generality, let X_1 be the location of the left node and X_2 that of the right node; that is, $X_1 \leq X_2$. The two-node network is connected if $X_2 - X_1 \leq r$. This is graphically shown in Figure 9.2. The set of values that (X_1, X_2) can take is denoted by the area OAB. The set of (X_1, X_2) that would result in a connected network is given by the shaded area S in the figure. S is the region satisfying $X_1 < X_2$ (by definition of X_1 and X_2) and $X_2 - X_1 < r$ (the connectivity requirement). Since the nodes are distributed uniformly in $[0, z]$, the probability that the network is connected is the ratio of the area of S to the area of OAB. The area of S is $\left(\frac{z^2}{2} - \frac{(z-r)^2}{2} \right)$ and that of OAB is $\frac{z^2}{2}$. Thus the probability that the network is connected is

$$\frac{z^2/2 - ((z-r)^2)/2}{z^2/2} = \frac{2zr - r^2}{z^2}$$

Let there be n nodes in the network and let the location of node i be denoted by X_i. X_i are i.i.d. with uniform distribution in $[0, z]$. Thus the random network is represented by a random vector $\mathbf{X} = [X_1, X_2, \ldots, X_n]$. Let $p_c(n, z, r)$

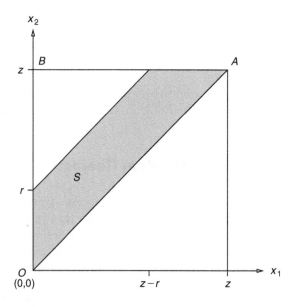

Figure 9.2 The feasible region of a random, connected 2-node, 1-dimensional ad hoc network.

be the probability that **X** represents a connected network when each node has a transmission range of r. Let $\hat{X} = [\hat{X}_1, \hat{X}_2, \ldots, \hat{X}_n]$ be the node locations ordered according to their positions on $[0, z]$; that is, $\hat{X}_1 < \hat{X}_2 < \ldots, < \hat{X}_n$. Define $\hat{X}_0 = 0$. The condition $\hat{X}_{i+1} - \hat{X}_i < r$ for $i = 1, \ldots, (n-1)$ needs to be satisfied for **X** to represent a connected network.

The set of all realizable networks is contained in the n-dimensional polytope A_n, defined by $0 = \hat{x}_0 \leq \hat{x}_1 \leq \hat{x}_2 \leq \ldots, \leq \hat{x}_n \leq z$. The set of connected networks is contained in the polytope $A_c(n, z, r)$, defined by $\hat{x}_{i+1} - \hat{x}_i < r$ for $i = 1, \ldots, (n-1)$. Let $V(n, z)$ and $V_c(n, z, r)$ be the volumes of the polytopes A_n and $A_c(n, z, r)$, respectively. Since we have assumed that the node locations are uniformly distributed in $[0, z]$, the probability that the network is connected, $p_c(n, z, r)$, will be $V_c(n, z, r)/V(n, z)$.

We obtain $p_c(n, z, r)$ as follows. Define $y_i = \hat{x}_{i+1} - \hat{x}_i$, $i = 0, \ldots n - 1$. Let B_n denote the polytope defined by

$$\left\{ y_0, y_1, \ldots y_{n-1} : y_i \geq 0 \text{ for } i \geq 0, \sum_0^{n-1} y_i \leq z \right\},$$

and let $U(n, z)$ be the volume of B_n. Let $B_c(n, z, r)$ be the polytope defined by

$$\left\{ y_0, y_1, \ldots y_{n-1} : y_0 \geq 0, \ 0 \leq y_i \leq r \text{ for } i \geq 1, \sum_0^{n-1} y_i \leq z \right\}$$

and let $U_c(n, z, r)$ be the volume of $B_c(n, z, r)$.

Note that the y are obtained from $\hat{x}$ using a linear transformation. Further this transformation is invertible because, given y, the $\hat{x}$ can be recovered. Since $\hat{x}$ and y are related by a linear invertible transformation, the polytope B_n is a scaled version of the polytope A_n and $B_c(n, z, r)$ is obtained from $A_c(n, z, r)$ using the same scaling. We thus have $U(n, z) = KV(n, z)$ and $U_c(n, z, r) = KV_c(n, z, r)$ for some constant $K > 0$. Thus

$$p_c(n, z, r) = \frac{U_c(n, z, r)}{U(n, z)}$$

It is easy to see that $U(1, z) = z$. To obtain $U(2, z)$, we need $0 \leq y_0 \leq z$ and $y_0 + y_1 \leq z$, and we have

$$U(2, z) = \int_0^z U(1, z - t) dt$$

Arguing along the same lines, $U(n, z)$ can be obtained using the recurrence relation

$$U(n, z) = \int_0^z U(n - 1, z - t) dt \tag{9.3}$$

Exercise 9.2

Evaluate the recursion in (9.3) to show that $U(n, z) = \frac{z^n}{n!}$.

We obtain a recursion for $U_c(n, z, r)$ as follows. The n-node network is connected if two conditions are satisfied: (1) the $(n-1)$-node network formed without the leftmost node is connected and (2) the leftmost node is within r of the node immediately to its right. This is saying that the $(n-1)$-node network formed by removing the segment between the first node and the second node is connected. Thus this connected network with $(n-1)$ nodes satisfies

$$\left\{ y_0, y_2, \ldots y_{n-1} : y_0 \geq 0, \quad 0 \leq y_i \leq r \text{ for } i \geq 2, \ y_0 + \sum_2^{n-1} y_i \leq (z-r) \right\}$$

Also, this $(n-1)$-node connected network will span over at least $(z-r)$. This is illustrated in Figure 9.3. We can therefore write the recursion for $U_c(n, z, r)$ as follows.

$$U_c(n, z, r) = \int_0^r U_c(n-1, z-t, r)dt \tag{9.4}$$

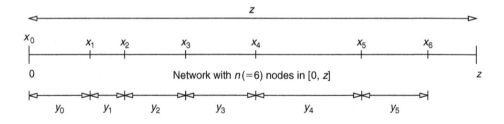

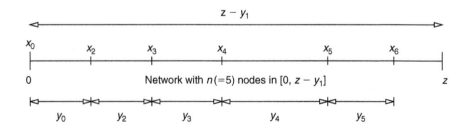

Figure 9.3 Illustrating the discussion leading to (9.4). For the n-node network to be connected, the $(n-1)$-node network without the node at x_1 should also be connected and y_1 should be at most r. Thus the $(n-1)$-node network will be spread over $(z-y_1)$ with y_1 taking values in $(0, r)$.

Defining

$$h(z) := \begin{cases} 1 & 0 \le z \le r \\ 0 & \text{otherwise} \end{cases},$$

we can write the right-hand side of (9.4) as the convolution of $U_c(n-1, z, r)$ and $h(z)$. Thus

$$U_c(n, z, r) = h(z) * U_c(n-1, z, r) = \cdots = h^{((n-1)*)}(z) * U_c(1, z, r)$$

where $h^{((n-1)*)}(\cdot)$ is the $(n-1)$-fold convolution of $h(\cdot)$ with itself. Note that $U_c(1, z, r) = zu(z)$, where $u(z)$ is the unit step function.

Let $\tilde{h}(s)$ and $\tilde{U}(n, s, r)$ denote the Laplace transform of $h(z)$ and $U_c(n, z, r)$, respectively.

$$\tilde{h}(s) = \frac{1 - e^{sr}}{s}$$

$$\tilde{U}_c(n, s, r) = \left(\frac{1 - e^{sr}}{s} \right)^{n-1} \frac{1}{s^2}$$

$$= \frac{(1 - e^{sr})^{n-1}}{s^{n+1}}$$

$$= \frac{\sum_{k=0}^{n-1} \binom{n-1}{k} (-1)^k e^{srk}}{s^{n+1}}$$

Taking the inverse Laplace transform, we get

$$U_c(n, z, r) = \sum_{k=0}^{n-1} \binom{n-1}{k} \frac{(-1)^k (z - kr)^n u(z - kr)}{n!} \tag{9.5}$$

Finally, the probability that the network is connected is

$$p_c(n, z, r) = \frac{U_c(n, z, r)}{U(n, z)} = \sum_{k=0}^{n-1} \binom{n-1}{k} (-1)^k \frac{(z - kr)^n}{z^n} u(z - kr)$$

Figure 9.4 plots $p_c(n, z, r)$ as a function of r for $z = 1$ and different values of n. Observe that, as n becomes large, for small values of r $p_c(n, 1, r)$ is close to zero and for large values of r it is close to one. The range of r over which $p_c(n, 1, r)$ takes intermediate values is very small. This indicates some kind of a *threshold* behavior for the connectivity probability: For large n, as r is increased from 0, the network is disconnected with probability very nearly 1 and at a threshold transmission range, it becomes connected with probability very nearly 1. There is no "intermediate value" for the probability of connectivity; either it is close to one or close to zero and the transition is sharp. We will now investigate this behavior for a network in two dimensions. The analysis extends easily to higher dimensions.

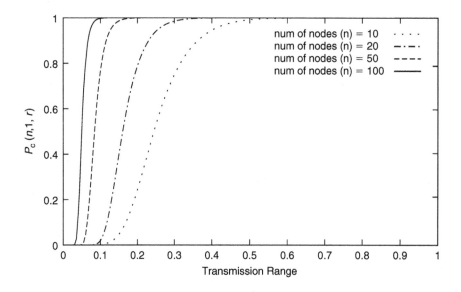

Figure 9.4   The probability of connectivity in a one-dimensional network as a function of r for different n, the number of nodes in the network. The operational area is [0, 1].

9.2.2   Networks in Two Dimensions: Asymptotic Results

To make the network connected, it is necessary that the network not have any isolated nodes (i.e., nodes that are not connected to any other node). Let us first find a sufficient condition on r_n, the transmission range as a function n, to make the isolated nodes disappear. In the following, we will always assume that $\pi r_n^2 < 1$.

Making Isolated Nodes Disappear

Let n nodes be distributed uniformly in the unit square. Let r_n be the transmission range of the nodes when there are n nodes in the network. Let $\mathcal{Z}_i$ be the event that node i is isolated (i.e., it is not in the transmission range of any other node), and let $\mathcal{Z} := \cup_i^n \mathcal{Z}_i$ be the event that there is at least one isolated node in the network. A node is isolated if the other $(n-1)$ nodes are not in the intersection of the circle of radius r_n centered at X_i (the location of node i) and the unit square. The part of the square that is not in this circle is at most $\left(1 - \frac{\pi r_n^2}{4}\right)$.

$$\Pr(\mathcal{Z}_i) \leq \left(1 - \frac{1}{4}\pi r_n^2\right)^{n-1}$$

From the union bound (see Appendix B), $\Pr(\cup_{i=1}^n \mathcal{Z}_i) \leq \sum_{i=1}^n \Pr(\mathcal{Z}_i)$ and we can write

$$\Pr(\mathcal{Z}) \le n \left(1 - \frac{1}{4}\pi r_n^2 \right)^{n-1}$$

$$= e^{\left(\log n + (n-1) \log\left(1 - \frac{1}{4}\pi r_n^2 \right) \right)}$$

$$\le e^{\left(\log n - (n-1)\left(\frac{1}{4}\pi r_n^2 \right) \right)}$$

$$= e^{\log n \left(1 - \frac{(n-1)}{\log n}\left(\frac{1}{4}\pi r_n^2 \right) \right)}$$

$$= e^{\log n \left(1 - \frac{\pi}{4}\left(\frac{r_n}{\sqrt{(\log n)/(n-1)}} \right)^2 \right)}$$

where, in writing the second inequality, we have used the fact that $\log(1+x) \le x$. To make $\Pr(\mathcal{Z}) \to 0$ as $n \to \infty$, from the last equation, it is sufficient to have $r_n^2/((\log n)/n) \to \infty$; r_n^2 is made to decrease strictly slower than $(\log n)/n$ with the ratio going to ∞ as $n \to \infty$. For example, using $r_n^2 = \frac{\log n}{n} + c_n$ where $c_n = o(n)$ and $c_n \to \infty$ is sufficient to make isolated nodes disappear. The condition $c_n = o(n)$ is necessary to ensure that $\frac{1}{4}\pi r_n^2 < 1$.

Interestingly, we will show that making $r_n/\sqrt{(\log n)/n} \to \infty$ is necessary and sufficient to make the network connected.

Exercise 9.3

Using arguments similar to the preceding, show that in a one-dimensional network, isolated nodes disappear if $\frac{r_n}{(\log n)/n} \to \infty$.

Necessary Cutoff for Connectivity

Let us ignore edge-effects. This is not too bad because, as n becomes large, smaller values of r_n would suffice for connectivity and the edge effects become negligible.

Let P_d denote the probability that the network is disconnected. Clearly,

$$P_d := \Pr(\text{network is disconnected}) \ge \Pr\left(\cup_{i=1}^{n} \mathcal{Z}_i \right)$$

$$\ge \sum_{i=1}^{n} \Pr(\mathcal{Z}_i) - \sum_{1 \le i < j \le n} \Pr(\mathcal{Z}_i \cap \mathcal{Z}_j) \tag{9.6}$$

The last inequality follows from the "inclusion-exclusion" formula for the probability of the union of events (see Appendix B).

We have discussed $\Pr(\mathcal{Z}_i)$ earlier. Ignoring edge effects, we can assume that the entire circle of radius r_n around node i is inside the unit square. This is illustrated in Figure 9.5. Hence

$$\Pr(\mathcal{Z}_i) = (1 - \pi r_n^2)^{n-1} \tag{9.7}$$

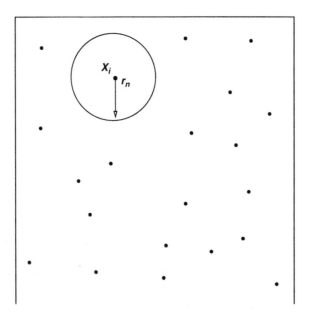

Figure 9.5 **For node *i* to be isolated, the other ($n-1$) nodes should not be in the circle of radius r_n around node X_i, the location of node *i*.**

Let us now consider the event $\{\mathcal{Z}_i \cap \mathcal{Z}_j\}$; that is, the event that nodes i and j are both isolated. For this to happen, the ($n-2$) nodes other than i and j should not be in the region determined by the union of circles of radius r_n centered at X_i and X_j. Also, X_j should not be inside the circle at X_i. For the following discussion, refer to Figure 9.6. If $\|X_i - X_j\| > 2r_n$, then the two circles centered at X_i and X_j are disjoint and the area not available to the ($n-2$) nodes is ($2\pi r_n^2$). This is shown in the left panel of Figure 9.6. In this case nodes i and j will be isolated with probability $(1 - 2\pi r_n^2)^{n-2}$. If $r_n \leq \|X_i - X_j\| \leq 2r_n$, then the area unavailable to the ($n-2$) nodes depends on the location of X_j. This is shown in the right panel of Figure 9.6. From elementary geometry, we can obtain the unavailable area to be

$$2\pi r_n^2 - r_n^2(2\phi - \sin 2\phi) = (2\pi - 2\phi + \sin 2\phi)r_n^2$$

where $a = \|X_i - X_j\|$ and $\phi = \arccos(a/2r_n)$. It is easy to see that $\pi/3 \leq \phi \leq 0$. Hence the area unavailable to the other ($n-2$) nodes when $r_n \leq \|X_i - X_j\| \leq 2r_n$ is given by $B\pi r_n^2$ where B is a random number satisfying

$$2 \geq B \geq \frac{8\pi + 3\sqrt{3}}{6\pi} > 1$$

Let b be the conditional expectation of B when $r_n \leq \|X_i - X_j\| \leq 2r_n$. Of course, $1 < b < 2$.

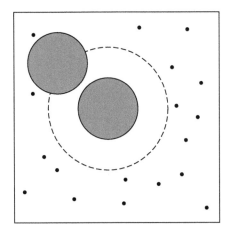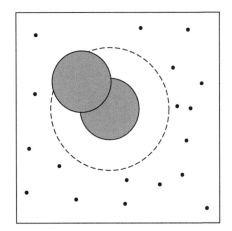

Figure 9.6 For nodes *i* and *j* to be isolated there are two cases. The left panel shows the case when $|X_i - X_j| \geq 2r_n$. In this case, the area unavailable to the other $(n-2)$ nodes to keep *i* and *j* isolated is the sum of the areas of the two shaded circles centered at X_i and X_j, $2\pi r_n^2$. The right panel shows the case when $r_n \leq |X_i - X_j| \leq 2r_n$. Here the area unavailable to the other $(n-2)$ nodes is less than $2\pi r_n^2$.

The probability that X_j is outside the outer dashed circle is $(1 - 4\pi r_n^2)$ and the probability that it is in the annulus between the dashed and solid circle is $3\pi r_n^2$. We therefore have

$$\Pr(Z_i \cap Z_j) = \left(1 - 4\pi r_n^2\right)\left(1 - 2\pi r_n^2\right)^{n-2} + \left(3\pi r_n^2\right)\left(1 - b\pi r_n^2\right)^{n-2} \tag{9.8}$$

To proceed with the analysis, we will need the following lemma.

Lemma 9.1

If $\pi r_n^2 = \frac{\log n + c}{n}$, $c \geq 0$, then, for fixed $\theta < 1$, and for all sufficiently large n,

$$n\left(1 - \pi r_n^2\right)^{n-1} \geq \theta e^{-c} \tag{9.9}$$

■

This is derived as follows. For $x \geq 0$,

$$e^{-x} \leq 1 - x + \frac{x^2}{2} \tag{9.10}$$

Also, since y^n is a convex function, for $a > 0$, and $0 < \epsilon < a$, we have

$$(y - \epsilon)^n \geq y^n - \epsilon n y^{n-1} \tag{9.11}$$

Let

$$x = \pi r_n^2 = \frac{\log n + c}{n} \quad \text{and} \quad \epsilon = \frac{x^2}{2}$$

Then

$$\left(1 - \pi r_n^2\right)^n = (1 - x)^n \geq \left(e^{-x} - \frac{x^2}{2}\right)^n \geq e^{-nx} - n \frac{x^2}{2} e^{-(n-1)x}$$

$$= e^{-nx} \left(1 - n \frac{x^2}{2} e^x\right)$$

$$= \frac{e^{-c}}{n} (1 - o(n))$$

The first inequality is obtained from (9.10) and the second inequality is obtained from (9.11). We thus have

$$n \left(1 - \pi r_n^2\right)^n \geq e^{-c} (1 - o(n)) \geq \theta e^{-c}$$

where θ is a constant.

Let us now go back to analyzing P_d. Applying (9.7) and (9.8) in (9.6), for large n, we get

$$P_d \geq n \left(1 - \pi r_n^2\right)^{n-1} - \frac{n(n-1)}{2} \left(3\pi r_n^2 \left(1 - b\pi r_n^2\right)^{n-2} + \left(1 - 4\pi r_n^2\right) \left(1 - 2\pi r_n^2\right)^{n-2}\right)$$

$$\geq \theta e^{-c} - \frac{n(n-1)}{2} \left(3\pi r_n^2 e^{-b(n-2)\pi r_n^2} + e^{-2(n-2)\pi r_n^2}\right) \tag{9.12}$$

In deriving the last inequality we have used the fact that $(1-x) \leq e^{-x}$ and $(1 - 4\pi r_n^2) \leq 1$. Let us now consider each of the terms separately.

$$e^{-b(n-2)\pi r_n^2} = e^{-bn\pi r_n^2} e^{2b\pi r_n^2}$$

$$= e^{-bn \frac{\log n + c}{n}} e^{2b\pi r_n^2}$$

$$= \frac{e^{-bc}}{n^b} e^{2b\pi r_n^2}$$

$$\frac{n(n-1)}{2} \left(3\pi r_n^2\right) e^{-b(n-2)\pi r_n^2} = \frac{n(n-1)}{2} \left(3 \frac{\log n + c}{n}\right) e^{-b(n-2)\pi r_n^2}$$

$$= \frac{3}{2} \left(\frac{n-1}{n^b}(\log n + c)\right) \left(e^{-bc} e^{2b\pi r_n^2}\right) \to 0$$

The last assertion is true because $b > 1$ and hence $\frac{\log n + c}{n^b/(n-1)} \to 0$ as $n \to \infty$. Let us now look at the second term in (9.12).

$$\frac{n(n-1)}{2} e^{-2n\pi r_n^2} e^{4\pi r_n^2} = \frac{n(n-1)}{2} e^{-2(\log n + c)} e^{4\pi r_n^2}$$

$$= \frac{n(n-1)}{2} \frac{1}{n^2} e^{-2c} e^{4\pi r_n^2}$$

$$= e^{-2c} \left(\frac{n-1}{2n} e^{4\pi r_n^2} \right) \le e^{-2c}(1 + \epsilon) \qquad (9.13)$$

for any ϵ for sufficiently large n. This is true because as $n \to \infty$, the term in the parenthesis goes to $1/2$.

Returning to (9.12), we have

$$P_d \ge \theta e^{-c} - (1 + \epsilon)e^{-2c} \qquad (9.14)$$

for all n sufficiently large, say $n \ge N_1(\epsilon, \theta, c)$.

Replace c by a sequence c_n. We will characterize P_d in terms of the properties of c_n. Also, since P_d is a function of n, we will make the relation explicit by denoting it as $P_d(n, r_n)$.

Let $\limsup c_n = \bar{c}$. Then, for any $\epsilon > 0$, $c_n \le \bar{c} + \epsilon$ for all $n \ge N_2(\epsilon)$. For any realization of the node locations, the number of edges can only increase when c is increased and hence P_d decreases monotonically with increasing c. Let $P_{d,\epsilon}(n, r_n)$ be the P_d with $c = \bar{c} + \epsilon$. Therefore

$$P_d(n, r_n) \ge P_{d,\epsilon}(n, r_n) \ge \theta e^{-(\bar{c}+\epsilon)} - (1 + \epsilon)e^{-2(\bar{c}+\epsilon)}$$

for $n > \max(N_2(\epsilon), N_1(\epsilon, \theta, c))$. Taking limits

$$\lim_{n \to \infty} \inf P_d(n, r_n) \ge \theta e^{-(\bar{c}+\epsilon)} - (1 + \epsilon)e^{-2(\bar{c}+\epsilon)}$$

This holds for all $\epsilon > 0$ and $\theta < 1$. Therefore, we can say that if $\pi r_n^2 = \frac{\log n + c_n}{n}$, then

$$\lim_{n \to \infty} \inf P_d(n, r_n) \ge e^{-c}(1 - e^{-c}) \qquad (9.15)$$

where $c = \lim_{n \to \infty} c_n$. This means that if $\pi r_n^2 = \frac{\log n + c_n}{n}$ and $\limsup_{n \to \infty} c_n < \infty$, then there is a nonzero probability that the network is disconnected. This in turn means that it is necessary to have

$$\pi r_n^2 = \frac{\log n + c_n}{n} \qquad c_n \to \infty$$

for the network to be asymptotically almost surely connected.

We next investigate a sufficient condition for connectivity.

Sufficient Transmission Range for Connectivity

To obtain a sufficient transmission range, we make the following construction. Tessellate the unit square into square "cells" each of side s_n, (there are $\frac{1}{s_n^2}$ cells). The subscript n is used to indicate that the size of the cell will be a function of n. Figure 9.7 illustrates this tessellation. We will first find an s_n for which all cells will contain one or more nodes with high probability. We will then choose r_n such that the nodes in a cell are connected to all nodes in the cells that are above and below it and on the right and left of it. It is easy to see that this can be achieved if we choose $r_n = \sqrt{5}s_n$.

Now consider an arbitrary cell in the tessellation, which we will call the tagged cell. When a node is randomly placed in the square, the probability that it will be in the tagged cell is s_n^2. We therefore have

$$\Pr\left(\text{tagged cell is empty}\right) = (1 - s_n^2)^n$$

Since there are $\frac{1}{s_n^2}$ cells, using the union bound, we see that

$$\alpha_n := \Pr\left(\text{one or more cells are empty}\right) \leq \frac{(1 - s_n^2)^n}{s_n^2}$$

$$\leq \frac{\left(e^{-s_n^2}\right)^n}{s_n^2}$$

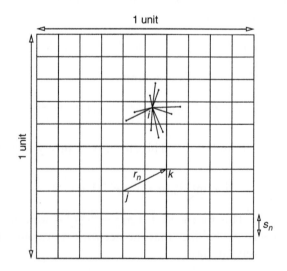

Figure 9.7   The unit square is tessellated into square cells of side s_n. Each node should have an edge to all the nodes in the neighboring cells as shown for node i. The maximum distance between two nodes in adjacent cells is when they are located like j and k. Thus it suffices to have $r_n = \sqrt{5}s_n$ for asymptotic connectivity.

The last inequality is true because $(1 - x) \leq e^{-x}$. Now, let $s_n = \sqrt{\frac{K \log n}{n}}$. We then have

$$\alpha_n \leq \frac{e^{-\frac{nK \log n}{n}}}{\frac{K \log n}{n}}$$

$$= \frac{ne^{-K \log n}}{K \log n}$$

$$= \frac{1}{n^{K-1}} \frac{1}{K \log n}$$

If we choose $K > 1$, then $\alpha_n \to 0$; if $s_n = \sqrt{\frac{K \log n}{n}}$, then no cell is empty with high probability.

Thus we see that $r_n = (\sqrt{5}K \log n)/n$, $K > 1$, is sufficient to make the network connected with high probability. Compare this with the necessary condition: $r_n = \sqrt{(\log n + c_n)/(\pi n)}$ with $c_n \to \infty$. Thus $\sqrt{\frac{\log n}{n}}$ is a *threshold function*. This means that if r_n decreases faster than $\sqrt{\frac{\log n}{n}}$ then the probability that the network is connected goes to zero. On the other hand, if it decreases slower, then the probability that the network is connected goes to one. For large n and a given deployment, if we construct random geometric graphs with increasing r_n, there is a *phase transition* of the connectedness property of the graph around $r_n = \sqrt{\frac{\log n}{n}}$. With high probability, it is disconnected below this value and is connected above this value.

We can also prove a slightly stronger result on connectivity. If we choose $K > 2$, we see that

$$\sum_{n=1}^{\infty} \alpha_n < \infty \tag{9.16}$$

Now consider the sequence of random vectors $\{\mathbf{X}_n\}$ with $\mathbf{X}_n$ denoting an n-node deployment.

$$\sum_{n=1}^{\infty} \Pr(\mathbf{X}_n \text{ has empty cells}) = \sum_{n=1}^{\infty} \alpha_n < \infty$$

From the Borel-Cantelli lemma (see Section B.3.3), this means $\mathbf{X}_n$ has no empty cells infinitely often and the network is asymptotically almost surely connected.

9.3 Connectivity in the Interference Model

Let us now consider connectivity in STIRGs. We will assume that the node locations form a homogeneous Poisson point process in $\mathbb{R}^2$ of intensity λ. This means that the number of nodes in two nonoverlapping areas is independent. Further, the number of nodes in an area A has a Poisson distribution with mean λA. If $\lambda > 0$, almost surely, there are an infinite number of nodes in the network. As we have mentioned earlier, the graph from a realization of the node locations depends on the orthogonality factor γ, SINR threshold β, the transmission powers P_i, the receiver noise $W N_0$, and the path loss function $L(\cdot)$. To simplify the analysis, let us assume that all nodes transmit with the same power P. Also, all nodes transmit in every slot. We will also choose a path loss model and fix the SINR threshold β. We will denote the random graph by $G(\lambda, \gamma)$ to make the dependence of a random graph on λ and γ explicit.

For a given realization of the node locations, decreasing γ increases the number of edges. However, for a given γ, increasing λ will increase the expected number of nodes near the tagged node and can increase the expected number of edges in the network. Increasing λ can also contribute to increased interference because of the increase in the expected number of nodes in the network. Thus the dependence of connectivity on λ is not straightforward.

Since there are an infinite number of nodes in the network, we usually investigate a property related to connectivity. For a given λ and γ, we look for the existence of a *giant component,* a connected component with an infinite number of nodes. The giant component is a subgraph that typically extends to infinity in all directions. When there is a giant component in the graph, we say that the network *percolates.* To understand this terminology, it will help to

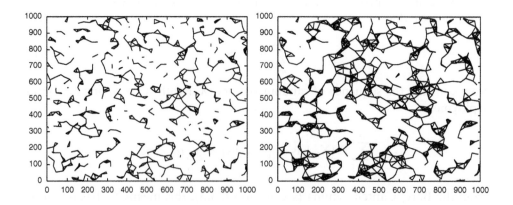

Figure 9.8 For both figures, we used $P = 100{,}000$, $\beta < 1$, and $N_0W = 1$. The node locations were generated using $\lambda = 10^4$. The same node locations are used in the two figures. The path loss function $L(d) = d^{-3}$. For the left figure $\gamma = 0.03$ and for the right figure $\gamma = 0.003$. It is easy to see that there is percolation in the right figure and not in the left.

visualize the links as representing open pipes. Consider a sufficiently large finite set in $\mathbb{R}^2$, say a square. If there is a giant component in the network, then a part of the giant component will be inside this square and there will be pathways connecting the sides of the square. In Figure 9.8, we show two examples, one in which there is percolation and the other in which there is no percolation. Note that the existence of a giant component does not imply that there is only one component.

Let us first examine useful forms of the path loss functions $L(\cdot)$. When $\gamma > 0$, the total interference at a receiver is the sum of an infinite number of terms and can become arbitrarily large. Further, observe that the interference is actually a random variable and depends on the node locations. If the node locations form a Poisson process, all nodes transmit with the same power, and the path loss function depends only on the transmitter-receiver distance, then we can show that $\sum_{k \text{ transmitting}} L(d_{k,j})$ almost surely converges if $L(\cdot)$ satisfies

$$\int_y^\infty L(t)\, dt < \infty,$$

for sufficiently large y, or, more conservatively, if $L(\cdot) \geq 0$ and is absolutely integrable,

$$\int_0^\infty L(t)\, dt < \infty$$

The first condition is satisfied for $L(d) = d^{-\alpha}$, $\alpha > 2$ and in the rest of the section we will use this form for the path loss function. Convergence of $\left(\sum_{k \text{ transmitting}} L(d_{k,j})\right)$ implies that the interference at a node is finite. This, in turn, means that edges can possibly exist in the network.

Consider a tagged node, say Node 0. Let us characterize the degree of Node 0 in the STIRG. Let v_0 be the number of neighbors of Node 0. Consider a neighbor of Node 0 from whom the received power at Node 0 is the least among all the neighbors. Call this neighbor Node 1 and number the other neighbors $2, 3, \ldots, v_0$. Let the other nodes in the network, those that are not neighbors of N_0, be numbered $v_0 + 1, v_0 + 2, \ldots$. By definition

$$P\, L(\|X_1 - X_0\|) \leq P\, L(\|X_k - X_0\|)$$

for $k = 2, \ldots, v_0$. Since Node 1 is a neighbor of Node 0, it satisfies the SINR constraint, and we can derive the following.

$$\frac{P\, L(\|X_1 - X_0\|)}{N_0 W + \gamma \sum_{k=2}^\infty P\, L(\|X_k - X_0\|)} \geq \beta$$

$$P L(\|X_1 - X_0\|) \geq \beta N_0 W + \beta \gamma \sum_{k=2}^{\infty} P L(\|X_k - X_0\|)$$

$$\geq \beta N_0 W + \beta \gamma (v_0 - 1) \, P L(\|X_1 - X_0\|)$$

$$+ \beta \gamma \sum_{k=v_0+1}^{\infty} P L(\|X_k - X_0\|) \geq \beta \gamma (v_0 - 1) P L(\|X_1 - X_0\|)$$

$$v_0 \leq 1 + \frac{1}{\beta \gamma}$$

Thus the number of neighbors for any node is upper bounded when $\gamma > 0$. Further, note that when $\gamma > \frac{1}{\beta}$ then each node has at most one neighbor. We are now ready to analyze percolation behavior as a function of γ. We remark here that in a narrowband system $\gamma = 1$ and $\beta > 1$. Thus the above analysis implies that each node can decode at most one transmission in a slot.

Observe that for $\gamma = 0$, we get the Boolean model of the previous section. For this case, with sufficiently large λ, say $\lambda \geq \lambda^*$, we should get a giant connected component. Since increasing γ can only decrease the number of edges, we can be sure that there is no percolation for $\lambda < \lambda^*$ for any value of γ. For $\lambda > \lambda^*$, we know that if $\gamma > \frac{1}{\beta}$, then each node has only one neighbor and there is no percolation. Therefore, we can argue that for every $\lambda \geq \lambda^*$, there exists $0 < \gamma < \frac{1}{\beta}$ at which the STIRG percolates. This is illustrated in Figure 9.9.

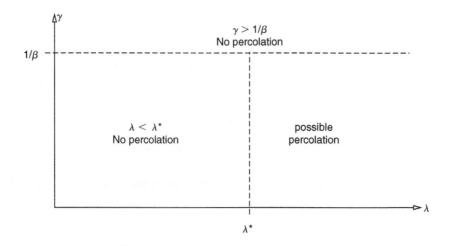

Figure 9.9 The values of γ and λ for which there may be percolation in STIRG.

Let $\gamma^*(\lambda)$ be the *critical* (largest) value of γ that achieves percolation for node density λ,

$$\gamma^*(\lambda) := \sup\{\gamma \geq 0 : G(\lambda, \gamma)\text{percolates }\}$$

In general, it is not easy to obtain the value of $\gamma^*(\lambda)$ and only some structural results, existence results, and bounds can be obtained. We now show that for the power law path loss model, the value of γ at which the network percolates is an increasing function of λ.

Let $\lambda_1 < \lambda_2$ and $a := \sqrt{\lambda_1/\lambda_2} < 1$. Consider a realization of node locations with node density λ_1. Recall that all nodes transmit in every slot. Let $G_1(\lambda_1, \gamma)$ represent the STIRG corresponding to this realization. Let (i, j) be an edge in $G_1(\lambda_1, \gamma)$ between Node i and Node j and let $d_{i,j}$ be the distance between them. The SINR at j is given by

$$\text{SINR}_{i,j} = \frac{Pd_{i,j}^{-\alpha}}{N_0 W + \gamma \sum_{k \neq \{i,j\}} Pd_{k,j}^{-\alpha}}$$

$$= \frac{Pa^{-\alpha}d_{i,j}^{-\alpha}}{a^{-\alpha}N_0 W + \gamma \sum_{k \neq \{i,j\}} Pa^{-\alpha}d_{k,j}^{-\alpha}}$$

$$\leq \frac{P\left(ad_{i,j}\right)^{-\alpha}}{N_0 W + \gamma \sum_{k \neq \{i,j\}} P\left(ad_{k,j}\right)^{-\alpha}}$$

The last inequality is obtained by observing that $\alpha > 1$ and $a < 1$. In this realization of node locations, scale the node locations by a so that nodes located at x are moved to ax. This means that the nodes in a unit square are now contained in a square of side a; the mean density of nodes in the scaled realization is $\frac{\lambda_1}{a^2} = \lambda_2$. Also, the distances in the scaled network are scaled by a; the distance between i and j in the scaled model is $ad_{i,j}$. Let $SINR'_{i,j}$ be the SINR for the transmission from Node i at Node j in the scaled network.

$$\text{SINR}'_{i,j} = \frac{P\left(ad_{i,j}\right)^{-\alpha}}{N_0 W + \gamma \sum_{k \neq j} P\left(ad_{k,j}\right)^{-\alpha}} > \text{SINR}_{i,j} \geq \beta$$

This means that all the links in $G_1(\lambda_1, \gamma)$ are also present in the STIRG when the distances are scaled to achieve a density of $\lambda_2 > \lambda_1$. Thus if there is percolation in the STIRG with density λ_1 there is also a percolation in the network when the density is λ_2 with the same γ. This in turn means that $\gamma^*(\lambda_1) \leq \gamma^*(\lambda_2)$; when the density of nodes is increased, a higher value of γ can achieve percolation.

Discussion

A closed form expression for $\gamma^*(\lambda)$ is not known and we take recourse to estimation via simulation. Figure 9.10 plots $\gamma^*(\lambda)$ as a function of λ as obtained from simulation for $L(d_{i,j}) = d_{i,j}^{-3}$. Observe that $\gamma^*(\lambda)$ seems to saturate with increasing λ. Let us now analyze this behavior and see if we can obtain some insight into the effect of interference on the connectivity.

When λ is increased, it increases the number of nodes in the neighborhood of a receiver. This in turn should increase the interference at the receiver. Thus it seems surprising that with increased λ, the network can tolerate a higher γ; that is, a higher value of interference from other transmissions. One can quickly see that this is an artifact of the path loss function. Notice that $L(d_{i,j}) \to \infty$ as $d_{i,j} \to 0$; the model assumes that there is amplification of the transmitted signal by the channel at nodes close to the transmitter! We need to use a more reasonable path loss function.

The power law path loss function usually is used for the far field, when the receiver is so far from the transmitter that the antennas do not get electrically coupled. In the near field, we should upper bound the path loss. One such function would be to ignore the path loss near the transmitter but use a power law for locations far from the transmitter; for example,

$$L(d_{i,j}) = \min\{1, d_{i,j}^{-\alpha}\}$$

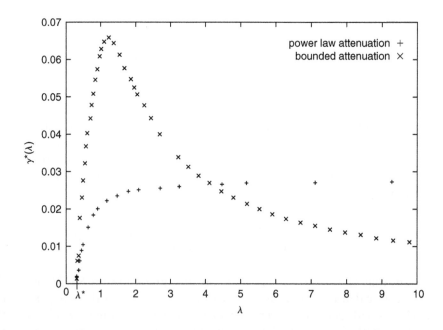

Figure 9.10 Critical γ for percolation as a function of the node density λ. Transmission power $P = 100{,}000$, $\beta = 1$, and the receiver noise $N_0 W = 10{,}000$. The path loss function is $L(d_{i,j}) = d_{i,j}^{-3}$. λ^* is also indicated. Adapted from [30].

It turns out that for $\alpha = 3$, this model does behave as expected—tolerates less interference as λ increases, $\gamma^*(\lambda) \to 0$ as $\lambda \to \infty$. This is shown in Figure 9.10.

9.4 Capacity and Spatial Reuse Models

Recall that the capacity of a wireless network is determined by the spatial reuse that can be achieved and increases with increasing spatial reuse. Thus, capacity analysis essentially determines the spatial reuse possible in the network. Hence, determining transport capacity (in bit-meters per second) boils down to obtaining the sum of the distances traveled by these simultaneous transmissions. The per-node capacity can then be obtained from the source-destination distances. In Chapter 8, we had introduced link activation constraints that determine the spatial reuse in the network. These were obtained from the protocol model, the interference model, and network graph-based constraints. We will first develop some connections between the first two models. Then, for the graph-based constraints, we will analyze the possible spatial reuse in a random geometric graph when r_n is chosen such that the network is connected.

Consider a network in which each node is transmitting with power P. Consider a tagged receiver, say Node 0, and a transmitter that is at a distance r_n from it. Let $d_n = r_n(1 + \Delta)$, $\Delta > 0$, be an exclusion zone around the receiver in which no other transmitter is located. We need to choose d_n to satisfy the SINR requirement at the receiver. Let us calculate the interference at the receiver of Node 0 with a tight packing of the transmitter-receiver pairs around Node 0 as shown in Figure 9.11. The dashed circles have radius $(2k + 1)d_n$ for $k = 0, 1, 2, \ldots$. In the annulus between the k-th and the $(k + 1)$-th dashed circles, there are at most

$$\frac{\pi((2k+3)d_n)^2 - \pi((2k+1)d_n)^2}{\pi d_n^2} = (2k+3)^2 - (2k+1)^2 = 8k+8$$

transmitters. The numerator in the first expression is the area of the annulus and the denominator is the area of an exclusion circle. Each of these transmitters is at a distance of at least $(2k + 1)d_n$ from the receiver. For the power law path loss function with $\alpha > 2$, an upper bound on the total interference power at the receiver is

$$\sum_{k=0}^{\infty} P((2k+1)d_n)^{-\alpha} 8(k+1) = 8Pd_n^{-\alpha} \sum_{k=0}^{\infty} \frac{k+1}{(2k+1)^\alpha}$$

$$= a_0 P d_n^{-\alpha}$$

where $a_0 = 8 \sum_{k=0}^{\infty} \frac{k+1}{(2k+1)^\alpha}$. Since $\alpha > 2$, the series converges and a_0 is a finite constant. Ignoring the receiver noise, the SIR at the receiver at Node 0 is

$$\text{SIR} \geq \frac{Pr_n^{-\alpha}}{N_0 W + a_0 P d_n^{-\alpha}} \approx \frac{1}{a_0}(1 + \Delta)^\alpha \tag{9.17}$$

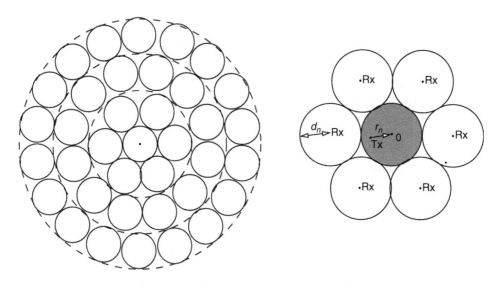

Figure 9.11 Arrangement of tightly packed transmitter-receiver pairs around Node 0 to obtain the interference at the receiver of the transmission from Node 0. The exclusion zone around each receiver is shown using solid lines. The transmitter for each receiver could be anywhere in the corresponding solid circle.

In making the approximation, we have assumed that the receiver noise is significantly smaller than the overestimation of the interference. For any SINR threshold β, we can obtain Δ from (9.17). We have thus obtained the protocol model from the interference calculations with the power law path loss model. However, as we have seen in the previous section, the power law path loss model needs to be used with care. We will discuss this in more detail in the next section.

Let us now consider the graph-based constraints. Consider a realization of an n-node random geometric graph with transmission range r_n. Consider a tagged node, say Node 0. Let us investigate the degree of this node, i.e., the number of neighbors of Node 0. Let $\mathcal{D}_0(n)$ be the degree of Node 0 when there are n nodes in the network. Ignoring edge effects, we see that this is the same as the number of nodes that are in a circle of radius r_n around Node 0;

$$\Pr(\mathcal{D}_0(n) = k) = \binom{n-1}{k} \left(\pi r_n^2\right)^k \left(1 - \pi r_n^2\right)^{n-1-k}$$

for $0 \leq k \leq n - 1$. Let δ_n be the minimum degree and D_n the maximum degree in $G_n(r_n)$. From Section 9.2.2 we know that if $r_n^2 = \frac{5K \log n}{n}$ with $K > 1$, then $\delta_n \geq 1$ with high probability. To obtain D_n we use the Chernoff bound as follows. First, let us evaluate $\mathsf{E}\left(e^{\theta \mathcal{D}_0(n)}\right)$.

$$\mathsf{E}\left(e^{\theta \mathcal{D}_0(n)}\right) = \left(1 + (e^\theta - 1)\pi r_n^2\right)^{n-1}$$

$$\leq e^{(n-1)\ (e^\theta - 1)\pi r_n^2}$$

$$= e^{(n-1)\ (e^\theta - 1)\pi (5K\frac{\log n}{n})}$$

$$= n^{\frac{n-1}{n}\ (e^\theta - 1)5\pi K}$$

The inequality on the second line is true because $1 + x \leq e^x$. For $\theta \geq 0$ and for any b, the Chernoff bound on the degree of Node 0 is

$$\Pr\left(\mathcal{D}_0(n) > b\right) \leq \frac{\mathsf{E}\left(e^{\theta \mathcal{D}_0(n)}\right)}{e^{\theta b}}$$

$$\leq \frac{n^{\frac{n-1}{n}\ (e^\theta - 1)5\pi K}}{e^{\theta b}}$$

Choosing $\theta = 1$ and $b = 28K \log n$, we get, for large n,

$$\Pr\left(\mathcal{D}_0(n) > 28K \log n\right) \leq \frac{n^{\frac{n-1}{n}\ (e-1)5\pi K}}{e^{28K \log n}}$$

$$= \frac{n^{\frac{n-1}{n}\ (e-1)5\pi K}}{n^{28K}}$$

$$= n^{K\left(\left(\frac{n-1}{n}\ (e-1)5\pi\right) - 28\right)}$$

$$\leq n^{-K}$$

The last inequality is true because for all $n \geq 1$, $\left(\left(\frac{n-1}{n}(e-1)5\pi\right) - 28\right) \leq -1$.

Our interest is in obtaining the maximum degree, D_n, in the n-node network. Applying the union bound, and noting that $K > 2$

$$\Pr\left(D_n > 28K \log n\right) \leq n^{-(K-1)}$$

$$\sum_{n=1}^{\infty} \Pr\left(D_n > 28K \log n\right) \leq \sum_{n=1}^{\infty} n^{-(K-1)} < \infty$$

The last inequality means that from the Borel-Cantelli lemma (like in Section 9.2.2),

$$\Pr\left(D_n > 28K \log n \ i.o.\right) = 0$$

This implies that asymptotically almost surely, the maximum degree D_n is less than $28K \log n$.

Recall that we had three graph theoretic constraints—primary conflict, receiver conflict, and transmitter-receiver conflict. As in Chapter 8, consider the primary conflict constraint. To determine the spatial reuse, we will color the edges with the minimum number of colors $1, 2, \ldots$ such that adjacent edges do not have the same color. If $\chi(G_n(r_n))$ colors are used then the number of simultaneous transmitters is at least $\frac{n}{\chi(G_n(r_n))}$. Recall from (8.14)

$$D_n \le \chi(G_n(r_n)) \le \frac{3D_n}{2}$$

Thus,

$$\chi(G_n(r_n)) \le 42K \log n$$

asymptotically almost surely. This means that there is a transmission on every edge at least once in $(42K \log n)$ slots.

In the next two sections we will analyze the transport capacity of wireless networks. We begin by analyzing the transport capacity of arbitrary networks for which no assumptions are made on the distribution of the nodes. To obtain the maximum possible capacity, we are allowed to choose the best position of the nodes that maximizes the capacity. Since there is an equivalence between the protocol and the STIRG with $\gamma = 0$, we will consider only the protocol model. We then provide a constructive lower bound on the capacity of randomly deployed networks. Once again we will primarily use the protocol model in this analysis.

9.5 Transport Capacity of Arbitrary Networks

In this section we will not consider random placement. Rather, we will assume that the node placements can be chosen arbitrarily. We will obtain upper and lower bounds on the achievable capacity.

Let n nodes be deployed in the unit square $[0, 1]^2$. Let (i, j) and (k, l) be two active transmitter-receiver pairs. Recall that the protocol model places the following constraints on the relative locations of these nodes.

$$\|X_k - X_j\| \ge (1 + \Delta)\|X_i - X_j\| \qquad \|X_i - X_l\| \ge (1 + \Delta)\|X_k - X_l\| \qquad (9.18)$$

Using the triangle inequality first, we can derive the following relation between the locations of the node pairs (i, j) and (k, l).

$$\|X_j - X_l\| + \|X_l - X_k\| \ge \|X_j - X_k\|$$

$$\|X_j - X_l\| \ge \|X_j - X_k\| - \|X_l - X_k\|$$

$$\ge (1 + \Delta)\|X_i - X_j\| - \|X_l - X_k\| \qquad (9.19)$$

The last relation is obtained by applying (9.18) to the second relation. Similarly,

$$\|X_l - X_j\| + \|X_j - X_i\| \geq \|X_l - X_i\|$$

$$\|X_l - X_j\| \geq \|X_l - X_i\| - \|X_j - X_i\|$$

$$\geq (1 + \Delta)\|X_k - X_l\| - \|X_j - X_i\| \tag{9.20}$$

Adding the inequalities in (9.19) and (9.20), we get

$$\|X_l - X_j\| \geq \frac{\Delta}{2}\left(\|X_k - X_l\| + \|X_j - X_i\|\right) \tag{9.21}$$

Note that j and l are the receivers and we can interpret this inequality to say that for (i, j) and (k, l) to be simultaneously active, a disk of radius $\frac{\Delta}{2}\left(\|X_i - X_j\|\right)$ centered at X_j and a disk of radius $\frac{\Delta}{2}\left(\|X_k - X_l\|\right)$ centered at X_l should not overlap. These disks can be called *exclusion disks*. This is illustrated in Figure 9.12. Of course, (9.21) is just a necessary condition on the relative locations and not sufficient to allow two node pairs to transmit.

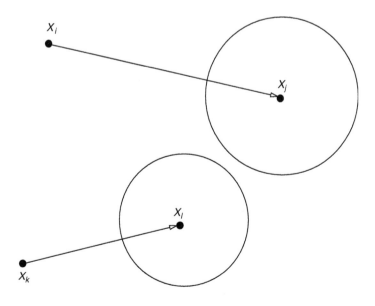

Figure 9.12 Illustrating the exclusion disk. (i,j) and (k,l) **are transmitter-receiver pairs that are active in the same slot. The circles centered at** X_j **and** X_l **have radii** $\frac{\Delta}{2}d_{i,j}$ **and** $\frac{\Delta}{2}d_{k,l}$, **respectively.**

Exercise 9.4
Show that (9.21) is not sufficient for two node pairs to transmit simultaneously in a slot by providing a counterexample.

To obtain the transport capacity, consider the set of all transmitter-receiver pairs (i, j) and designate this set by $\mathcal{S}$. Let $d_{i,j} := \|X_i - X_j\|$ be the Euclidean distance between the transmitter at X_i and the receiver at X_j. As usual, assume that the network operates in a time slotted manner and that the nodes are synchronized. Without loss of generality, we will also assume that in each slot one bit is transmitted. Then the transport capacity of the network is $\mathcal{C}_T := \sum_{(i,j)\in\mathcal{S}} d_{i,j}$ bit-meters in this slot. We will now obtain an upper bound on $\mathcal{C}_T$.

Since the exclusion disks are to be nonoverlapping, the sum of the areas of the exclusion disks corresponding to elements of $\mathcal{S}$ should be less than 1, the area of the unit square. For a given $d_{i,j}$, the minimum area of the exclusion disk inside the square will be when the receiver is at one of the corners of the square, in which case a quarter of the exclusion disk is inside the unit square. See Figure 9.13 for an illustration of this minimum case. The radius of the exclusion disk for (i, j)

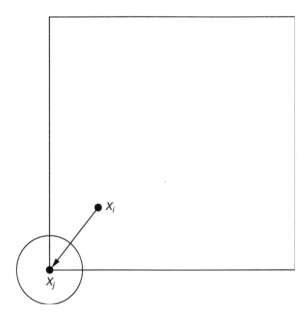

Figure 9.13 **The minimum area of the exclusion disk that is inside the operational area defined by the unit square.**

is $\frac{\Delta}{2}d_{i,j}$. The total area of the exclusion disks must satisfy

$$\sum_{(i,j)\in S} \frac{1}{4}\frac{\pi\Delta^2 d_{i,j}^2}{4} = \sum_{(i,j)\in S}\frac{\pi\Delta^2 d_{i,j}^2}{16} \leq 1$$

$$\sum_{(i,j)\in S} d_{i,j}^2 \leq \frac{16}{\pi\Delta^2}$$

Since there are n nodes in the network, $|S| \leq \frac{n}{2}$. By the Cauchy-Schwarz inequality, we can write

$$\sum_{(i,j)\in S} d_{i,j} \leq \sqrt{\sum_{(i,j)\in S}d_{i,j}^2 \sum_{(i,j)\in S}1}$$

$$\leq \sqrt{\frac{n}{2}\sum_{(i,j)\in S}d_{i,j}^2}$$

$$\leq \sqrt{\frac{n}{2}\frac{16}{\pi\Delta^2}}$$

$$= \sqrt{\frac{8}{\pi}}\frac{\sqrt{n}}{\Delta} \qquad (9.22)$$

Thus we see that under the protocol model, the transport capacity is upper-bounded by $c\frac{\sqrt{n}}{\Delta}$, where c is a constant. Using the order notation we say that the transport capacity is $O(\sqrt{n})$.

We now want to know if this capacity is realizable—can we arrange the nodes in such a manner that this transport capacity is indeed achieved? If we can find such an arrangement, then we can say that the transport capacity is also $\Omega(\sqrt{n})$. Since we have already showed that the transport capacity is $O(\sqrt{n})$, this implies that the transport capacity is $\Theta(\sqrt{n})$.

Consider the following construction. Let $k = \frac{\sqrt{n+1}-1}{2}$. Tessellate the unit square into cells of side $\frac{1}{k}$. Let

$$r = \frac{1}{k(1+2\Delta)} = \frac{2}{(\sqrt{n+1}-1)(1+2\Delta)}$$

Place the $\frac{n}{2}$ transmitter-receiver pairs as shown in Figure 9.14. For this construction, the following are easy to verify.

- The exclusion disks of radius $\Delta r/2$ centered at the receivers do not overlap.

- The two closest interfering transmitters are at distances $(r + 2\Delta r)$ and $\sqrt{(r + \Delta r)^2 + (\Delta r)^2}$. Both of these distances satisfy the interference constraint of the protocol model.

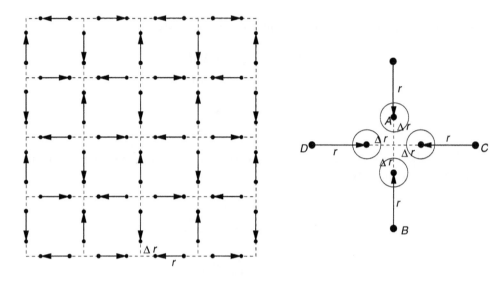

Figure 9.14   Figure on the left shows the arrangement of the transmitter-receiver pairs. Distance between every transmitter-receiver pair is _r_ and each pair is placed on the grid lines at an offset of Δ_r_ from the grid point. Figure on the right shows the exclusion disks for the four receivers around a grid point. The closest interfering transmitters to receiver _A_ are _B_ and _C_ and _D._ _B_ is at a distance $(r + 2\Delta r)$ and _C_ and _D_ are at $\sqrt{(r + \Delta r)^2 + (\Delta r)^2}$.

In this arrangement all the $\frac{n}{2}$ transmitter-receiver pairs can communicate simultaneously. Thus the transport capacity achieved in this arrangement is $\frac{n}{2}r$ bit-meters per slot or

$$\frac{n}{(\sqrt{n+1}-1)(1+2\Delta)} = O\left(\sqrt{n}\right)$$

Thus the transport capacity of an arbitrary network in which the spatial reuse can be described by the protocol model is $\Theta(\sqrt{n})$ bit-meters per second.

9.6   Transport Capacity of Randomly Deployed Networks

9.6.1   Protocol Model

We have argued before that having a low transmission range (above the threshold at which the network is connected) increases spatial reuse and can increase capacity when the nodes are uniformly distributed in the operational area and when the packets choose their destinations randomly. However, reducing the transmission radius increases the number of hops for each flow, and the number of times a packet has to be transmitted increases in inverse proportion to the transmission range. This offsets some of the gains made by spatial reuse. Also, the transmission range for connectivity is $\sqrt{\log n}$ times that for the arbitrary network. This reduces the per-node throughput compared to the arbitrary network derived in the previous section. We will investigate these issues and describe a scaling law for the per-node capacity of the network.

Informal Arguments: A Bound from Spatial Reuse

Let r_n be the transmission range of all the nodes in an n-node network, h_n the average number of hops that a packet has to make before reaching its destination, and λ_n the arrival rate of packets per node. Hence, the total required rate of transmissions in the network will be $n\lambda_n h_n$ packets per slot. We observe that decreasing r_n increases spatial reuse by a factor proportional to the area covered by a transmission and hence the spatial reuse is $O(1/r_n^2)$. It follows that we must have

$$n\lambda_n h_n = O\left(\frac{1}{r_n^2}\right)$$

Hence

$$n\lambda_n = \frac{1}{h_n}\,O\left(\frac{1}{r_n^2}\right)$$

Now the number of hops is directly proportional to the distance between the source and destination nodes, and a packet can cover a distance proportional to r_n in one hop. Therefore, h_n scales inversely with r_n. It then follows that

$$n\lambda_n = O\left(\frac{1}{r_n}\right)$$

However, we have seen that for the network to remain connected, r_n should decrease slower than $\sqrt{\frac{\log n}{n}}$. Thus, we obtain

$$\lambda_n = O\left(\frac{1}{\sqrt{n\log n}}\right)$$

This means that as the number of nodes increases, the per-node capacity—the maximum rate at which a node can generate and transmit data—decreases at least as fast as $\frac{1}{\sqrt{n\log n}}$. Intuitively, this is because as the number of nodes increases, each node spends more and more time relaying packets from other nodes and the benefits of reduced transmission range and the consequent increased spatial reuse are not quite realized.

It should be clear that the "derivation" that we have just provided is informal. Since we have randomly distributed nodes, and packets choose random destinations, a precise statement of the result should be a probabilistic statement. Further, in deriving the result we should carefully model the conditions under which packet transmissions are successful. We will now do just that.

Analysis

Before we analyze the capacity of a randomly deployed network, we will discuss a problem from distributed computation whose solution will be used in the analysis.

Consider n^2 processors connected in an $n \times n$ wired mesh network as shown in Figure 9.15. Let each processor have k packets that it needs to send to another arbitrary processor in the mesh. Also, each processor is a receiver of exactly k packets from another arbitrary processor in the network. The packets from the source to the destination are transported over a multihop path. For example, a packet from $(3,2)$ to $(1,4)$ can take the path $(3,2) \rightarrow (2,2) \rightarrow (2,3) \rightarrow (2,4) \rightarrow (1,4)$. Many other paths are also possible. We assume that the network of processors has a time-slotted operation. In each slot, one packet can be transferred in each direction on each of the links. This means that each processor can receive up to four packets in a slot and also transmit up to four packets in a slot. The packets reach the destination in a store-and-forward manner; each packet is completely stored at a hop before being forwarded on the next hop toward the destination. This means that a packet that is being received cannot be forwarded toward the destination in the same slot. For the preceding example, the packet from Node $(3,2)$ would require four slots to reach Node $(1,4)$. Our interest is

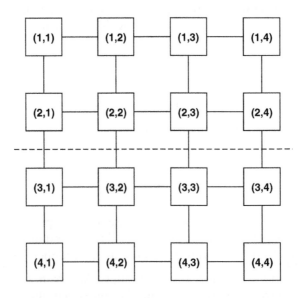

Figure 9.15   16 processors are connected in a 4 x 4 wired mesh network. The links are bidirectional, or full duplex, and are activated in a time slotted manner. In each slot, each processor can receive and transmit one unit of data on each of the links that is connected to it. The nodes above and below the dashed line form a bisection of the graph.

in designing an efficient routing algorithm that will route these kn^2 packets in a minimum number of slots.

The *bisection bound* is a lower bound on the minimum number of slots that is needed to complete the transport of the packets as before. Consider the network graph shown in Figure 9.15. A *bisection* of this graph is a partition of the nodes into disjoint sets with the same number of nodes in each partition. This is illustrated in Figure 9.15. When the transmitters choose their destinations arbitrarily, the worst-case choice is one in which every node of one set of a bisection chooses a destination in the other set of the bisection. Thus in the worst case, the maximum number of packets will flow across the edges connecting the two sets of nodes in the bisection. This means that half of kn^2 packets have to cross the n edges connecting the two partitions in one direction and the other half in the other direction. Since there are n edges connecting the two sets of the partition, at least $\frac{kn^2/2}{n} = \frac{kn}{2}$ slots would be needed to complete the transport for an arbitrary choice of destinations by the nodes.

There are many algorithms for routing of the packets that very nearly match the bisection bound. One such randomized algorithm is described in Algorithm 9.1.

---

assign color brown or black with equal probability to each of the kn^2 packets
for black-colored packets from (i, j) to (s, t) **do**
    phase 1: choose $i' \in [1, \ldots, n]$ randomly; transport packet to (i', j);
    phase 2: transport packet from (i', j) to (i', t);
    phase 3: transport packet from (i', t) to (s, t)
end for
for brown-colored packets from (i, j) to (s, t) **do**
    phase 1: choose $j' \in [1, \ldots, n]$ randomly; transport packet to (i, j');
    phase 2: transport packet from (i, j') to (s, j');
    phase 3: transport packet from (s, j') to (s, t)
end for

Algorithm 9.1 $k \times k$ routing on an $n \times n$ mesh.

The black-colored packets first move to a random node in the same column, and then reach the destination by first reaching the destination column. The brown-colored packets are in phase-quadrature. They first move to a random node in the same row and then reach the destination by first reaching the destination row. The randomization in Phase 1 allows the redistribution of the packets to avoid bad patterns. The random splitting of packets at the beginning of the algorithm allows the vertical and horizontal movements to be executed simultaneously. This is illustrated in Figure 9.16.

It can be shown that, when $k \geq 8$, Algorithm 9.1 can accomplish the routing in $\left(\frac{3}{4}kn + \tilde{o}(kn)\right)$ slots. This means that the routing can be accomplished in $\frac{3}{4}kn + o(kn)$ slots with high probability.

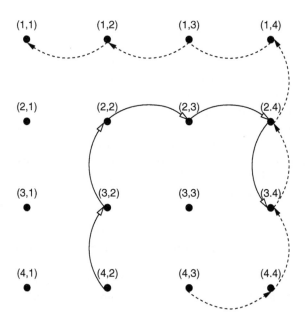

Figure 9.16 The solid arrows show the path of the black packets and the dashed arrows that of the brown packets. The packet from (4, 2) randomly chooses (2, 2) for its first phase. It then goes to (2, 4) and then to its destination (3, 4) in a total of five slots. The packet from (4, 3) randomly chooses (4, 4) for its first phase and then reaches its destination (1, 1) via (1, 4) in seven slots.

Let us now see how we can use this $k \times k$ routing algorithm to bound the capacity of the random wireless network under the protocol model. To do that let us revisit the tessellation of the unit square into square cells of side s_n in Section 9.2.2. Before proceeding with the analysis of the randomly deployed network, let us consider a deterministic deployment of n nodes in the unit square. Let the number of nodes in a cell be uniformly bounded by c_n. Assume that Node i, $i = 1, \ldots, n$, has m packets that it needs to transport to a destination at distance d_i meters from it. If all the nodes can complete the exchange of their packets in T slots, then the transport capacity is $\frac{1}{T} \sum_{i=1}^{n} m d_i$ packet-meters per slot.

We will now construct the routing and scheduling scheme to transport these kn packets. Each cell is analogous to the processors in the mesh network of processors that we discussed earlier. The routing is performed as in Algorithm 9.1 with the nodes routing their packets via an arbitrary node in a neighboring cell. Now consider a slot in which Node i in Cell 0 is transmitting to a node in a neighboring cell as illustrated in Figure 9.17. This receiver could be at a distance of up to $\sqrt{5}s_n$ meters from the transmitter. For this transmission to be successfully

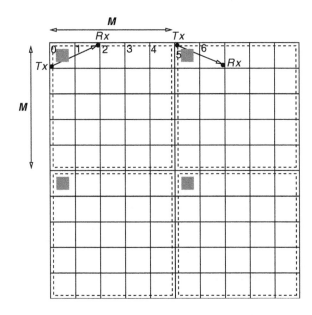

Figure 9.17 Each cell has side s_n. Let $\Delta = 0.34$. Node i in Cell 0 is transmitting to node j. The closest transmitter, other than to Node j, other than i, is in cell 5 at a distance of $3s_n$. In any slot, at most one node in each of the dashed squares may be transmitted. Each dashed square is of size $M \times M$ cells.

decoded, no other node in cells in an $M \times M$ square located as shown in Figure 9.17 can transmit in the same slot. M is obtained from the protocol model as follows. The closest transmitter will be at least $(M-2)s_n$ meters from the receiver. We can thus obtain M as follows.

$$(M-2)s_n \geq (1+\Delta)\sqrt{5}s_n$$

$$M = \lceil (1+\Delta)\sqrt{5} + 2 \rceil$$

M is independent of n and is a constant. This is illustrated in Figure 9.17, where we have used $\Delta = 0.34$, giving us $M = 5$.

Let us now calculate the time that it would take to implement Algorithm 9.1 in a wireless network. There are up to M^2 cells in each square and a node in only one of these may transmit in a slot. However, in the mesh network using Algorithm 9.1, each node could receive up to four packets and simultaneously transmit up to four packets in a slot. Thus the exchanges in one slot of Algorithm 9.1 can be accomplished in at most $8M^2$ slots in the wireless network. We have a grid of $\left(\frac{1}{s_n}\right) \times \left(\frac{1}{s_n}\right)$ cells with at most mc_n packets in each cell. Applying the analysis of Algorithm 9.1, we see that the transport of all the packets to their respective destinations can be accomplished in at most

$$\left(8M^2\right)\left(\frac{3}{4}mc_n\frac{1}{s_n}\right) + \tilde{o}\left(\frac{mc_n}{s_n}\right) = a\left(\frac{mc_n}{s_n}\right) + \tilde{o}\left(\frac{mc_n}{s_n}\right)$$

slots. Here $a = 6\left(\lceil(1+\Delta)\sqrt{5}+2\rceil\right)^2$ is a constant independent of n.

Now we will move to the capacity analysis of the randomly deployed network. Recall that for $s_n = \sqrt{\frac{K\log n}{n}}$, $K > 2$, every cell almost surely contains one or more nodes. Let us now obtain an upper bound on the maximum number of nodes in a cell. To do that, consider a tagged cell, say Cell i. Following the arguments in the derivation of the maximum degree of a node in Section 9.4, we can obtain an upper bound on the number of nodes in Cell i as follows. We will assume $s_n^2 = \frac{K\log n}{n}$.

Let $\mathcal{N}_i$ be the number of nodes in Cell i. From the Chernoff bound,

$$\Pr(\mathcal{N}_i > b) \leq \frac{\mathsf{E}\left(e^{\mathcal{N}_i\theta}\right)}{e^{\theta b}}$$

$$= \frac{\left(1 + (e^\theta - 1)s_n^2\right)^n}{e^{\theta b}}$$

$$= \frac{\left(1 + (e^\theta - 1)\frac{K\log n}{n}\right)^n}{e^{\theta b}}$$

$$\leq \frac{e^{n\left((e^\theta-1)\frac{K\log n}{n}\right)}}{e^{\theta b}}$$

$$= \frac{\left(e^{\log n}\right)^{K(e^\theta-1)}}{e^{\theta b}}$$

$$= \frac{n^{K(e^\theta-1)}}{e^{\theta b}}$$

The second inequality in the preceding follows because $(1 + x) \leq e^x$. Choosing $\theta = 1$ and $b = Ke\log n$, we have

$$\Pr(\mathcal{N}_i > Ke\log n) \leq \frac{n^{K(e-1)}}{e^{Ke\log n}}$$

$$= n^{K(e-1)-Ke} = n^{-K}$$

Now let us obtain the maximum number of nodes in any cell in the network. We derive this by examining the probability that there exists a cell with more than $(Ke \log n)$ nodes.

$$\mathsf{Pr}\big(\mathcal{N}_i > (Ke \log n) \text{ for some } i\big) = \mathsf{Pr}\bigg(\cup_{i=1}^{1/s_n^2} \big(\mathcal{N}_i > (Ke \log n)\big)\bigg)$$

$$\leq \sum_{i=1}^{1/s_n^2} \mathsf{Pr}\big(\mathcal{N}_i > (Ke \log n)\big)$$

$$\leq \sum_{i=1}^{1/s_n^2} n^{-K} = \frac{n^{-K}}{s_n^2}$$

$$= \frac{n^{-K}}{(K \log n)/n} = \frac{n^{-(K-1)}}{K \log n}$$

The second line is from the union bound. For $K > 2$,

$$\sum_{n=1}^{\infty} \frac{n^{-(K-1)}}{K \log n} < \infty$$

and from the Borel-Cantelli lemma, this implies that the probability that there exists a cell with more than $(Ke \log n)$ nodes infinitely often is zero. Thus we can say that asymptotically almost surely none of the cells has more than $(Ke \log n)$ nodes. This gives us $c_n = Ke \log n$ when $s_n^2 = \frac{K \log n}{n}$.

There is just one thing left. Since our interest is in the transport capacity, we need to now analyze $\sum_{i=1}^{n} d_i$ because that determines the transport capacity. Let us evaluate an upper bound on this quantity. Assume that all the source destination locations are identically and independently chosen. Consider a tagged flow from a source at $X_j = (X_{j,1}, X_{j,2})$ to a destination at $X_k = (X_{k,1}, X_{k,2})$. If the source and destination are randomly chosen, $X_{j,1}, X_{j,2}, X_{k,1}, X_{k,2}$ are all uniformly distributed in $(0, 1)$.

Since $d_i \leq \sqrt{2}$, we have $\sum_{i=1}^{n} d_i^2 \leq \sqrt{2} \sum_{i=1}^{n} d_i$. Further, $\sum_{i=1}^{n} d_i^2$ has the same distribution as the sum of n i.i.d. random variables of the form $((X_{j,1} - X_{k,1})^2 + (X_{j,2} - X_{k,2})^2)$ where $X_{j,1}, X_{k,1}, X_{j,2}$ and $X_{k,2}$ are i.i.d. uniform in $(0, 1)$. Thus

$$\mathsf{E}\big(e^{-\theta d_i^2}\big) = \Big(\mathsf{E}\big(e^{-\theta(V-W)^2}\big)\Big)^2$$

where V and W are i.i.d. uniform in $(0, 1)$. We therefore have

$$\Pr\left(\sum_{i=1}^{n} d_i < \frac{n}{a}\right) \leq \Pr\left(\sum_{i=1}^{n} d_i^2 < \frac{\sqrt{2}n}{a}\right)$$

$$= \Pr\left(\frac{\sqrt{2}n}{a} - \sum_{i=1}^{n} d_i^2 > 0\right)$$

$$\leq E\left(e^{\theta\left(\frac{\sqrt{2}n}{a} - \sum_{i=1}^{n} d_i^2\right)}\right)$$

$$= \frac{\left(E\left(e^{-\theta(V-W)^2}\right)\right)^{2n}}{e^{-\frac{\theta\sqrt{2}n}{a}}}$$

The first inequality follows from the preceding discussion because $\sum_{i=1}^{n} d_i \leq \alpha$ implies $\sum_{i=1}^{n} d_i^2 \leq \sqrt{2}\alpha$. The second inequality follows from the Chernoff bound. $E\left(e^{-\theta(V-W)^2}\right)$ is obtained as follows.

$$E\left(e^{-\theta(V-W)^2}\right) = \int_0^1 \int_0^1 e^{-\theta(v-w)^2} \, dw \, dv$$

$$= 2 \int_0^1 \int_0^v e^{-\theta(v-w)^2} \, dw \, dv$$

$$= 2 \int_0^1 \int_0^v e^{-\theta u^2} \, du \, dv$$

$$= 2 \int_0^1 \int_u^1 e^{-\theta u^2} \, dv \, du$$

$$= 2 \int_0^1 (1 - u)e^{-\theta u^2} \, du$$

$$\leq 2 \int_0^\infty e^{-\theta u^2} \, du = \sqrt{\frac{\pi}{\theta}}$$

Choosing $a = \sqrt{2}\theta$ and $\theta = 2\pi e$, we have

$$\Pr\left(\sum_{i=1}^{n} d_i < \frac{n}{2\sqrt{2\pi e}}\right) < \frac{1}{2^n}$$

Since $\sum_{n=1}^{\infty} \frac{1}{2^n} < \infty$, from the Borel-Cantelli lemma, asymptotically almost surely,

$$\sum_{i=1}^{n} d_i \geq \frac{n}{2\sqrt{2\pi e}}$$

This implies that asymptotically almost surely, $\sum_{i=1}^{n} d_i = \Omega(n)$. To summarize the preceding discussion, asymptotically almost surely,

- Each cell contains at most $Ke \log n$ nodes. This means that the routing of the mn packets can be accomplished in

$$a\left(\frac{mc_n}{s_n} + \tilde{o}\left(\frac{mc_n}{s_n}\right)\right) = a\left(\frac{mKe\log n}{\sqrt{\frac{K\log n}{n}}} + \tilde{o}\left(\frac{mKe\log n}{\sqrt{\frac{K\log n}{n}}}\right)\right)$$

$$= a\left(me\sqrt{nK\log n} + \tilde{o}\left(me\sqrt{nK\log n}\right)\right)$$

$$= O\left(m\sqrt{n\log n}\right)$$

  slots.

- The total distance traveled by the mn packets is $m\sum_{i=1}^{n} d_i = \Omega(mn)$ meters.

- Thus the transport capacity is $\Omega\left(\frac{mn}{m\sqrt{n\log n}}\right) = \Omega\left(\sqrt{\frac{n}{\log n}}\right)$ packet-meters per slot.

It can also be shown that the transport capacity is indeed $\Theta\left(\sqrt{\frac{n}{\log n}}\right)$ packet-meters per slot. We omit that proof here.

9.6.2 Discussion
Effect of Path Loss Model

Recall that for the power law path loss model, we can choose $\Delta = \frac{r_n}{d_n}$ such that the SINR at a receiver is greater than a threshold β. This allows us to specify an exclusion ratio of radius d_n around the receiver. This in turn means that spatial reuse can be $O\left(\frac{1}{r_n^2}\right) = O\left(\frac{n}{\log n}\right)$. Like in the STIRGs (see Section 9.3), when n is increased we expect the interference to increase and hence possibly reduce the spatial reuse. That spatial reuse continues to increase with n is an artifact of the path loss model—it permits power amplification in the near field. However, r_n should be very small, or equivalently n should be very large before the amplification effect of the path loss model begins. Thus there is a limit to the achievable spatial reuse imposed by the transmission range requirement to keep the network connected. This suggests that very dense wireless networks should not be used for communication between arbitrary pairs of nodes (as is the case in a general internet); instead such networks may be more useful for nearest neighbor communication as would be required for distributed instrumentation applications over ad hoc wireless sensor networks. This is the subject of the next chapter.

The previous discussion, and also our discussion on STIRGs under different path loss models, suggests that the network behavior is very sensitive to the path loss model. It can shown that with a slight change in the path loss model, the spatial reuse can become O(1).

It must also be noted that in much of the analysis under the protocol model, the interference is ignored. For example, in Figure 9.14, we have ignored the fact that the interference at A when B, C, and D transmit simultaneously can be significant.

Penalty of Randomness

It is interesting to compare the effect of randomness on the transport capacity. With arbitrary placement of the nodes, we get a transport capacity of $\Theta\left(\sqrt{n}\right)$ and for random deployment it is $\Theta\left(\sqrt{\frac{n}{\log n}}\right)$. Thus there is the $O\left(\frac{1}{\sqrt{\log n}}\right)$ penalty for random deployment. This comes about because of the transmission range required to make the network connected. Whereas n nodes can be placed in a $\sqrt{n} \times \sqrt{n}$ grid in the unit square and achieve connectivity with a range of $r_n = \frac{1}{\sqrt{n}-1}$, for random deployment, we need $r_n^2 = \frac{K \log n}{n}$, $K > 1$ to ensure connectivity with high probability.

Exploiting Mobility

The transport capacity scales sublinearly with n. This means that the per-node throughput decreases as n increases. If the source and the destination of the flows are randomly selected, we have seen that the sum of the source destination distances is $\Omega(n)$. Hence, for an arbitrary network, the throughput per node is $O\left(\frac{1}{\sqrt{n}}\right)$ and for the random network it is $O\left(\frac{1}{\sqrt{n \log n}}\right)$. The decrease with increasing n is due to the increasing number of hops, which in turn means that most nodes essentially spend most of the time relaying traffic from other nodes. We can exploit mobility of the nodes to increase the per node capacity. An informal introduction is as follows. In a slot $O\left(\frac{1}{r_n^2}\right)$ nodes can transmit simultaneously. We would like to obtain a throughput of $O\left(\frac{1}{r_n^2}\right)$ packets per slot while reducing the average hop length. Each source transmits its packet to a randomly selected one-hop neighbor that will be the relay node for the packet. Each node maintains a separate packet queue for each of the destinations. Since the nodes are mobile, either the relay nodes or the source will eventually be near the destination. The packet then makes one more hop to the destination. Thus each packet makes exactly two hops to the destination, $h_n = O(1)$. Following the informal arguments on the transport capacity, we have for this model

$$n\lambda_n \leq \frac{1}{h_n} O\left(\frac{1}{r_n^2}\right)$$

$$n\lambda_n \leq O\left(\frac{1}{r_n^2}\right) = O\left(\frac{n}{\log n}\right)$$

$$\lambda_n \leq \frac{1}{n} O\left(\frac{1}{r_n^2}\right) = O\left(\frac{1}{\log n}\right),$$

the per-node throughput is $O\left(\frac{1}{\log n}\right)$. This increased throughput comes at the cost of delay. The delay will depend on the time that it takes for either the relay node or the source node to be near the destination and also get the opportunity to transmit. Thus the delay will depend on the mobility characteristics of the nodes.

9.7 Notes on the Literature

The classic reference for Erdös-Renyi random graphs is the book by Bollobas [14] and for random geometric graphs it is the book by Penrose [109].

Connectivity analysis of the one-dimensional network is adapted from the work of Desai and Manjunath [26]. Godehardt and Jaworski [42] obtain more results on networks in one dimension; for example, the probability of there being k components in the network is shown to be

$$\sum_{k=m-1}^{n-1} \binom{n-1}{k} \binom{k}{m-1} (-1)^{k+m-1} (1 - kr)^n$$

One-dimensional networks in which node locations are i.i.d. exponential random variables have been analyzed by Gupta, Iyer, and Manjunath [48] and by Karamchandani, Manjunath, and Iyer [67]. A recent survey of Iyer and Manjunath [63] has some of the newer results on random geometric graphs and also some new proofs.

The necessary condition for connectivity is adapted from the work of Gupta and Kumar [49] and the sufficient condition is based on that of Xue and Kumar [137] and Kulkarni and Viswanath [84]. Onkar Dabeer gave the simpler proof of Lemma 9.1. The early days of packet radio networks also had inspired many connectivity results, notably the works of Cheng and Robertazzi [22]; Philips, Panwar, and Tantawi [110]; and Piret [111]. The discussion on STIRGs is based on the work of Dousse, Baccelli, and Thiran [29] and Dousse and Thiran [30]. The plot from Figure 9.10 is reconstructed from the plots in [30]. The capacity of arbitrary networks is based on [137] which in turn is based on the classic paper by

Gupta and Kumar [50]. This paper has led to a veritable explosion on the study of capacity under different models or derive the same results again, at least in the order sense, using different approaches. The randomized algorithm for $k \times k$ routing was based on the work of Kaufmann and Rajasekaran and Sibeyn [71]. Most recently, Franceschetti et al. [36] have showed that the capacity in STIRG can be $O(\sqrt{n})$ bit-meters per second. Exploiting mobility to increase the capacity to $O(1)$ is formally derived by Grossglauser and Tse [47]. Recently, Bansal and Liu [5], El Gamal et al. [37], Sharma and Mazumdar [122], and others have studied delay capacity tradeoffs for different mobility models. This being an active area of research, we are able to provide only a sample of the literature.

Problems

9.1 Consider a one-dimensional network in $[0, 1]$ with cutoff r. Let $X_1 = 0$ and X_2 and X_3 be randomly located nodes in $[0, 1]$. Conditioned on $(1, 2)$ and $(2, 3)$ being edges, find the probability that $(1, 3)$ is an edge. Repeat the problem for the case when X_1 is arbitrarily located.

9.2 Let X_1, X_2 and X_3 be three randomly located nodes in $[0, 1]^2$. Conditioned on $(1, 2)$ and $(2, 3)$ being edges, find the probability that $(1, 3)$ is an edge. Ignore edge effects. Let r be the cutoff.

9.3 Consider an n-node network on the positive x-axis with Node i located at X_i. X_i are i.i.d. exponential with mean $1/\lambda$. For a transmission range of r, find $P(\lambda, n, r)$, the probability that the network is connected. Obtain $\lim_{n \to \infty} P(n, \lambda, r)$. Show that the limit is strictly less than one for $\lambda r < \infty$.

9.4 Consider a one-dimensional $(n + 1)$-node network with Node 0 located at 0 and the other n nodes uniformly distributed in $(0, 1)$. Find the probability that the network is connected. This corresponds to a *hybrid network* where the node at 0 corresponds to a base station.

9.5 In a network in $\mathbb{R}^2$, Node i has a neighbor Node r that is a possible next hop node for transmissions from i. For all transmissions from i to j we wish to use the minimum energy path. Thus for all possible receivers j in the network, there will be a one-hop path from i and a multihop path through r. Define $R_i(r)$ to be set of the locations of j such that the two-hop path from i through r is more energy efficient than the one-hop path from i directly to j. r is located at the origin and i is at $-a$ on the y-axis. Assume power law path loss (path loss exponent $\alpha > 0$) and transmission power that just satisfies the SNR threshold β. Draw the boundary of $R_i(r)$ for different α.

9.6 Consider a network with nodes distributed according to a spatial Poisson process of intensity λ. A node at the origin wants to send a packet to a node on the x-axis through a relay node. It looks for a relay node in an sector of angle $(-\alpha, \alpha)$. Assume that the next-hop node is the k-th nearest node to the origin in this sector. Find the distribution of the distance to this relay node. Find the expectation of the horizontal distance (distance along the x-axis) covered by the hop through this relay node.

9.7 As a model of a wireless ad hoc network consisting of a large number of randomly strewn devices, consider a Poisson field of points of intensity λ per m^2 on the plane, with each point denoting the location of a wireless transceiver. All transceivers transmit on a common frequency with unit power using omnidirectional antennas. For the propagation model, assume only path loss with exponent η. Time is divided into fixed-length slots. In each slot, a device decides to transmit with probability α and decides to receive with probability $(1 - \alpha)$ independent of past transmissions and other devices. Find the distribution of interference power received at a randomly selected device in a slot. Show that the mean interference power is finite only if $\eta > 2$. Show also that, when $\eta > 2$, the mean is linear in λ and α.

9.8 Consider a network in $\mathbb{R}^2$ with nodes distributed according to a spatial process of intensity λ. Consider a unit area A. Let $N(A)$ be the number of nodes in A. For any $0 < \epsilon < 1$, show that $\lim_{\lambda \to \infty}$ $\Pr(N(A) \le (1 - \epsilon)\lambda A) = 0$.

9.9 Consider a network in $\mathbb{R}^2$. Assume that all nodes transmit in every slot with power P. Let the path loss function be a decreasing function with $L(0) = M$ and $L(a) \ge m$. Consider a finite square of side a. For a SINR threshold of β, show that if the number of nodes in the square is greater than $\frac{(1 + 2\beta\gamma)M}{\beta\gamma m}$, then none of the nodes in the square has a neighbor.

9.10 Using the $k \times k$ *permutation on an $n \times n$ mesh* analogy obtain the transport capacity of an arbitrary network.

9.11 Consider a 9-node network with nodes arranged on a 3×3 grid. The decode and interference region of every node just covers the four nodes on the neighboring grid points. The network uses s-Aloha protocol with each node transmitting in a slot with probability p. Find the bit-transport capacity of the network as a function of p. Find the optimal p.

CHAPTER 10

Ad Hoc Wireless Sensor Networks (WSNs)

Advances in microelectronics technology have made it possible to build inexpensive, low-power, miniature sensing devices. Equipped with a microprocessor, memory, radio, and battery, such devices can now combine the functions of sensing, computing, and wireless communication into miniature *smart sensor nodes*, also called *motes*. Since smart sensors need not be tethered to any infrastructure because of on-board radio and battery, their main utility lies in being *ad hoc*, in the sense that they can be rapidly deployed by randomly strewing them over a region of interest. Several applications of such wireless sensor networks have been proposed, and there have also been several experimental deployments. Example applications are:

- Ecological monitoring: wild-life in conservation areas, remote lakes, forest fires

- Monitoring of large structures: bridges, buildings, ships, and large machinery, such as turbines

- Industrial measurement and control: measurement of various environment and process parameters in very large factories, such as continuous process chemical plants

- Assistance in navigation and guidance through the geographical area where the sensor network is deployed

- Defense applications: monitoring of intrusion into remote border areas; detection, identification, and tracking of intruding personnel or vehicles

The ad hoc nature of these wireless sensor networks means that the devices and the wireless links will not be laid out to achieve a planned topology. During the operation, sensors would be difficult or even impossible to access and hence their network needs to operate autonomously. Moreover, with time it is possible that sensors fail (one reason being battery drain) and cannot be replaced. It is, therefore, essential that sensors *learn about each other* and *organize into a network* on their own. Another crucial requirement is that since sensors may often be deployed randomly (e.g., simply strewn from an aircraft), in order to be useful, the devices need to determine their locations. In the absence of a centralized control, this whole process of self-organization needs to be carried out in a distributed fashion.

In a sensor network, there is usually a *single, global objective* to be achieved. For example, in a surveillance application, a sensor network may be deployed to detect intruders. The global objective here is intrusion detection. This can be contrasted with multihop wireless *mesh networks*, where we have a collection of source-destination pairs, and each pair is interested in optimizing its *individual* performance metric. Another characteristic feature of sensor networks appears in the *packet scheduling* algorithms used. Sensor nodes are battery-powered, and the batteries cannot be replaced. Hence, energy-aware packet scheduling is of crucial importance.

A smart sensor may have only modest computing power, but the ability to communicate allows a group of sensors to collaborate to execute tasks more complex than just sensing and forwarding the information, as in traditional sensor arrays. Hence, they may be involved in online processing of sensed data in a distributed fashion so as to yield partial or even complete results to an observer, thereby facilitating control applications, interactive computing, and querying. A distributed computing approach will also be energy efficient as compared to mere data dissemination since it will avoid energy consumption in long haul transport of the measurements to the observer; this is of particular importance since sensors could be used in large numbers due to their low cost, yielding very high resolutions and large volumes of sensed data. Further, by arranging computations among only the neighboring sensors the number of transmissions is reduced, thereby saving transmission energy. A simple class of distributed computing algorithms would require each sensor to periodically exchange the results of local computation with the neighboring sensors. Thus the design of distributed signal processing and computation algorithms, and the mapping of these algorithms onto a network, is an important aspect of sensor network design.

Design and analysis of sensor networks must take into account the native capabilities of the nodes, as well as architectural features of the network. We assume that the sensor nodes are *not mobile*. Further, nodes are *not equipped with position-sensing technology*, like the Global Positioning System (GPS). However, each node can set its transmit power at an appropriate level—each node can exercise *power control*. Further, each node has an associated *sensing radius*; events occurring within a circle of this radius centered at the sensor can be detected.

In general, a sensor network can have multiple sinks, where the traffic generated by the sensor sources leaves the network. We consider networks in which only a *single sink* is present. Further, we will be concerned with situations in which sensors are *randomly deployed*. In many scenarios of practical interest, preplanned placing of sensors is infeasible, leaving random deployment as the only practical alternative; for example, consider a large terrain that is to be populated with sensors for surveillance purposes. In addition, random deployment is a convenient assumption for analytical tractability in models. Our study will also assume a simple *path loss model*, with no shadowing and no fading in the environment.

Overview

With the sensor node capabilities and sensor network characteristics mentioned earlier, we will begin by recapitulating some results from Chapter 9. We will be concerned with the question of how sensor nodes should set their transmit powers; specifically, how should transmit powers be set so that the randomly deployed network is *connected* with high probability? After this brief look at *communication coverage*, we will consider the problem of *sensing coverage*. Each sensor can sense events within a certain radius of itself. All points within the disk of this radius are said to be covered by the sensor. If the sensor deployment is random, it is not clear that every point within the deployment region can be covered by at least one sensor. We are interested in finding the *density* of deployment that ensures complete sensing coverage with high probability.

The next problem we consider is that of *localization*. A group of sensors called *anchors* are aware of their own positions and transmit this information to others via beacons. The problem is for the nonanchor nodes to estimate their own locations utilizing the information provided by the anchors. Next, we turn to the problem of *routing* in the sensor network. We discuss *face routing*, where the estimated node location information is used, and also *attribute-based routing*, which does not depend on the knowledge of node locations. *Directed Diffusion* is a prominent example of attribute-based routing. Sensor networks are deployed with specific objectives and usually, some kind of inference about a phenomenon is desired. The inference is based on measurements and subsequent computation of some function of the measurements.

We consider the generic problem of *function computation* next. Our interest is in understanding the maximum rate at which a particular type of function computation can be carried out. Lastly, we briefly describe two MAC *scheduling algorithms* that have been designed, keeping the resource-constrained nature of the sensors in mind.

10.1 Communication Coverage

Formally, we will view the network as a graph, with the motes being the vertices of the graph. If two motes can hear each other *in the absence of interference from other nodes*, then there will be an edge between the corresponding vertices. Essentially, this corresponds to the receiver being within the *decode region* of the transmitter, as discussed in Chapter 7.

In this graph model, which is obtained when only the decode regions are considered, it is desirable that each vertex have a path to the vertex corresponding to the sink. This assures us that there *is* a way for a sensor node i to communicate its measurements to the sink. This is because one can think of a strategy in which i is the *only* node that transmits in a time slot, thereby passing its information to a neighbor within its decode region. Similarly, in the next slot, the neighbor is the only node that transmits. This naive strategy, albeit inefficient, will succeed in transferring information from i to the sink, over several time slots, *if* there is a path in the graph model from i to the sink.

Let us enlarge the requirement slightly and ask that there be a path between *any* pair of nodes. Thus, we are asking the question: What is the minimum power at which the nodes should transmit so that the graph obtained is *connected*?

In passing, we recall from our discussions in Chapter 9 that *not all* the edges in a path from a vertex to the sink can be active *simultaneously*. The question of the set of links that can be activated simultaneously is discussed extensively in Chapters 8 and 9; recall the notion of Activation Sets in Chapter 8, as well as the Protocol model and the Physical model in Chapter 9. However, in this section, we will consider only the question of connectivity of the graph obtained by considering just the decode regions.

Now for a random placement of nodes, the right question to ask is: What is the minimum power at which the nodes should transmit so that the graph is connected *with a given high probability*? For a given number of motes N, this question is hard to answer. Rather, answers have been found in the asymptotic regime where N tends to ∞.

Suppose that N sensors are deployed in a square region of unit area. Each sensor is located independently of any other, and the location is chosen by sampling the uniform distribution. Further, let $r_c(N)$ be the *range* of each of the nodes, that is, if nodes i and j are separated by a distance less than or equal to $r_c(N)$, then they can decode each other's transmission. We note that $r_c(N)$ is being regarded as a function of the total number of nodes N; this suggests that the range changes as N varies. In fact, we would be interested in understanding how to set $r_c(N)$, for a given N, so that the sensor network remains connected.

As N increases, it is expected that the range required to maintain connectivity decreases; $r_c(N)$ is a decreasing function of N. Suppose we consider a range such that

$$\pi r_c^2(N) = \frac{\ln N + c(N)}{N}$$

where $c(N)$ is some function of N that we will discuss later. Note that this range assignment essentially means that a disk of area $\frac{\ln N + c(N)}{N}$ is within reach of a node. Let $P_d(N, r_c(N))$ be the probability that, with this $r_c(N)$, the graph $\mathcal{G}(N, r_c(N))$ is disconnected.

The following has been shown:

$$\liminf_{N \to \infty} P_d(N, r_c(N)) \geq e^{-c}(1 - e^{-c})$$

where $c = \limsup_{N \to \infty} c(N)$. Also,

$$\limsup_{N \to \infty} P_d(N, r_c(N)) \le 2e^{-c}$$

Discussion

Let us consider the implication of these results. Suppose we set the range $r_c(N)$ such that $\pi r_c^2(N) = \frac{\ln N + c(N)}{N}$. The first result says: As $N \to \infty$, suppose $c = \limsup_{N \to \infty} c(N)$ is finite; then, the probability that the network is disconnected is positive. The second result says: As $N \to \infty$, suppose $c = \limsup_{N \to \infty} c(N)$ is infinite; then, the probability that the network is disconnected goes to zero.

Together, the two results provide a necessary and sufficient condition: As $N \to \infty$, with the range assignment as shown, the probability that the network remains connected approaches 1 if and only if $\limsup_{N \to \infty} c(N) = \infty$.

The significance of this result is that if we simply set the range such that $\pi r_c^2(N) = \frac{\ln N}{N}$, then, *with positive probability*, we would get a disconnected network as N increases. This range assignment decreases too rapidly as N increases. It is necessary to ensure that the decrease is not so rapid; this can be done, for example, by adding a term $c(N) = \sqrt{N}$ to the numerator, so that we have $\pi r_c^2(N) = \frac{\ln N + \sqrt{N}}{N}$. Another example is given by $c(N) = \epsilon \ln N$, so that we get $\pi r_c^2(N) = \frac{(1+\epsilon) \ln N}{N}$. Even $c(N) = \ln(\ln N)$ suffices to ensure connectivity with high probability as $N \to \infty$. In all cases, it is still true that the range $r_c(N)$ decreases as N increases; but the decrease is slow enough to ensure connectedness with probability approaching 1.

10.2 Sensing Coverage

Next, let us turn to the question of sensing coverage. We recall that the question here is essentially this: Given an area to be monitored and given a sensing disk around each sensor, how many sensors are required? Now as the node deployment process is random, as a first step, we assume that the nodes are deployed as a two-dimensional spatial Poisson process, of intensity λ points per unit area. The significance of the Poisson assumption is that in two nonoverlapping areas, the numbers of sensors are independent random variables. Further, in an area $\mathcal{A}$, the number of sensors is Poisson-distributed with parameter λA, where A is the area of $\mathcal{A}$.

This question must be refined as follows: What is the minimum intensity λ such that the probability that every point in the monitoring region is covered by at least k nodes is close to 1?

Let r_s denote the *sensing radius* of each disk. Let us choose the unit of area such that each sensor covers unit area: $\pi r_s^2 = 1$. Let us define V_k to be the total area that is *not* k-covered. This means that each point in the area V_k is at most $(k-1)$-covered. V_k is referred to as the *k-vacancy value*. Clearly, V_k is a nonnegative random variable that depends on the particular instance of the Poisson deployment process.

First, it can be shown that no finite λ, no matter how large, can ensure that each point in the monitoring area is covered by at least k nodes. To see this, let $I_k(x)$ denote the indicator function corresponding to k-vacancy at location x. That is,

$$I_k(x) = \begin{cases} 1 & \text{if at most } k-1 \text{ nodes cover point } x \\ 0 & \text{else} \end{cases}$$

If the point x is covered by at most $(k-1)$ sensors, then it is within the sensing distance r_s from at most that many sensors. Equivalently, if we draw a circle of radius r_s centered at x, then there are at most $(k-1)$ sensors within it. Recalling that the deployment process is Poisson with intensity λ and that r_s has been chosen such that the area of a circle with radius r_s is unity, we have

$$\Pr(I_k(x) = 1) = e^{-\lambda} \sum_{i=0}^{k-1} \frac{\lambda^i}{i!} \tag{10.1}$$

Now V_k can be written as

$$V_k = \int_{\mathcal{A}} I_k(x)\, dx$$

Then

$$E(V_k) = \int_{\mathcal{A}} E(I_k(x))\, dx$$

$$= A\Pr(I_k(x) = 1)$$

$$= a^2 e^{-\lambda} \sum_{i=0}^{k-1} \frac{\lambda^i}{i!}$$

$$> 0$$

where we have assumed that $\mathcal{A}$ is a square region with each side of length a. In arriving at the second line, we have used the fact that $\Pr(I_k(x) = 1)$ does not depend on x. We note that $E(V_k) > 0$ for any finite λ, no matter how large it is. But $E(V_k) > 0$ implies that $\Pr(V_k = 0)$ cannot be 1. Thus, for any finite λ no matter how large, we see that $\Pr(V_k > 0) > 0$; we cannot ensure that each point in the area is covered by at least k nodes.

Exercise 10.1

Show that as $\lambda \to \infty$, $E(V_k) \to 0$.

From this exercise, as λ increases, $E(V_k)$ goes to zero. In other words, $\Pr(V_k = 0) \to 1$ as $\lambda \to \infty$. This agrees with intuition: Given a finite monitoring area, as the intensity of the Poisson process increases, it is expected that the probability of the whole area being k-covered (i.e., covered by at least k or more sensors) will increase to 1.

However, it turns out that even more can be shown. Consider the square area $\mathcal{A}$, with sides of length a. Let $a \to \infty$ and along with this, let $\lambda \to \infty$ in a certain way to be discussed later. It can then be shown that even when the monitoring area grows to infinity (i.e., becomes the whole first quadrant), $\Pr(V_k > 0) \to 0$ as $\lambda \to \infty$ in that particular way.

We will say that a point in $\mathcal{A}$ is covered by a sensor if it lies *strictly* within the sensing circle of the sensor, which is the circle of radius r_s centered at it. Consider the sensing circles around the Poisson-distributed sensors in $\mathcal{A}$. Let us define a *crossing* as an intersection point of the boundaries of two or more sensing circles, or an intersection of the boundary of a circle with the boundary of $\mathcal{A}$.

Lemma 10.1

If all crossings in $\mathcal{A}$ are k-covered, then $\mathcal{A}$ is k-covered.   ■

Figure 10.1 shows a square monitoring region $\mathcal{A}$ with the sensing circles of several sensors. It can be seen that the set of sensing circles *partitions* $\mathcal{A}$ into several coverage patches. Each patch is bounded by the arcs of sensing circles and/or the boundary of $\mathcal{A}$. Some patches are 1-covered, some are 2-covered and some are 0-covered. If a patch is k-covered, we will also say that the *coverage degree* of the patch is k. It can be seen that *all* points in a patch have the same coverage degree.

Now suppose that all crossings are k-covered. Let us recall that each sensing circle is *open*. Consider a point x whose coverage degree is the *least* in $\mathcal{A}$, say m. If possible, let $m < k$. Let us now consider the patch S within which x lies. One can claim that the boundary of S cannot be that of a sensing circle.

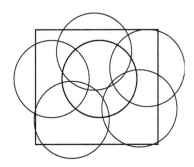

Figure 10.1 Sensor nodes are distributed randomly in a square area. The radius of each circle is r_s, the sensing radius. Several crossings can be seen. The circles define a *partition* of the area.

Suppose that the boundary of S, the patch with the least coverage degree, is a circle. Then no other sensing circle can overlap with any part of S since that would break up S into smaller patches. So, other sensing circles can, at best, touch S at some points on its circumference.

Thus, we can at best have S surrounded by some other sensing circles; see Figure 10.2. The dark circle in the center represents S (if possible). In such a situation, geometry shows us that there are always points like x, y, in the interstitial spaces, where the coverage degree is lower than that in S. But this leads to a contradiction, because the coverage degree cannot be lower than that in S, where it is lowest. Hence, the boundary of S cannot be that of a sensing circle.

The possibilities that remain are shown in Figure 10.3. It can be seen that when S is *not* circular, at least one point on the boundary of S *is* a crossing point.

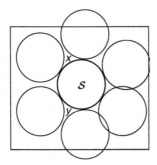

Figure 10.2 The dark circle in the center represents S (if possible). Other sensing circles cannot overlap with any part of S; at best, the other circles can touch S's circumference. In such a situation, there are points like x, y where the coverage degree is lower than that in S.

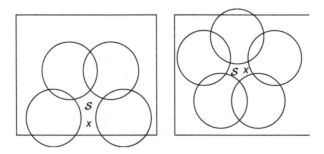

Figure 10.3 Different possibilities for the coverage patch S. When S is not circular, points on its boundary *can* be crossing points. In these examples, S is 0-covered. The crossings are 0-covered because they are not *strictly* inside any sensing circle.

Let us now go back and show by contradiction that if all crossings are k-covered, then $\mathcal{A}$ is also k-covered. As before, let x be a point in $\mathcal{S}$, the coverage patch with the least coverage degree m, with $m < k$. Now, recalling that sensing circles are *open*, a crossing on the boundary of $\mathcal{S}$ cannot have the same coverage degree as the sensing circle on whose circumference it lies; in other words, the crossing is not covered by the circle since it is not strictly inside the circle. Then the coverage degree of the crossing must be the same as that of $\mathcal{S}$, *viz.*, m. But $m < k$, and this means that not all crossings are k-covered—a contradiction. Therefore, if all crossings are k-covered, it must be true that $\mathcal{A}$ is also k-covered.

Equivalently, Lemma 10.1 states that the event $\{\mathcal{A}$ is not k-covered$\}$ implies the event $\{$there is at least one disk with two or more less-than-k-covered crossings on its boundary$\}$. We use this to proceed with the analysis. Let M_k denote the total number of less-than-k-covered crossings in $\mathcal{A}$. Recalling that V_k is the random variable representing the total area that is not k-covered, we have

$$\mathsf{Pr}(V_k > 0) \leq \mathsf{Pr}(M_k \geq 2)$$

Exercise 10.2
Show that

$$\mathsf{Pr}(M_k \geq 2) \leq \frac{\mathsf{E}(M_k)}{2}$$

Using this, we have $\mathsf{Pr}(V_k > 0) \leq \frac{\mathsf{E}(M_k)}{2}$, and we will now proceed to find an upper bound on $\mathsf{E}(M_k)/2$.

Let us consider first the crossings created by two disks intersecting. If two nodes are within a distance of $2r_s$ from each other, their coverage disks intersect. So, given a particular node, the expected number of crossings due to it is twice the expected number of nodes within $2r_s$ of this node, and hence, is given by

$$2\lambda\pi(2r_s)^2$$

where we have used the fact that sensors are distributed according to a Poisson process. The factor 2 in the expression arises because two disks intersect at two points.

Exercise 10.3
Show that if N_1 represents the total number of crossings created by two disks intersecting in the square area of side a and $\pi r_s^2 = 1$ (as assumed before), then

$$\mathsf{E}(N_1) = 4\lambda^2 a^2$$

Next, we consider the crossings created by coverage disks intersecting the boundary of the deployment area. If a node is within a distance r_s from the boundary, then two crossings are created. Let N_2 denote the number of such crossings in the area.

Exercise 10.4

Show that

$$E(N_2) \leq 4\lambda a r_s$$

Let us recall from Equation 10.1 that the probability that a given crossing is not k-covered is

$$e^{-\lambda} \sum_{i=0}^{k-1} \frac{\lambda^i}{i!}$$

Here, we recall that a crossing is *not* covered by any of the circles that intersect at the crossing. So, as far as k-coverage is concerned, a crossing is just like any other location x in the area. Then we have

$$E(M_k) = \left(E(N_1) + E(N_2)\right)e^{-\lambda} \sum_{i=0}^{k-1} \frac{\lambda^i}{i!}$$

$$\leq (4\lambda^2 a^2 + 4\lambda a r_s)e^{-\lambda} \sum_{i=0}^{k-1} \frac{\lambda^i}{i!}$$

$$\leq 4\lambda^2 a^2 (1 + o(1))e^{-\lambda} \sum_{i=0}^{k-1} \frac{\lambda^i}{i!}$$

where, as usual, $o(1)$ indicates a function $f(\lambda)$ such that $\lim_{\lambda \to \infty} f(\lambda) = 0$ (see Appendix A).

Exercise 10.5

Show that

$$\frac{E(M_k)}{2} \leq 2a^2 e^{-\lambda} \frac{\lambda^{k+1}}{(k-1)!}(1 + o(1))$$

Using these results, we arrive at the following inequality:

$$\Pr(V_k > 0) \leq 2a^2 e^{-\lambda} \frac{\lambda^{k+1}}{(k-1)!}(1 + o(1))$$

Now our task is to see how λ should increase with a so that the right-hand side of the expression goes to zero as $a \to \infty$. To this end, consider

$$\lambda = \ln a^2 + (k+1)\ln(\ln a^2) + c(a) \qquad (10.2)$$

where $c(a) = o(\ln a^2)$. This means that $\lambda = \ln a^2 + o(\ln a^2)$. Then we have

$$e^{-\lambda} = e^{-\ln a^2} e^{-(k+1)\ln(\ln a^2)} e^{-c(a)}$$

$$= \frac{1}{a^2} \frac{1}{(\ln a^2)^{k+1}} \frac{1}{e^{c(a)}}$$

$$\therefore \frac{\mathsf{E}(M_k)}{2} \le \frac{2}{(k-1)!} \left(\frac{\lambda}{\ln a^2}\right)^{k+1} \frac{(1+o(1))}{e^{c(a)}}$$

Now $\lambda = \ln a^2 + o(\ln a^2)$ implies that

$$\frac{\lambda}{\ln a^2} \to 1$$

as $a \to \infty$, and we can write $\frac{\lambda}{\ln a^2} = 1 + o(1)$. Therefore, we get

$$\frac{\mathsf{E}(M_k)}{2} \le \frac{2(1+o(1))}{e^{c(a)}(k-1)!}$$

From this expression, we can see that if $c(a) \to \infty$ as $a \to \infty$, then the upper bound on $\Pr(V_k > 0)$ goes to zero as the length of the square goes to ∞. If we did not have the term $c(a)$ in the expression for λ (10.2), then we would not be able to assert that $\Pr(V_k > 0) \to 0$ as $a \to \infty$. Thus, a sufficient condition on λ, to ensure that $\Pr(V_k > 0) \to 0$ as $a \to \infty$, *demands* that the term $c(a)$ be present, and increase to infinity. As an example, we could have $c(a) = \sqrt{\ln a}$.

Actually, it can be shown that $c(a) \to \infty$ as $a \to \infty$ is not only a sufficient condition for $\Pr(V_k > 0) \to 0$ as $a \to \infty$, but also *necessary*. This means that if $c(a)$ is bounded above as $a \to \infty$, then $\Pr(V_k > 0)$ remains strictly positive.

Discussion

We began this section with the question: How dense should the deployment of sensors be so that an entire area $\mathcal{A}$ is k-covered? To get quantitative answers, we modelled the distribution process as a spatial Poisson process with rate λ. Our first observation was that in a finite area, we cannot ensure that each point is k-covered, no matter how large λ is. However, it *is* true that as λ increases, the probability of a nonzero less-than-k-covered area tends to zero; with high probability, the entire area becomes k-covered.

Next, we considered A to be a square area of side a, and allowed a to increase to infinity. The question was: How should λ change so that, as in the finite case, the entire first quadrant is k-covered with arbitrarily high probability? It is, of course, expected that λ should increase to infinity; however, the increase has to satisfy some criterion, as (10.2) shows. Notice the similarity of this result with that related to the critical range for connectivity, due to Gupta and Kumar, discussed in Chapter 9.

10.3   Localization

In many situations of practical interest, sensor nodes are strewn randomly over the deployment area. Consequently, the position of each sensor node is not known *a priori*. Position information, however, is crucial in many situations; for example, to report *where* an event has occurred. Moreover, knowledge of node positions can be exploited in routing also, as in *geographic routing*. Hence, localization is an important problem in sensor networks.

Let us consider several sensors distributed over an area. A small fraction of these are *anchor* devices, that know their own positions. They could be GPS-enabled, or they could have been placed precisely at particular positions, with the position information being programmed into them. The problem is to localize the other sensors with help from these anchors.

A crude idea of the distance from an anchor node can be obtained by noting the strength of the signal received from the anchor, and the transmit power of the anchor. The quality of a distance estimate obtained in this way depends on the accuracy of the model of signal attenuation used. Further, the transmit power used by an anchor may not be easily available to a sensor. For this reason, let us consider "range-free" localization, where we do not calculate distances from anchors based on the received signal strength.

Suppose each anchor sends out messages including its own position, and including a hop count parameter. The anchor initializes the hop count to 1. A sensor (i.e., nonanchor node) receiving the message notes down the anchor's position and the hop count contained in the packet. Next, it increments the hop count value and broadcasts the packet again.

In this way, a wave of packets originates from an anchor and spreads outward. If a sensor receives a packet with a hop count value that is greater than the one stored locally, it ignores the received packet.

The hop count from the i-th anchor, stored at a sensor, is a crude measure of its distance from the anchor. As the density of sensor deployment increases, the distance estimate indicated by the hop count becomes more reliable. As the density increases, sensors at the same hop count from an anchor tend to form concentric rings, of annular width approximately r_c, where r_c is the communication range of a sensor. Thus, if h_i is the hop count from anchor i, then the sensor is at a distance approximately $h_i r_c$ from anchor i.

After obtaining several node-anchor distance estimates as before, nodes follow the multilateration technique. Suppose a node has heard from M anchors, and the anchors' positions are, respectively, (x_i, y_i), $1 \leq i \leq M$. The sensor node j is located at position (x_j, y_j), and this information is not available to it. The actual distance between node j and anchor i is given by

$$d_{j,i} = \sqrt{(x_j - x_i)^2 + (y_j - y_i)^2}$$

which, of course, is unknown to j. The estimate of this distance that *is* available to j is $\hat{d}_{j,i} := h_i r_c$. Then, a natural criterion that can be used to determine the unknown (x_j, y_j) is the total *localization error* E_j, defined as

$$E_j = \sum_{i=1}^{M} (d_{j,i} - \hat{d}_{j,i})^2$$

$$= \sum_{i=1}^{M} \left(\sqrt{(x_j - x_i)^2 + (y_j - y_i)^2} - \hat{d}_{j,i} \right)^2$$

Exercise 10.6

Show that the partial derivatives of E_j with respect to x_j and y_j are given by

$$\frac{\partial E_j}{\partial x_j} = 2 \sum_{i=1}^{M} (x_j - x_i) \left(1 - \frac{\hat{d}_{j,i}}{d_{j,i}} \right) \quad \text{and} \quad \frac{\partial E_j}{\partial y_j} = 2 \sum_{i=1}^{M} (y_j - y_i) \left(1 - \frac{\hat{d}_{j,i}}{d_{j,i}} \right)$$

Using these expressions, we can get an iterative procedure to obtain x_j and y_j. We start with some initial guess of the position of j: $(x_j^{(0)}, y_j^{(0)})$. This allows us to calculate $d_{j,i}$ approximately, and also evaluate $\frac{\partial E_j}{\partial x_j}$ and $\frac{\partial E_j}{\partial y_j}$. Then, *updates* to the initial guessed position can be obtained as

$$\Delta x_j = -\alpha \frac{\partial E_j}{\partial x_j} \quad \text{and} \quad \Delta y_j = -\alpha \frac{\partial E_j}{\partial y_j}$$

where α is a small positive fraction. It can be seen that the sign of the update Δx_j is always opposite of that of $\frac{\partial E_j}{\partial x_j}$; a similar conclusion follows for the update Δy_j. The updated position is obtained by taking a small step in the direction of the *negative gradient* of error with respect to the current position. Therefore, the iterative process of updating positions is such that the error tends to decrease.

Essentially, this method estimates distance from an anchor by the hop count from that anchor. In a dense deployment of sensors, this estimate is reasonable. Thus, it can be expected that the quality of the localization obtained from this method is critically dependent on the density of sensor deployment. Further, the method also assumes that the communication range r_c is known.

Convex Position Estimation

We now discuss an alternative approach to sensor localization in which a *convex position estimation* problem is formulated and solved. As before, it is assumed that the positions of anchors are known. A sensor node wishing to localize itself notes the identities of the anchors it can hear, and computes its position as follows.

If a sensor node j can hear an anchor i, then j must be within a distance r_c from i. In other words, j can be localized to within a circle of radius r_c around i. Let us assume that the *boundary* of the circle of radius r_c is out of bounds, and the distance between i and j should be *strictly* less than r_c. Formally, we have

$$\|i - j\|_2 < r_c \tag{10.3}$$

where $i = (x_i, y_i)$ and $j = (x_j, y_j)$ are the positions of anchor i and node j, respectively, in the two-dimensional plane, and $\|i - j\|_2$ represents the Euclidean norm of $(i - j)$, *i.e.*, $\|i - j\|_2 = \sqrt{(x_i - x_j)^2 + (y_i - y_j)^2}$.

This constraint can be represented in terms of a *Linear Matrix Inequality* (LMI), as we discuss now. The motivation for formulating the constraint in these terms comes from the availability of powerful numerical methods for solving such problems.

Let us recall how a *positive definite* matrix F is defined. Suppose F is a real and symmetric $N \times N$ matrix. Then F is positive definite if for every *nonzero* N-vector $u \in \mathbb{R}^N$, $u^T F u > 0$, where u^T denotes the transpose of u.

We note the following facts. Let

$$G = \begin{bmatrix} G_1 & G_2 \\ G_3 & G_4 \end{bmatrix} \tag{10.4}$$

be a positive definite $N \times N$ matrix, where G_1 is an $M \times M$ matrix, G_4 is an $(N - M) \times (N - M)$ matrix, and the dimensions of G_2 and G_3 are evident (G_2 and G_3 are not necessarily square matrices). Then it is known that

- G_1 and G_4 are both positive definite.

- $G_4 - G_3 G_1^{-1} G_2$ is also positive definite. $(G_4 - G_3 G_1^{-1} G_2)$ is called the *Schur complement* of G_1 in G.

To see how this is used, consider the real, symmetric matrix

$$F = \begin{bmatrix} r_c I_2 & i - j \\ (i - j)^T & r_c \end{bmatrix}$$

where I_2 is the 2×2 identity matrix. By correspondence with (10.4), $N = 3$ and $M = 2$ here. Suppose that F is positive definite. Then, considering the Schur complement of $r_c I_2$ in F, we have

$$G_4 - G_3 G_1^{-1} G_2 = r_c - (i - j)^T \frac{1}{r_c} I_2^{-1} (i - j)$$

$$= r_c - \frac{1}{r_c} \|i - j\|_2^2$$

As **F** is positive definite, so is the Schur complement, and therefore we have $(r_c - \frac{1}{r_c} \|i - j\|_2^2) > 0$, as positive definiteness reduces to simple positivity for a 1×1 matrix. Thus, we get

$$\|i - j\|_2 < r_c$$

as in (10.3).

i represents the position of the anchor, which is known. The position of the node is unknown, which means that j is unknown. The matrix **F** can be regarded as a function of the unknowns x_j, y_j; thus, $\mathbf{F} = \mathbf{F}(x_j, y_j)$. What we saw earlier can be rephrased as follows: If j is such that $\mathbf{F}(x_j, y_j)$ is positive definite, then $\|i - j\|_2 < r_c$. Thus, if we *define* the set of *feasible positions* for node j as the set

$$\{j : \mathbf{F}(x_j, y_j) \text{ is positive definite}\}$$

then we are assured that the constraint $\|i - j\|_2 < r_c$ is respected. It may be noted that this definition is *sufficient* for the condition in (10.3) to hold. Hence, by this definition, we get a *smaller* feasible set than that indicated by (10.3). The smaller feasible set is the price we pay when we formulate the problem in terms of a Linear Matrix Inequality.

Exercise 10.7
Show that the preceding set, *viz.*,

$$\{j : \mathbf{F}(x_j, y_j) \text{ is positive definite}\}$$

is convex; that is, the set of feasible positions for node *j* is a convex set. This is why this approach is referred to as convex position estimation.

In the previous discussion, we have considered the situation when node *j* hears only one anchor, *viz.*, anchor *i*. What happens when *j* hears from *M* anchors? The LMI approach readily extends to cover this situation. For this, we have to define a number of matrices $\mathbf{F}^{(i)}(x_j, y_j)$, $i = 1, 2, \ldots, M$, as follows:

$$\mathbf{F}^{(i)}(x_j, y_j) = \begin{bmatrix} r_c I_2 & i - j \\ (i - j)^T & r_c \end{bmatrix}$$

For each $i = 1, 2, \ldots, M$, the arguments of $\mathbf{F}^{(i)}$ are the same: the unknowns (x_j, y_j). One can arrange these matrices in block-diagonal form to get a large block-diagonal matrix $\mathbf{F}(x_j, y_j)$:

$$\mathbf{F}(x_j, y_j) = \begin{bmatrix} \mathbf{F}^{(1)} & 0 & \cdots & 0 & 0 \\ 0 & \mathbf{F}^{(2)} & 0 & \cdots & 0 \\ & & \cdots & & \\ 0 & 0 & \cdots & 0 & \mathbf{F}^{(M)} \end{bmatrix}$$

Exercise 10.8

Show that this block-diagonal matrix $\mathbf{F}(x_j, y_j)$ is positive definite if and only if each block matrix $\mathbf{F}^{(i)}(x_j, y_j)$, $i = 1, 2, \ldots, M$, is also positive definite.

Suppose we require $\mathbf{F}(x_j, y_j)$ to be positive definite. Then, by virtue of the preceding exercise, $\mathbf{F}^{(i)}(x_j, y_j)$ is also positive definite for $i = 1, 2, \ldots, M$. This, in turn, allows us to conclude that if $\mathbf{j} = (x_j, y_j)$ is feasible, then $\|\mathbf{i} - \mathbf{j}\| < r_c$ for each anchor $i = 1, 2, \ldots, M$. Thus, if a feasible point exists, it is guaranteed to be in the intersection of the circular discs of radius r_c around each anchor.

Thus, for each sensor j that needs to be localized, we can pose the problem: Find $\{\mathbf{j} : \mathbf{F}(x_j, y_j)$ is positive definite$\}$, where M represents the number of anchors that j hears from. As we saw, this will give a convex feasible set to which j can be localized.

Further, the feasible set obtained can be bounded within a rectangle. For this, consider a two-element vector $\mathbf{c}$, and consider the *semidefinite* program (SDP)

$$\min \mathbf{c}^T \mathbf{j}$$

$$\text{subject to } \mathbf{F}(x_j, y_j) \succ 0$$

where $\mathbf{F}(x_j, y_j) \succ 0$ means $\mathbf{F}(x_j, y_j)$ is positive definite. A semidefinite program is a generalization of a linear program in which the objective function is linear in the unknowns, but the constraints are expressed in terms of a positive/negative definite/semidefinite matrix. As mentioned before, efficient computational methods for solving SDPs are available.

In this formulation, $\mathbf{c}$ represents the cost of the position estimation (x_j, y_j) that we are interested in. Suppose we choose $\mathbf{c} = (1, 0)^T$. Then, the SDP corresponds to finding the *smallest* x_j that is consistent with the feasibility constraint. On the other hand, consider $\mathbf{c} = (-1, 0)^T$; now the SDP corresponds to finding the *largest* x_j that is consistent with the feasibility constraint. Similarly, by choosing appropriate values of $\mathbf{c}$, we can obtain bounds on y_j, too.

Discussion

This approach leads to the conclusion that it is possible to obtain a rectangular bounding box within which each sensor j can be localized (see Figure 10.4). In this

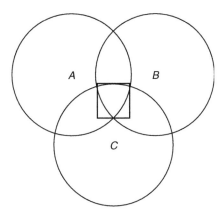

Figure 10.4 A sensor node *j* which hears from three anchor nodes will lie in the intersection of the three circles. Thus, the closed region *ABC* is the feasible set. The dark rectangle around *ABC* gives the bounding box within which the feasible region must lie.

way, each sensor can be localized to a rectangular box within which it must lie. As long as the feasible set is nonempty, the dimensions of the box provide estimates of locationing errors along the x and y axes.

The convex position estimation approach discussed previously also requires knowledge of the communication radius r_c, just as the first method did. However, in contrast to the first method, the convex position estimation method does not demand dense sensor deployment. Further, the first method was iterative, but the second method discussed is not. On the other hand, the LMI-based approach can place a significant computational burden on the sensors.

10.4 Routing

Standard table-driven routing approaches are often not attractive in the sensor network context. Route discovery and route maintenance are periodic and energy-intensive tasks, and typical sensor nodes are severely constrained in energy, memory, and computing power. Thus, alternate approaches have been considered in the literature.

Let us consider routing ideas referred to as geographic or geometric or position-based routing. Its characteristic features are: (1) every node knows its own position, and the positions of its neighbors; (2) the source knows the position of the destination; (3) there are no routing tables stored in the nodes; (4) the additional information stored in a packet is bounded above by a constant times the number of nodes in the graph; the additional information is $O(|\mathcal{V}|)$, where $\mathcal{V}$ is the vertex set of the graph $\mathcal{G}(\mathcal{V}, \mathcal{E})$ representing the sensor network. Because of the memory restrictions, geometric routing is known also as $O(1)$-memory routing.

Evidently, geometric routing is based entirely on local information. We will assume that nodes have *acquired knowledge of their own positions.*

The simplest approach is for a source node *s* to forward data to a neighbor who is closer to *t*. Basically, this is a *greedy* approach—a packet is passed to a neighbor who is closest to the destination.

As can be expected, however, the greedy approach does not always work: What if none of the neighbors of a node *i* is closer to the destination *t* than *i* itself? An example is shown in Figure 10.5.

As mentioned before, we will consider the sensor network as a graph, with the node positions being the vertices and a link between two nodes being an edge. For simplicity, we will assume that all nodes transmit at the same power, so that the communication radius r_c is the same for all nodes. This means that any two nodes at a distance less than or equal to r_c can communicate directly with each other. If r_c is defined as the unit of distance, then we have what is called a *Unit Disk Graph (UDG)* (see Chapter 9).

In a dense network, it is clear that a UDG will give rise to numerous edges. Typically, in a graph with a large number of edges, significant computational effort is required to find routes between pairs of nodes. In a resource-constrained sensor network, one can ill-afford this energy expenditure.

This suggests that *removing* some edges from the UDG, while retaining graph connectivity, is an option worth exploring. However, removing edges also means that path lengths between pairs of nodes increase; for example, if the direct link between *i* and *j* is removed, then evidently, these nodes are connected through a multihop path instead of the earlier single-hop one. Now higher bit rates may be achievable over paths with fewer hops. Clearly, we are trading off bit rate for computational efficiency here.

A *planar* graph is one that *can* be drawn such that no edges intersect on the plane. When it *is* drawn in such a way, what we get is a plane graph. Planar graphs are of interest because they are usually sparse, as we will see later. The energy and

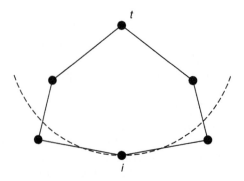

Figure 10.5   Scenario where greedy routing fails. Neither of the neighbors of *i* is closer to the destination *t* than *i* itself.

computation overhead of finding routes can be expected to be significantly less on a sparse graph.

It is noteworthy that a given graph may *appear* to have edge intersections, but it may be possible to distort the given graph, while ensuring that each edge connects the same pair of nodes as before, so that in the new drawing, there are no edge crossings on the plane. For example, the graph on the left in Figure 10.6 can be redrawn as shown on the right, keeping all edges between the same pair of nodes in both cases.

How would one obtain a planar graph from a UDG? One possibility is to start with the node positions of the original UDG, eliminate all edges in the UDG, and then reintroduce some edges appropriately. Several standard geometric constructions are used to get planar graphs in this way. The basic idea is to introduce an edge between nodes i and j, say $i \rightarrow j$, if a suitable region around $i \rightarrow j$ (called the *witness region*) is free of other nodes.

Let $d(i, j)$ be the geometric or Euclidean distance between nodes i and j. Consider two circles of radius $d(i, j)$ centered around the nodes i and j (see Figure 10.7). The intersection of the two circles is called the *lune*. Suppose we introduce the edge $i \rightarrow j$ if the lune is free of other nodes. If we do this for every pair of nodes, we get the *Relative Neighborhood Graph* (RNG). On the other hand, suppose we introduce the edge $i \rightarrow j$ if the circle of diameter $d(i, j)$ within the lune is free of other nodes. Doing this for every pair of nodes gives us the *Gabriel Graph* (GG). For the RNG, the witness region is the lune; for the GG, the witness region is the circle within the lune.

It is known that both the RNG and the GG are planar graphs. Further, it is also evident that the RNG is a subgraph of the GG. It turns out that both of these planar graphs can be computed using local algorithms, involving exchange of information among a node and its neighbors only.

We observe that a planar graph induces a partitioning of the plane into a set of regions with disjoint interiors. Each such region is called a face. The outer unbounded region is also regarded as a face.

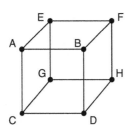 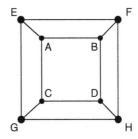

Figure 10.6 The graph on the left is planar because it can be redrawn as shown on the right, keeping all node adjacencies unchanged.

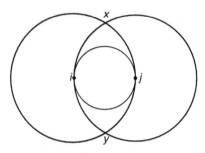

Figure 10.7 Obtaining the Relative Neighborhood Graph (RNG) and the Gabriel Graph (GG) from the node positions. To get the RNG, we introduce edge $i \to j$ if the lune $ixjy$ is free of other nodes. To get the GG, we introduce edge $i \to j$ if the circle of diameter $d(i,j)$ within the lune is free of other nodes.

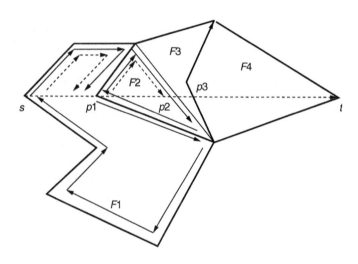

Figure 10.8 A planar graph with four faces—$F1$, $F2$, $F3$, and $F4$—is shown. s and t are the source and the destination, respectively. $p1$, $p2$, and $p3$ are the points at which the straight line from s to t intersects the edges between the corresponding faces. The solid lines with arrows indicate the first circumnavigation around a face, and the dashed lines indicate the second traversal until the switch point. After the switch point is reached, the packet retraces its path and starts moving around an adjacent face.

To get around the problem encountered with greedy routing, consider a strategy that routes along the boundaries of the faces that are crossed by the straight line between the source and the destination. Figure 10.8 shows an example.

In Figure 10.8, we see four faces: $F1$, $F2$, $F3$, and $F4$ in the planar graph. The source s and the destination t are also marked. $F1$ is the face that contains s and the line s, t intersects $F1$. To begin with, the boundary of $F1$ is explored, in the clockwise direction (say). This is indicated by the thin solid arrows in Figure 10.8. As the boundary is traversed, the algorithm notes all points where the line s, t intersects the boundary, and the point closest to t is stored for future use. In Figure 10.8, there is only one such point, *viz.*, $p1$. After traversing the whole boundary and reaching s again, a second traversal (along the dashed line) is started. This time, when $p1$ is reached, the face $F2$ is explored in a similar manner. We may refer to the point $p1$ as a switch point.

In this way, faces are explored successively, with a new face being taken up at switch points closest to t. It can be shown that on simple planar graphs, face routing terminates in $O(N)$ steps, where N is the number of nodes. Essentially, the idea is to show that in the course of the execution of the face routing algorithm, each edge is traversed *at most* a constant number of times. Then, a theorem about planar graphs is applied; this theorem says that the number of edges $|\mathcal{E}|$ in a connected plane graph with at least three vertices satisfies

$$|\mathcal{E}| \leq 3|\mathcal{V}| - 6$$

This relation is significant, because it indicates that planar graphs are basically sparse graphs. In a fully connected graph, the number of edges is $O(|\mathcal{V}|^2)$, whereas the relation asserts that in a planar graph, the upper bound is only a linear function of $|\mathcal{V}|$.

Using this relationship, it follows that face routing terminates in $O(N)$ steps, as $|\mathcal{V}| = N$. Thus, we are assured that face routing works correctly. Nevertheless, face routing may select a path that is considerably longer than the shortest path between a pair of nodes.

This description provides the principle underlying the face routing algorithm. In practice, nodes will need to deduce the locations of the switch points p_1, p_2, and p_3. This is possible because each node knows its own position and the positions of its neighbors, as well as the positions of the source s and destination t, because the latter are carried in the packet header.

In summary, we note that face routing requires very little memory and computation, but the price to be paid is the knowledge of nodes' own positions, as well as the positions of their neighbors. Not unexpectedly, face routing is sensitive to the accuracy of position information, and large errors in position may cause face routing to fail.

Attribute-based Routing

In the previous section, we considered a routing approach that relied upon nodes knowing their own positions. In this section, we consider a routing strategy that is oblivious to the nodes' positions, and even addresses. In fact, the routing scheme to be discussed here does not even *attempt* to reach a specific node from another; it uses a completely different philosophy.

In a deployed sensor network, a specific type of event may be of interest. For example, in a sensor network used for environment monitoring, observations of a particular type of animal may be of interest. The strategy in *attribute-based* routing schemes is to launch a query that *describes the data of interest.* As this query propagates through the network, it encounters nodes that have observed the event of interest. Such nodes now provide replies that move back toward the original node issuing the query. Thus, attribute-based routing provides a way for information seekers and information gatherers to "meet," *without* knowing one another beforehand.

Data are described by *attribute-value* pairs that characterize the information that is of interest. For example, a query can be expressed as a *record* consisting of multiple attribute-value pairs:

```
type = animal
instance = leopard
rectangle = [0, 400, 0, 400]
```

In the first line, "type" is the attribute and "animal" is the value; similarly, "instance" is the attribute and "leopard" is the value. The third line specifies an area (in some coordinate system) within which the observation is sought; a rectangle of size 400×400 is shown here, with the ranges along the x and y axes specified.

A prominent example of attribute-based routing is provided by a scheme known as *Directed Diffusion* (DD). In DD terminology, the node issuing the query is called a *sink*, while the nodes providing the observation are called *sources*. The query itself is referred to as the *interest*. In the following paragraphs, we provide a high-level description of how DD finds paths between sources and sinks.

To begin with, the sink broadcasts the interest. As shown in the previous example, the interest is a collection of several attribute-value pairs, among which are the attributes *duration, interval,* and *update*. The duration attribute specifies the period of time for which the interest is valid. The interval attribute specifies the intervals at which the observation is to be reported. Implicit in this attribute is the characteristic of DD that *repeated observations* of the event of interest are important. Further, the duration attribute indicates that the interest persists for some time. Both attributes suggest that the sink is not satisfied with just one observation of the event; rather observations extended over time are important. This allows DD to amortize the cost of finding paths between sources and sinks over the duration of the communication. The use of the update attribute is discussed shortly.

The interest generated originally by the sink passes through nodes in the network. A node that receives the interest checks if it has any event record that matches the interest. If it does, then it becomes a source and proceeds to relay the information back to the sink; we will see this in more detail shortly. If it does not, then it forwards the received interest to its neighbors.

Each node maintains an *interest cache*, where valid interests are stored. Along with each interest, the node notes down the neighbor from which the interest was received. It is noteworthy that this is strictly *local information*; the sink that generated the interest in the first place is not tracked.

A particular interest may be received at a node from a number of neighbors because, initially, the interest is flooded through the network and may arrive at the node via multiple paths. All such neighbors are stored in the interest cache. How often should each neighbor receive a report of the observed event? This is determined by the *update* attribute of the interest received from that neighbor. The smaller the value of update, the more frequently the neighbor receives a report. In fact, the update attribute determines what is called a *gradient* toward a neighbor, where gradient is defined as the reciprocal of the update value.

In this way, utilizing neighbors and the gradient toward each neighbor, an observed event record makes its way back toward the sink. As in the forward pass when the interest was making its way into the network, the requested information may arrive back at the sink over multiple paths. In the DD scheme, the update attribute is now used in a clever way to reinforce good quality paths.

To do this, the sink observes the returned reports received from various neighbors. The neighbor from which the *first* report was received is likely to lie on the least-delay path from the source to the sink. The sink now resends the interest to *this* particular neighbor, with a smaller update attribute. This leads to higher gradients being set up all along the backward path from the source to the sink. In this way, DD provides a way of adaptively selecting preferred paths. Similarly, the update interval toward a nonpreferred neighbor can be increased, so that ultimately, that particular path from source to sink is suppressed.

It is worth noting that DD is inherently robust to node failures, because if the currently preferred path becomes unavailable (due to node failure, say), then an alternate path (which was currently not preferred) may be picked up and reinforced by update attribute manipulation.

10.5 Function Computation

Let us imagine a situation where N sensors have been distributed uniformly and independently in a square area $\mathcal{A}$. These sensors have self-organized to form a network. In particular, the transmission range of each has been set to a value $r_c(N)$ such that the network $\mathcal{G}(N, r_c(N))$ is *connected* asymptotically; the probability that $\mathcal{G}(N, r_c(N))$ is connected goes to 1 as N goes to infinity. As we have seen before, this will happen when

$$r_c(N) = \sqrt{\frac{\ln N + c(N)}{\pi N}} \tag{10.5}$$

with $\limsup_{N \to \infty} c(N) = \infty$.

The sensors make periodic measurements. There is a single sink in the network where some function of the sensors' measurements is to be computed.

The function, in general, would depend on the inference problem that the sensor network has been designed to solve. For example, in an event detection application, the function could be the conditional probability of the sensor output being in a certain range, given that there has been no event (the *null* hypothesis). In statistics, such a function is referred to as a *likelihood* function, and it is extensively used in event detection.

One naive way to compute the function is to forward *all* the measured data to the sink, which then computes the function. This is a centralized model of computation. However, this fails to take advantage of the processing capability of the sensors. An alternative approach is *in-network processing*, where the sensors compute intermediate results and forward these to the sink. This aggregation helps in reducing the amount of data to be forwarded to the sink, and thus helps in easing congestion as well as prolonging battery life.

For example, consider a linear network of $(N + 1)$ sensors as shown in Figure 10.9, with s denoting the sink. Suppose that each sensor measures the temperature in its neighborhood and the objective is to compute the maximum temperature. If all measurements are simply forwarded to the sink, then the communication effort is $O(N^2)$; sensor 1 data requires N hops to reach the sink, sensor 2 data requires $(N - 1)$ hops, and so on. On the other hand, an alternate strategy is one in which node i compares received data with its own measurement, and forwards the maximum of the two. In this strategy, in-network processing is being done, and the communication effort drops to just $O(N)$.

Suppose that the N sensors make periodic measurements. Let each sensor reading belong to a discrete set $\mathcal{X}$. Let time be slotted. $\mathbf{X}^{(t)}$ denotes the vector of N sensor readings at discrete time-slot t. Let us also assume that readings over a period $t = 1, 2, \ldots, T$ are available with each sensor; here T is the block length over which measurements have been collected. The $N \times T$ matrix $\mathbf{X}$ represents the complete data set, across sensors as well as across the block length, that is available. $\mathbf{X}_i$, the i-th row of the matrix, represents the readings of the i-th sensor over the block. Correspondingly, the t-th column $\mathbf{X}^{(t)}$ represents the readings across the sensors at time t. The objective is to compute the function $f(\mathbf{X}^{(t)})$, for every $t \in \{1, 2, \ldots, T\}$.

Generally, if $\mathcal{C}$ is a subset of sensors, then $f(\mathbf{X}_{\mathcal{C}}^{(t)})$ is the function computed by taking the readings of sensors in the set $\mathcal{C}$ at time t.

We have tacitly assumed that the function to be computed admits distributed computation in a divide-and-conquer fashion, in which the result of a partial

Figure 10.9 A linear network of _(N+1)_ sensors is shown. Each sensor makes measurements, and the maximum of all measurements is desired at the sink _s_.

computation by some sensors is forwarded to others, which then repeat the process. To formally state the property that we assumed, we introduce the notion of *divisible functions*.

Let C be a subset of $\{1,2,\ldots,N\}$, and let $\pi := \{C_1, C_2, \ldots, C_s\}$ be a *partition* of C. The function $f(.)$ is said to be divisible, if for any $C \subset \{1,2,\ldots,N\}$ and any partition $\pi = \{C_1, C_2, \ldots, C_s\}$ of C, there exists a function $g^{(\pi)}(.)$ such that

$$f(\mathbf{X}_C^{(t)}) = g^{(\pi)}(f(\mathbf{X}_{C_1}^{(t)}), f(\mathbf{X}_{C_2}^{(t)}), \ldots, f(\mathbf{X}_{C_s}^{(t)}))$$

This says that if we know the values of the function $f(.)$ evaluated over the sets in any partition π of C, then we can combine these values, using the function $g^{(\pi)}(.)$, to obtain $f(.)$ evaluated over C. Thus, it is possible to compute $f(\mathbf{X}_C^{(t)})$ in a divide-and-conquer fashion.

Exercise 10.9

Let each sensor reading belong to a discrete set $\mathcal{X}$. Consider the function $\tau(\mathbf{X}^{(t)})$ that gives the frequency histogram or "type vector" corresponding to the sensor readings $\mathbf{X}^{(t)}$ at time t. This function is a vector with $|\mathcal{X}|$ elements, where $|\mathcal{X}|$ denotes the size of the set $\mathcal{X}$:

$$\tau : \mathcal{X}^N \to \{0,1,2,\ldots,N\}^{|\mathcal{X}|}$$

Show that $\tau(.)$ is a divisible function.

Exercise 10.10

Show that the function that provides the *second largest value* in a set of sensor measurements is *not* divisible.

Let $\mathcal{R}(f)$ denote the range of the function; let us recall that the range of a function $f : \mathcal{X} \to \mathcal{Y}$ is defined as

$$f(\mathcal{X}) := \{y \in \mathcal{Y} : \exists x \in \mathcal{X} \text{ with } f(x) = y\}$$

Exercise 10.11

In the previous exercise, is $\mathcal{R}(\tau) = \{0,1,2,\ldots,N\}^{|\mathcal{X}|}$?

Now the sensors together compute $f(\mathbf{X}^{(t)})$ for $t = 1,2,\ldots,T$, by passing messages among one another according to some scheme. Let $\mathcal{U}_{N,T}$ denote such a scheme. Further, let $\mathcal{T}(\mathcal{U}_{N,T})$ denote the *maximum* time (in slots) taken to complete the computation of the function for all times in $t = 1,2,\ldots,T$, where the maximum is taken over all possible values of $\mathbf{X}^{(t)}, t = 1,2,\ldots,T$. This is the time at which the *sink* in the network is able to obtain the values $f(\mathbf{X}^{(t)}), t = 1,2,\ldots,T$.

With the previous notation, the *rate* of function computation when scheme $\mathcal{U}_{N,T}$ is followed is defined, in computations/slot, as

$$R(\mathcal{U}_{N,T}) = \frac{T}{\mathcal{T}(\mathcal{U}_{N,T})}$$

Considering the maximum of $R(\mathcal{U}_{N,T})$ over all possible schemes and all possible block lengths T, we obtain the *maximum rate of function computation* for a given divisible function, written as $R_{\max}^{(N)}$. The value of $R_{\max}^{(N)}$ is specific to the N-sensor network; the superscript in $R_{\max}^{(N)}$ is intended to serve as a reminder of this fact.

How large can $R_{\max}^{(N)}$ be? To compute $f(\mathbf{X}^{(t)}), t = 1, 2, \ldots, T$, the sink must receive the results of the partial computations carried out by outlying sensors and complete the task using its own data. Considering the protocol model (see Chapter 9), it is clear that only one link terminating at the sink can be active at any instant. If the maximum possible bit rate on a link is W bits per slot, then the sink cannot receive more than W bits in a slot.

On the other hand, to identify the specific function value $f(\mathbf{X}^{(t)})$, we would need $\log_2 |\mathcal{R}(f)|$ bits, where we have assumed that the discrete set $\mathcal{R}(f)$ has 2^m elements for some $m \geq 1$. This leads to the following upper bound on $R_{\max}^{(N)}$:

$$R_{\max}^{(N)} \leq \frac{W}{\log_2 |\mathcal{R}(f)|}$$

Before proceeding further, we recall the following observations from Chapter 9. Suppose N nodes are placed uniformly and independently in the unit square. Let us denote by $\mathcal{G}(N, \phi_N)$ the graph that results when each node is connected to its ϕ_N *nearest neighbors*. Then, $\mathcal{G}(N, \phi_N)$ is connected with high probability if and only if $\phi_N = \Theta(\ln N)$. Specifically, there are two constants $0 < c_1 < c_2$ such that

$$\lim_{N \to \infty} \Pr\big(\mathcal{G}(N, c_1 \ln N) \text{ is disconnected}\big) = 1$$

$$\lim_{N \to \infty} \Pr\big(\mathcal{G}(N, c_2 \ln N) \text{ is connected}\big) = 1$$

Thus, by selecting a transmission range such that a node connects to at least $c_2 \ln N$ nearest neighbors, we are assured of getting a connected graph with high probability. Also, we note that the *degree* of the resulting graph is $O(\ln N)$ with high probability.

Let us now consider the N-sensor network with the transmission range $r_c(N)$ set appropriately (as before), so that each node has enough number of neighbors for the graph to be connected with high probability. Consider a tessellation of the unit square into small squares (called cells) of side $r_c(N)/\sqrt{2}$. This implies that nodes within a cell are always within range of one another.

Let us now define a *cell graph*. Each nonempty cell of the tessellation (see Figure 10.10) is a vertex in the cell graph. Further, two vertices c_1 and c_2 of the cell graph are defined to be adjacent (i.e., to have an edge between them), if we can find a node inside cell c_1 and a node inside cell c_2 such that the nodes are neighbors.

Figure 10.10 shows a cell graph corresponding to a number of sensors deployed randomly in a unit square. The small rectangle inside the cell in the middle represents the sink of the network. Corresponding to each nonempty cell in this tessellation, we have a vertex in the cell graph. Further, as shown in Figure 10.10, we can find a node in cell c_1 and a node in cell c_2 such that these nodes are neighbors (indicated by the line joining them); hence, vertices c_1 and c_2 are adjacent in the cell graph. The figure also shows relay nodes in a cell and relay parents; we define these later.

Let us consider a spanning tree on the cell graph, rooted at the cell containing the sink. It is possible to obtain such a spanning tree because the underlying network of nodes is connected, and, therefore, so is the cell graph.

Consider a cell c and its parent cell in the spanning tree. A node i in cell c will be called the *relay node* in that cell if (1) it has at least one neighbor in cell c, and (2) it collects data from all neighbors in cell c, runs a partial computation on the data and forwards the result to a node j in an adjacent cell. This node j will be called the *relay parent* of relay i. By definition, a relay node in cell c cannot serve as a relay parent for any other cell.

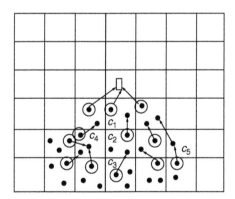
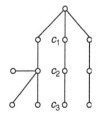

Figure 10.10   On the left panel, a random deployment of sensors is shown. The nodes with a circle around them are the relay nodes. The arrow pointing away from a relay node identifies the corresponding relay parent, located in a parent cell. Cell c_4 has *two* relay parents and one relay, and cell c_2 has one relay, one relay parent, and also a node that is neither a relay nor a relay parent. The cell graph for this deployment would be obtained by considering a vertex for every nonempty cell; two vertices would be adjacent if each of the corresponding cells has a node within communication range of the other. The right panel shows a spanning tree on the cell graph.

This definition implies that a cell has a relay node in it only if there are two or more nodes in it. If a cell has only one node, then we will *not* consider that node to be a relay node. However, it will, of course, make measurements and pass them to some other node in an adjacent cell. It can also *receive* measurements from nodes in adjacent cells and pass them on. Thus, the single node *may* behave as a relay parent for relays in adjacent cells. In Figure 10.10, the cell c_5 has a single node in it, and this node behaves as a relay parent; however, it is not considered a relay node.

In summary, a relay node collects data from other nodes in its *own* cell only, whereas a relay parent collects data from relay nodes in *other* cells only. A node can be either a relay or a relay parent, but not both. Of course, it is possible that a node is neither a relay nor a relay parent.

In Figure 10.10, the node with a circle around it is the relay node in that cell, and the arrow indicates its relay parent. As Figure 10.10 shows, there is one relay node in each cell and possibly several relay parents. A node that has neither a circle around it nor an arrow pointing to it is neither a relay node nor a relay parent.

For computing function values, data need to be collected at the sensors and sent toward the sink. On the way, partial function computations can be carried out. In Figure 10.10, the process would start with the sensor nodes in the leaf cells. Next, we outline an approach that makes distributed function computation possible.

Suppose a divisible function $f(.)$ is to be computed at T epochs. Let us divide time into T rounds. For $n = 1, 2, \ldots, T$, round n consists of T_1 slots, where T_1 will be specified later.

In round 1, all *nonrelay* nodes in the leaf cells in Figure 10.10 transmit to the relay node the result of a partial computation of $f(.)$, where the function is evaluated at a node's *own* sensor measurements. For example, in cell c_3 in Figure 10.10, each of the two nonrelay nodes transmits the result of its computation to the relay node in c_3. No other transmission occurs in c_3 or in any other cell.

In round 2, the *relay node* of cell c_3 carries out a partial computation based on the values it received in the previous round, as well as its own sensor reading. The result is transmitted to its *relay parent* in cell c_2. Further, in round 2, the nonrelay nodes in cell c_3 again transmit computed function values to their relay node. No transmissions occur in other cells.

Consider the situation in round 3. Before round 3 begins, the cells that are one hop closer to the sink than the leaf-cells, like cell c_2, have received the results of some particular partial computations. Specifically, before round 3 begins, a relay parent in c_2 already has the result of a partial computation based on the measurements made by a subset of sensors below it. Hence, during round 3, the relay parent is in a position to include its own measurement, carry out the function evaluation, and transmit the result to the relay in c_2; at the end of round 3, we have the result of a partial computation based on the measurements of sensors constituting a *subtree* rooted at the relay parent in c_2. Further, during round 3, the nonrelay nodes in cell c_3 are again occupied in transmitting their results to the relay node in c_3. Similarly, in round 4, it is the relay node in c_2 that carries out a computation and transmits the result to *its* relay parent in c_1.

We can see that intermediate results progress toward the sink along the spanning tree on the cell graph in a *pipeline*. Nodes occurring lower in the tree deposit their computed results with a node that is higher up. In turn, the node that is higher up in the tree picks up the result, includes its own measurement, carries out a fresh partial computation, and passes the result upward. Moreover, as we go up the tree, partial computation and transmission are initiated *later*, once all the necessary inputs have arrived.

Our discussion is summarized in Figure 10.11, for an example cell graph of depth 2. It can be seen that several transmissions need to occur in the same round. How can we be sure that these transmissions can be arranged in time and space such that they do not interfere? In other words, how do we know that a *feasible schedule* exists?

The following observation is critical in showing the existence of a feasible schedule:

Each cell has a bounded number of *interfering cell-neighbors* (say k_2), where two cells c_1 and c_2 are interfering cell-neighbors if there exist a node in c_1 and a node in c_2 separated by a distance less than the bound imposed by the Protocol Model (see Chapter 9). This means that interfering cell-neighbors cannot support simultaneous transmissions.

According to the Protocol Model, if node i is transmitting to node j, then other transmitters that can *interfere* with successful reception at node j must be located within a disc of radius $(1 + \Delta)d_{i,j}$ around node j, where $d_{i,j}$ is the distance between nodes i and j. Here, Δ is a parameter used in the Protocol Model.

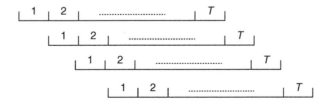

Figure 10.11 Sketch showing *T* rounds of computation and transmission at various nodes. The two rows at the top correspond to the activities of the nonrelay nodes and the relay node, respectively, at the leaf-cells of the cell graph. The two rows at the bottom refer to the actions of the nonrelay nodes and relay node, respectively, in a cell that is one hop closer to the sink. It may be noted that the rounds are *staggered* according the nodes' positions in the cell graph. Computation and transmission are pipelined.

As the number of nodes N increases, the communication range $r_c(N)$ decreases, and so the cells in the tessellation shrink, since the cell side is of length $\frac{r_c(N)}{\sqrt{2}}$. Therefore, interferers must be located within discs of smaller radii around the receivers. This argument leads to the conclusion that the number of interfering cell neighbors is *uniformly bounded*; the bound k_2 does not depend on N. With this, a graph coloring argument (see Problem 10.8) is used to show that there exists a schedule in which each cell receives 1 out of every $(1 + k_2)$ slots to transmit.

Hence, in an interval of T_1 slots constituting one round, each cell can be allotted $\frac{T_1}{1+k_2}$ slots. Each relay parent j in the cell requires at most $\log_2 |\mathcal{R}(f)|$ bits to communicate the result of the partial function computation based on data received from its children. Further, each cell in the cell graph has a uniformly bounded number of children, say k_3, and therefore, it follows that there are *at most* k_3 relay parents per cell.

Similarly, a relay node also requires at most $\log_2 |\mathcal{R}(f)|$ bits. A node that is neither a relay parent nor a relay node requires at most $\log_2 |\mathcal{X}|$ bits because it merely transmits its own readings.

To get an upper bound on the total number of bits that a cell needs to transmit, we need to bound the number of nodes in a cell. This is where the degree of the sensor node graph plays a role. Let us *assume* that the degree of the sensor node graph $d(\mathcal{G}(N, r_c(N)))$ is bounded above by

$$d(\mathcal{G}(N, r_c(N))) \leq k_1 \log_2 |\mathcal{R}(f)|$$

for some $k_1 > 0$. Clearly, then, the number of nodes in a single cell cannot be more than $k_1 \log_2 |\mathcal{R}(f)|$, because, by construction, all nodes in a cell are within communication range. Hence, the total number of bits that a cell needs to transmit per computation is bounded above by

$$k_3 \log_2 |\mathcal{R}(f)| + \log_2 |\mathcal{R}(f)| + \log_2 |\mathcal{X}| \times k_1 \log_2 |\mathcal{R}(f)|$$

$$= (k_3 + 1 + k_1 \log_2 |\mathcal{X}|) \log_2 |\mathcal{R}(f)|$$

Recalling that in the feasible schedule mentioned earlier, a cell gets $\frac{T_1}{1+k_2}$ slots in an interval of T_1 slots, and also that at most W bits can be sent in a slot, we can see that the transmissions can be feasibly scheduled if

$$\frac{T_1}{1 + k_2} = \frac{(k_3 + 1 + k_1 \log_2 |\mathcal{X}|) \log_2 |\mathcal{R}(f)|}{W}$$

We can now get an upper bound on the total time required for the function computation to be completed. Extending the idea depicted in Figure 10.11, it can be seen that for the scheme described, by time $(T + 2\delta_{\max})T_1$, the sink can complete

T computations (here, $\delta_{\max}$ is the maximum depth of the spanning tree on the cell graph). This implies that

$$R_{\max}^{(N)} \geq \lim_{T \to \infty} \left(\frac{T}{T + 2\delta_{\max}} \right) \frac{1}{T_1}$$

$$= \frac{1}{T_1}$$

$$= \frac{W}{(1 + k_2)(k_3 + 1 + k_1 \log_2 |\mathcal{X}|) \log_2 |\mathcal{R}(f)|}$$

Thus, we see that the maximum rate of function computation satisfies

$$\frac{W}{\{(1 + k_2)(k_3 + 1 + k_1 \log_2 |\mathcal{X}|)\} \log_2 |\mathcal{R}(f)|} \leq R_{\max}^{(N)} \leq \frac{W}{\log_2 |\mathcal{R}(f)|}$$

and hence

$$R_{\max}^{(N)} = \Theta \left(\frac{W}{\log_2 |\mathcal{R}(f)|} \right)$$

Discussion

We started by noting the communication range $r_c(N)$ that ensures, with high probability, that the graph $\mathcal{G}(N, r_c(N))$ is connected. Then we considered a divisible discrete-valued function $f(.)$, with $\mathcal{R}(f)$ denoting its range. What we found is that *if* the degree of the graph, $d(\mathcal{G}(N, r_c(N)))$, satisfies

$$d(\mathcal{G}(N, r_c(N))) \leq k_1 \log_2 |\mathcal{R}(f)|$$

then the maximum rate of function computation $R_{\max}^{(N)}$ satisfies

$$R_{\max}^{(N)} = \Theta \left(\frac{W}{\log_2 |\mathcal{R}(f)|} \right)$$

It is worth noting that the proof of the result is constructive, in that a scheme for function computation has been obtained. As the supportable bit rate W increases, it is expected that the rate of function computation will increase, because the network's communication capability has increased. Also, if the range of the function is larger, more bits will be required to specify its value at a particular argument, and hence the rate of computation would decrease. We see that both these aspects are captured in the expression $R_{\max}^{(N)} = \Theta \left(\frac{W}{\log_2 |\mathcal{R}(f)|} \right)$.

Consider the simple scenario where all measured data simply are uploaded to the sink. In this extreme case, there is no in-network processing at all. For this case, we can consider $f(X^{(t)}) = X^{(t)}$; $f(.)$ is just the *identity function*. Since each sensor measurement takes values in the discrete set $\mathcal{X}$, we have

$$\mathcal{R}(f) = \mathcal{X}^N$$

Thus, $\log_2 |\mathcal{R}(f)| = N \log_2 |\mathcal{X}|$, i.e., $\log_2 |\mathcal{R}(f)|$ is *linear* in N. Therefore, to apply the previous result, we need to check if the degree of the graph is bounded above by a linear function of N. But this is true for *any* connected graph. Hence, applying the result, we then conclude that there *is* a scheme that allows us to communicate $f(.)$, the identity function, at rate $O(\frac{1}{N})$. Thus, for large sensor networks, straightforward data uploading to the sink will lead to very low rates of extracting information.

10.6   Scheduling

Sensor nodes share the wireless medium. Therefore, they need a Medium Access Control (MAC) protocol to coordinate access. We discussed the IEEE 802.11 protocol in detail in Chapter 7. However, in a sensor network, energy efficient MACs are extremely important, and this forces us to look at MAC protocols closely.

Even when a sensor node is not transmitting but merely listening to the medium, significant energy is spent. This is because the electronic circuitry in the radio transceiver has to be kept on. Studies have shown that the ratio of energy spent in transmitting a packet to a receiver at unit distance, to that spent in receiving a packet and to that spent in *listening* for the same length of time, is 3:1.05:1. (Of course, when the receiver is far away, a transmitter would use more power and therefore, the energy spent would be more). The notable point here is that a sensor spends the same order of energy in simply listening on the medium as in actually receiving a packet.

Since saving energy is so important, we need to understand how energy can be *wasted*. The following causes can be discerned.

- **Idle listening:** If the medium is idle and yet a sensor node's radio transceiver is on, then it is spending energy unnecessarily.

- **Collision:** If transmissions collide, then all packets involved are garbled, leading to waste of energy all around.

- **Overhearing:** Overhearing occurs when a node receives a packet that is not addressed to it.

- **Control packet overhead:** From an application's point of view, energy spent in carrying information bits is energy usefully spent. It is desirable that the energy spent on control path activities, like channel reservation, acknowledgement, and route discovery, be as small as possible.

A good sensor MAC protocol leads to savings on all four of these fronts.

Sensor MAC protocols are significantly different from other wireless MAC protocols (like IEEE 802.11) because they can put the sensors into the *sleep state*. In this state, the radio transceiver is turned off completely. Nodes wake up periodically, listen on the medium for a short while, and then go back to sleep. This reduces idle listening drastically, and is a major reason for energy savings.

However, it is also immediate that as a result of the cycling between sleep and wake states, the latency in transferring information across nodes can be considerably increased. A transmitter has to hold on to the information it must send until it is certain that the receiver is ready and listening. Nevertheless, in many application scenarios, the increase in latency does not cause difficulties. For example, in an intrusion detection application, the speed at which the network transfers information is orders of magnitude higher than that of an intruder's movements, even when additional latency due to sleep-wake duty cycling is considered.

10.6.1 S-MAC

The protocol sensor-MAC (S-MAC) is one of the first to use the notion of sleep-wake duty cycles heavily. It aims to ensure a low duty cycle operation on the network. It introduces the notion of *coordinated sleeping*, in which clusters of nodes synchronize their sleep-schedules so that all of them sleep together. This ensures that when a node wakes up and wishes to transmit to a neighbor node in its cluster, the neighbor node will be awake to receive the transmission.

S-MAC reduces the energy wasted due to collisions by using the same approach as in IEEE 802.11, *viz.,* distributed channel reservation by RTS-CTS exchange (see Chapter 7). It is worth recalling that the exchange of RTS and CTS by the transmitter and receiver results in a silent neighborhood around each, allowing the transmission to complete successfully. This means that collisions involving long data packets are avoided at the small additional energy expense due to the short control packets.

Further, S-MAC utilizes the information available in the RTS and CTS packets to reduce overhearing by nodes. The RTS/CTS packet structure includes a duration field, which informs listeners of the interval for which the medium will be busy with the impending packet transfer. All nodes other than the transmitter and receiver can now afford to switch off their radios for this interval.

Finally, S-MAC reduces control packet overhead by resurrecting the old technique of *message passing*. Link layer frames normally have a maximum frame size, and a long message needs to be fragmented into pieces of the largest possible size. Now if each resulting fragment is transmitted as a separate entity, then each must be preceded by the RTS-CTS exchange. To reduce the control overhead, S-MAC proposed that the RTS-CTS exchange be carried out *only once* at the beginning, and the multiple fragments be sent in a burst, one after the other. The reservation interval indicated in the RTS-CTS packets corresponds to not just one fragment but the total time required to transmit *all* the fragments.

It is apparent that message passing allows one node to hog the channel and thereby cause unfairness in channel access opportunities among nodes. However, in a sensor network context, node-level unfairness over a *short time interval* is not a matter of concern. As mentioned before, a sensor network is not a collection of nodes that are interested in data transfer in a peer-to-peer fashion. Rather, the network has a single objective and all nodes collaborate toward achieving the same. However, over longer time intervals, we do need fairness because otherwise, distributed computation of functions can get held up.

S-MAC forms a flat, peer-to-peer topology. Thus, unlike clustering protocols, there is no cluster-head to coordinate channel access. We will see that some sensor MACs, like the IEEE 802.15.4 MAC, do require the presence of a coordinator. S-MAC also builds reliability into unicast data transfer by using explicit acknowledgements. Recall that we have seen the same idea before in the context of IEEE 802.11.

Because coordinated sleeping is so important in S-MAC, nodes need to exchange schedules before data transfer can begin. The *SYNC* packet is used for this purpose. The transmission time for a SYNC packet is called the *synchronization period*. Each node maintains a *schedule table* that stores the schedules of all its neighbors.

To choose a schedule, a node first listens for at least the synchronization period. If no SYNC packet is heard within this time, then the node chooses its own schedule and starts to follow it. It also broadcasts its schedule by transmitting its own SYNC packet.

If the node does receive a SYNC packet within the initial listen interval, then it sets its own schedule to the received one. Thus, synchronization with a neighbor is achieved. As before, it announces its schedule by transmitting its own SYNC packet later.

However, the following can also happen: After a node chooses and announces its own schedule, it receives a new and different schedule. What it does now depends on how many neighbors it had heard from. If the node had no neighbors, then it discards its original schedule and switches to the new schedule just received. If the node had one or more neighbors, it adopts *both* schedules, by waking up at the listen times of both. Such behavior typically is found among nodes that are located at the borders of two virtual clusters and facilitates communication between the two.

10.6.2　IEEE 802.15.4 (Zigbee)

The other sensor MAC protocol that has received wide attention is the IEEE 802.15.4 MAC. The protocol was introduced first in the context of Low-Rate Wireless Personal Area Networks (LR-WPANs). The PHY and MAC layers in LR-WPANs are defined by the IEEE 802.15.4 group, whereas the higher layers are defined by the Zigbee alliance.

IEEE 802.15.4 defines two types of devices: a *Full Function Device* (FFD) and a *Reduced Function Device* (RFD). The FFDs are capable of playing the role of a *network coordinator*, but RFDs are not. FFDs can talk to any other device, while RFDs can only talk to an FFD. Thus, one mode of operation of the IEEE 802.15.4 MAC is based on a hierarchy of nodes, with one FFD and several RFDs connected in a *star* topology (see Figure 10.12). The FFD at the hub, which is a network coordinator, plays the role of a cluster-head, and all communication is controlled by it. In the *peer-to-peer* topology, however, all nodes are equally capable; all are FFDs.

Figure 10.13 shows the superframe structure defined for IEEE 802.15.4. The superframe begins with a beacon. Nodes hearing the beacon can set their local

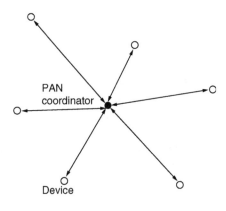

Figure 10.12 IEEE 802.15.4 nodes in a star topology.

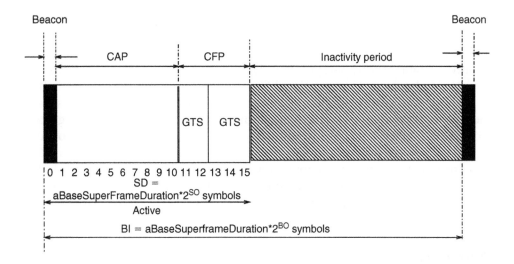

Figure 10.13 The Zigbee MAC superframe structure. CAP and CFP stand for Contention Access Period and Contention Free Period, respectively. GTS means Guaranteed Time Slot. The other parameters in the figure are defined in [61].

clocks appropriately, so that they go to sleep and wake up at the same time. This means *synchronized operation*.

The superframe is divided into an *active* and an *inactive* period. During the inactive period, nodes sleep. The active period consists of at most three parts—beacon transmission interval, the *Contention Access Period* (CAP) and an optional *Contention Free Period* (CFP). During the CAP, nodes contend using slotted CSMA/CA, as in IEEE 802.11 (see Chapter 7). In the CFP, a node can be allotted *Guaranteed Time Slots* (GTSs) by the network coordinator. Nodes request for GTS allocation by sending explicit GTS allocation *requests*. Transmitted frames are always followed by *Inter-Frame Spacings*.

10.7 Notes on the Literature

We began our discussion with a question about the transmission range to be used by sensor nodes such that the network is connected. As the number of nodes increases to infinity, we discussed a necessary and sufficient condition on the transmission range such that the probability that the network remains connected approaches 1. This result is due to Gupta and Kumar [49]. Our discussion of sensing coverage is based on the work of Zhang and Hou [143]. The notion of crossings that we used in analyzing the sensing coverage problem appeared in Wang et al. [134]. Our initial discussion of localization follows Nagpal et al. [106]; the convex position estimation approach follows Doherty et al. [27]. A detailed discussion of Linear Matrix Inequalities, with many applications, can be found in Boyd et al. [16].

The first geographic routing algorithms were greedy, as in Takagi and Kleinrock [124]. West [135] provides a very readable introduction to graph theory and associated algorithms. The notions of Unit Disk Graph, Gabriel Graph, Relative Neighborhood Graph, and Delaunay Triangulation are extensively discussed in books on Computational Geometry, for example, the one by de Berg et al. [24]. The survey article by Jaromczyk and Toussaint [64] also discusses these ideas. Face routing was introduced for the first time by Kranakis, Singh, and Urrutia [82], albeit under a different name. Several refinements are due to Kuhn et al. [83]. Directed Diffusion was proposed by Intanagonwiwat et al. in [62]. Our treatment of function computation follows Giridhar and Kumar [41]. The S-MAC protocol, due to Ye et al., appeared in [139]. Our brief discussion of the IEEE 802.15.4 standard uses [61].

Problems

10.1 Sensors are deployed according to a Poisson process of rate λ. Assume that each sensor covers a circular area of radius 1 unit. Consider two points x and y. Let C_x be the indicator variable that x is covered by at least one sensor. Find $\mathsf{E}(C_x C_y)$ and $\mathsf{E}((1 - C_x)(1 - C_y))$.

10.2 Sensors are deployed according to a Poisson process of rate λ. Assume that each sensor covers a circular area of radius 1 unit. Now consider an arbitrary straight line path through the sensor field. Find the distribution of the segment of the line covered by a sensor.

10.3 Consider a one-dimensional sensor field in which the sensors are deployed according to a Poisson process on a straight line. Each sensor covers a random segment whose mean is $\overline{X}$. What is the probability that a point is covered by exactly k sensors?

10.4 Consider the sensing coverage problem discussed in Section 10.2. Suppose that the coverage degree of a patch is k. Is the patch an open set or a closed set in $\mathbb{R}^2$?

10.5 Consider the first sensor localization method discussed in Section 10.3. Is it possible that the true distance of sensor j from anchor i, viz., $d_{j,i}$ is equal to the estimated distance $\hat{d}_{j,i}$, for $i = 1, 2, \ldots, M$, but the position of sensor j, as concluded by the algorithm, is incorrect?

10.6 Consider N nodes connected according to a complete graph; that is, all nodes can decode every other node's transmission. All communication channels are independent binary symmetric channels with error probability $0 \leq q \leq 0.5$. Each node has a bit and it is required that all the nodes know the parity of the N bits. Each node transmits its bit M times and every receiver uses the majority rule to decode; that is, a bit is decoded as 1 if more that $M/2$ transmissions are decoded as 1. Find $P_c(N, M)$, the probability that **all** the N nodes have the correct parity. Investigate the asymptotics of $P_c(N, M)$ when $M = \log_2 N$ and $N \to \infty$.

10.7 Consider N nodes connected according to a complete graph; that is, all nodes can decode every other node's transmission. Assume that the Aloha MAC protocol is used and each node transmits in a slot with probability p. Each node has a bit and it is required that all the nodes know the parity of the N bits. Assume that there is no channel error. If p is to be fixed, find the expectation and variance of the time at which all the nodes have transmitted once and all know the parity. If p can be dynamic, i.e., it can be changed in possibly every slot, design an algorithm that will minimize the expected time to complete the computation. Find the corresponding variance.

10.8 Consider the function computation problem discussed in Section 10.5. For a specific choice of $r_c(N)$, consider the cells of the cell graph as nodes of a new graph. Define two cells to be adjacent—having an edge between them—if they are interfering neighbors. The degree of such a graph is clearly $\leq k_2$, from Section 10.5. Our objective is to *color* the vertices of this graph using the *minimum* number of colors, subject to the constraint that no two adjacent vertices have the same color. Show that the minimum number of colors is $\leq (1 + k_2)$.

Note: After the coloring is done, nodes having the same color correspond to cells that can support simultaneous transmissions. Therefore, the result above says that within an interval of length $(1 + k_2)$ slots, *every* cell gets a transmission opportunity.

Appendices

Appendices

APPENDIX A

Notation and Terminology

In this appendix we have collected together, for ready reference, various notation and mathematical terminology. In some instances, we also briefly explain the concepts related to the notation.

A.1 Miscellaneous Operators and Mathematical Notation

$\mathbb{R}$	The set of real numbers
$\mathbb{Z}^+$	The set of nonnegative integers
x^+	For a real number x, $x^+ = \max\{0, x\}$
t_-, t_+	For t a point in time, t_- is interpreted as "just before t" and t_+ is interpreted as "just after t"; more formally, for $f(t)$ some function of time $f(t_-) = \lim_{\epsilon \downarrow 0} f(t - \epsilon)$, and $f(t_+) = \lim_{\epsilon \downarrow 0} f(t + \epsilon)$
t_{k-}, t_{k+}	The same interpretation as t_-, t_+ but for the indexed time instant t_k
$x \approx y$	Is read as x is approximately equal to y
$A \backslash B$	For sets A and B, $A \backslash B$ is the set difference, i.e., $A \cap B^c$
$|A|$	For a set A, $|A|$ is the *cardinality* of, or the number of element in A
I_A	Is the *indicator function* of the set A; if A is a subset of elements of Ω, whose elements are generically labelled by ω, then I_A is a function from Ω to $\{0, 1\}$, with $I_A(\omega) = 1$ if $\omega \in A$, and $I_A(\omega) = 0$ otherwise.

A.2 Vectors and Matrices

An element of (or, a *vector* in) $\mathbb{R}^n$ is denoted by a boldface lowercase symbol (e.g., **x**). Vectors are viewed as *column vectors*, unless stated otherwise. Matrices are denoted by bold upper case symbols (e.g., **A**). The element in row i and column j of the matrix **A** is denoted by $a_{i,j}$; the ith row of the matrix **A** is denoted by $\mathbf{A}_{i,\cdot}$, and is viewed as a row vector, by default; the jth column of the matrix **A** is denoted by $\mathbf{A}_{\cdot,j}$ and is viewed as a column vector. The transpose of a vector **x** is denoted by $\mathbf{x}^T$, and the transpose of the matrix **A** is denoted by $\mathbf{A}^T$.

A.3 Asymptotics: The *O*, *o*, and $\sim$ Notation

We often are interested in expressing compactly the behavior of a function $f(x)$, $x \in \mathbb{R}$, or $f(n), n \geq 1$, for large values of the argument, or for the argument going to zero. Some standard notation has been developed for this. A comprehensive reference for this material is the book on asymptotics by De Bruijn [25]; an extensive discussion on the subtleties of this notation is provided in [46].

The O, Ω, and Θ Notation:

We write

$$f(x) = O(g(x)) \text{ as } x \to \infty$$

if there is a positive number a, such that $|f(x)| \leq a|g(x)|$ for all large enough x. The same statement holds for $f(n) = O(g(n))$. We write $f(n) = \Omega(g(n))$ if and only if $g(n) = O(f(n))$, and we write $f(n) = \Theta(g(n))$ if and only if $f(x) = O(g(x))$ and $f(n) = \Omega(g(n))$.

Thus, for example, if $f(x) = x^9 + e^x$ then $f(x) = e^x(x^9 e^{-x} + 1)$, and hence there is an x_0 such that for all $x > x_0$, $f(x) \leq 2e^x$. Thus we can write $f(x) = O(e^x)$ as $x \to \infty$, which is to say that for large x, $f(x)$ grows exponentially. As another example, $f(n) = an^2 + bn^9 = O(n^9)$ as $n \to \infty$. In this example, if $b > 0$, it can also be seen that $f(n) = \Omega(n^9)$ and, therefore, $f(n) = \Theta(n^9)$.

Similarly, we write

$$f(x) = O(g(x)) \text{ as } x \to 0$$

to mean that $|f(x)| \leq a|g(x)|$ for all x close enough to zero. Thus, for example, if $f(x) = \ln(x + 1)$ then we can write $f(x) = O(x)$ as $x \to 0$. This follows from the Taylor's series expansion of $\ln(x + 1)$, and says that close to the origin $f(x)$ is linear.

The o ("little oh") Notation:

We write

$$f(x) = o(g(x)) \text{ as } x \to \infty$$

to mean $\lim_{x \to \infty} \frac{f(x)}{g(x)} = 0$, and similarly, we write

$$f(x) = o(g(x)) \text{ as } x \to 0$$

to mean $\lim_{x \to 0} \frac{f(x)}{g(x)} = 0$. For example, $f(n) = an^3 + bn^9 = o(n^{10})$ as $n \to \infty$, and $f(x) = \ln(x + 1) - x = o(x)$ as $x \to 0$; the error between $\ln(x + 1)$ and its approximation x, near 0, decreases strictly faster than x does near 0.

Asymptotic Equivalence; $\sim$ Notation:

We write

$$f(x) \sim g(x) \text{ as } x \to \infty$$

to mean $\lim_{x \to \infty} \frac{f(x)}{g(x)} = 1$, and similarly, we write

$$f(x) \sim g(x) \text{ as } x \to 0$$

to mean that $\lim_{x \to 0} \frac{f(x)}{g(x)} = 1$. For example, $f(n) = an^3 + bn^9 \sim n^9$ as $n \to \infty$, and $f(x) = \ln(x+1) \sim x$ as $x \to 0$. Note that the latter can be seen by writing $\ln(x+1) = x + o(x)$ as $x \to 0$. Divide the right-hand side by x, and use the definition of $o(x)$ to obtain the result.

A.4 Probability

i.i.d. A sequence of random variables, $X_n, n \in \{0, 1, 2, \ldots\}$, that are mutually independent and identically distributed is said to be independent and identically distributed, abbreviated as i.i.d.

⊔ A binary operator denoting statistical independence. So if A and B are random variables, then $A \sqcup B$ is to be read as A is independent of B.

$\overset{dist}{=}$ If two random vectors $\mathbf{X}$ and $\mathbf{Y}$ have the same joint distributions we write $\mathbf{X} \overset{dist}{=} \mathbf{Y}$.

w.p. 1 or a.s. With probability 1 or almost surely; thus, for example, saying that a random variable is nonnegative almost surely, or with probability 1, means $\Pr(X \geq 0) = 1$.

marginal If $(X_1, X_2, \ldots, X_n)$ is a random vector with some joint distribution, then the distribution of any of the component random variables $X_i, 1 \leq i \leq n$, is called a marginal distribution. The term also applies to random processes, and if, for example, the random process is stationary, then we can refer to *the* marginal distribution of the process.

A Review of Some Mathematical Concepts

B.1 Limits of Real Number Sequences

Limit and Limit Points

A sequence of real numbers $x_n, n \geq 1$, is said to converge to the *limit* $x \in \mathbb{R}$ if for every $\epsilon > 0$, there exists an n_ϵ such that for all $n > n_\epsilon$, $|x_n - x| < \epsilon$; i.e., no matter how small an $\epsilon > 0$ we take, there is a point in the sequence (denoted by n_ϵ) such that all elements of the sequence after this point are within ϵ of x (the proposed limit). A limit, if it exists, is clearly unique. This is written as

$$\lim_{n \to \infty} x_n = x$$

If $x_k, k \geq 1$, viewed as a set, is bounded above, and is such that $x_k \leq x_{k+1}$ (x_k is a nondecreasing sequence), then $\lim_{n \to \infty} x_n$ exists and is, in fact, the sup of the set of numbers $x_k, k \geq 1$. The corresponding result holds if the sequence is nonincreasing and bounded below.

   If the sequence $x_k, k \geq 1$, is bounded above and below, then define the sequence $a_k, k \geq 1$, as follows:

$$a_k = \inf\{x_n : n \geq k\}$$

Then each a_k exists, and is a bounded nondecreasing sequence (each bound of $x_k, k \geq 1$, is also a bound for $a_k, k \geq 1$), and hence $\lim_{k \to \infty} a_k$ exists. This limit, say a, is called the lim inf of the sequence $x_k, k \geq 1$, and we write

$$\liminf_{n \to \infty} x_n = a$$

A corresponding discussion holds for the sequence

$$b_k = \sup\{x_n : n \geq k\}$$

Here $b_k, k \geq 1$, is a nonincreasing sequence, bounded below, and hence it converges. Then $b = \lim_{k \to \infty} a_k$ is called the lim sup of the sequence $x_k, k \geq 1$, and we write

$$\limsup_{n \to \infty} x_n = b$$

We say that $x \in \mathbb{R}$ is a *limit point* of the sequence $x_k, k \geq 1$, if for all $\epsilon > 0$, and for every n, there exists an $m_{n,\epsilon} > n$, such that $|x_{m_{n,\epsilon}} - x| < \epsilon$. Thus for every $\epsilon > 0$ (no matter how small), the sequence comes within ϵ of x *infinitely often*; for if it did not, there would be some n at which x_n comes within ϵ of x for the last time, and there would be no $m_{n,\epsilon} > n$, with $|x_{m_{n,\epsilon}} - x| < \epsilon$. It can be seen that if the lim sup and the lim inf of a sequence are equal then there is only one limit point, which we call the limit of $x_n, n \geq 1$.

B.2   A Fixed Point Theorem

A function mapping a set C ($\subset \mathbb{R}^n$) into C (i.e., $f : C \rightarrow C$) is said to have a *fixed point* at $x \in C$ if $f(\mathbf{x}) = \mathbf{x}$. Given a function $f : C \rightarrow C$, the problem of determining if there exists an $\mathbf{x} \in C$ such that $f(\mathbf{x}) = \mathbf{x}$ is called a *fixed point problem*. When faced with such a problem we need to ask questions such as: Does a fixed point exist? If one exists is it unique?

Before we can state an important theorem we need to understand some elementary·concepts from real analysis. We say that a set $C \in \mathbb{R}^n$ is *closed*, if every convergent sequence $\mathbf{x}_n, n \geq 1$, of points in C has its limit point also in C. Thus for example the set $C = (0, 1]$ is not closed, since the limit of the sequence $\frac{1}{n}, n \geq 1$, (i.e., 0) is not in C. We say that a set $C \in \mathbb{R}^n$ is bounded if there is an n-dimensional ball centered at the origin and of radius large enough, but finite, such that C is entirely inside that ball.

We say that a set $C \in \mathbb{R}^n$ is *convex* if whenever $\mathbf{x}_1 \in C$, and $\mathbf{x}_2 \in C$, then for every $\lambda, 0 < \lambda < 1, \lambda \mathbf{x}_1 + (1 - \lambda)\mathbf{x}_2 \in C$; the entire line segment joining $\mathbf{x}_1$ and $\mathbf{x}_2$ is in C. Thus $C = [0, 1] \cup [2, 3]$ is a closed but nonconvex set.

Theorem B.1   Brouwer's Fixed Point Theorem

Let $C \subset \mathbb{R}^n$ be a closed, bounded, and convex set. Then a continuous function $f : C \rightarrow C$ has a fixed point in C.   ■

Figure B.1 illustrates the theorem for $n = 1$ and $C = [0, 1]$.

B.3   Probability and Random Processes
B.3.1   Useful Inequalities and Bounds

The Union Bound:
Consider the events, $A_i, i = 1, \ldots, n$. The union bound, also called Boole's inequality, on the probability of $\cup_{i=1}^n A_i$ is given by

$$\Pr\left(\cup_{i=1}^n A_i\right) \leq \sum_{i=1}^n \Pr(A_i)$$

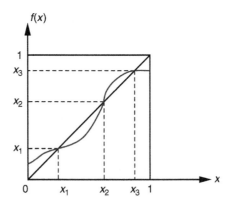

Figure B.1 Illustration of Brouwer's Fixed Point Theorem in one dimension. x_1, x_2, and x_3 are the fixed points.

A more general bounding method is as follows. Define

$$B_1 = \sum_{i=1}^{n} \Pr(A_i)$$

$$B_2 = \sum_{1 \le i < j \le n} \Pr(A_i \cap A_j)$$

$$B_k = \sum \Pr(A_{i_1} \cap A_{i_2} \dots \cap A_{i_k})$$

$i_1, \dots i_k$ are k distinct indices, $1 \le i_1, \dots, i_k \le n$, and the summation is over all combinations of these distinct integers. Then the *inclusion-exclusion principle* says that

$$\Pr(\cup_{i=1}^{n} A_i) \le \sum_{i=1}^{k} (-1)^{i+1} B_i \quad \text{for odd } k$$

$$\Pr(\cup_{i=1}^{n} A_i) \ge \sum_{i=1}^{k} (-1)^{i+1} B_i \quad \text{for even } k$$

These are also called the Bonferroni inequalities.

Jensen's Inequality:
For a convex function $f(x)$, and a random variable X with finite expectation,

$$E(f(X)) \ge f(E(X))$$

with equality if $f(\cdot)$ is a linear function. Equality will also occur for a general convex function if X is constant with probability 1.

Chernoff's Bound:
For a random variable X, if the moment generating function $E(e^{\theta X})$ exists for some $\theta \geq 0$, then

$$\Pr(X \geq 0) \leq E\left(e^{\theta X}\right)$$

This is a simple consequence of the fact that $I_{\{X \geq 0\}} \leq e^{\theta X}$ for every $\theta \geq 0$. Taking expectation on both sides yields Chernoff's bound.

Hölder's Inequality:
For $p > 1$ and $\frac{1}{p} + \frac{1}{q} = 1$,

$$E(|XY|) \leq \left(E(|X|^p)\right)^{\frac{1}{p}} \left(E(|Y|)^q\right)^{\frac{1}{q}}$$

As an example, take $p = q = 2$, and Hölder's inequality yields

$$\left(E(|XY|)\right)^2 \leq E\left(|X|^2\right) E\left(|Y|^2\right)$$

which is called Schwarz's inequality.

B.3.2 Convergence Concepts

Consider a sequence of real valued random variables $X_n, n \geq 1$. For each sample path ω, we obtain the real number sequence $X_n(\omega), n \geq 1$, which may or may not converge. We would like to talk about the convergence behavior of the sequence of random variables without necessarily requiring convergence for each sample path. Several useful notions of convergence of a sequence of random variables have been defined.

Convergence in Probability
The sequence of random variables $X_n, n \geq 1$, is said to converge *in probability* to a random variable X, if for each $\epsilon > 0$,

$$\lim_{n \to \infty} \Pr(|X_n - X| > \epsilon) = 0$$

which says that the probability that X_n differs from X by more than ϵ converges to 0 as $n \to \infty$. This is written as $X_n \xrightarrow{p} X$.

Convergence with Probability One
The sequence of random variables $X_n, n \geq 1$, is said to converge *almost surely* or *with probability one* to a random variable X, if

$$\Pr\left(\lim_{n \to \infty} X_n = X\right) = 1$$

or, more explicitly, $\Pr(\{\omega : \lim_{n\to\infty} X_n(\omega) = X(\omega)\}) = 1$. This says that the set of sample paths along which $X_n(\omega), n \geq 1$, converges to $X(\omega)$ has probability 1. This is denoted as $X_n \overset{a.s.}{\to} X$, or as $X_n \overset{w.p.1}{\to} X$.

Convergence in Distribution

A sequence of random variables $X_n, n \geq 1$, with distributions $F_n(\cdot), n \geq 1$, is said to converge in *distribution* to the random variable X, with distribution $F(\cdot)$ if

$$\lim_{n\to\infty} F_n(x) = F(x)$$

whenever x is not a point of discontinuity (i.e., a "jump" point) of $F(\cdot)$. This is denoted as $X_n \overset{dist}{\to} X$.

B.3.3 The Borel-Cantelli Lemma

The Borel-Cantelli lemma is a useful tool in proving almost sure convergence. Let $A_n, n \geq 1$, be a sequence of events. The event $\cap_{m=1}^{\infty} \cup_{n=m}^{\infty} A_n$ is called $\{A_n\}$ *infinitely often* or $\{A_n\}i.o.$, because an ω is in this event if and only if it belongs to an infinite number of events in the sequence $\{A_n\}$. Notice that the complement of the event $\{A_n\}i.o.$ comprises those ωs that belong to the events in the sequence $\{A_n\}$ only for finitely many n.

Lemma B.1   Borel-Cantelli

$$\sum_{n=1}^{\infty} \Pr(A_n) < \infty \Rightarrow \Pr(\{A_n\}i.o.) = 0$$

$\blacksquare$

There is also a complementary result to the Borel-Cantelli Lemma for the case when $\{A_n\}$ is a sequence of independent events. In such a case

$$\sum_{n=1}^{\infty} \Pr(A_n) = \infty \quad \Rightarrow \quad \Pr(\{A_n\}i.o.) = 1$$

B.3.4 Laws of Large Numbers and Central Limit Theorem

Theorem B.2   Weak Law of Large Numbers

$X_n, n \geq 1$, is a sequence of identically distributed uncorrelated random variables with finite mean μ and finite variance then

$$\lim_{n\to\infty} \frac{1}{n} \sum_{k=1}^{n} X_k \overset{p}{\to} \mu$$

$\blacksquare$

Theorem B.3   Kolmogorov's Strong Law of Large Numbers

$X_n, n \geq 1$, is a sequence of i.i.d. random variables with finite mean μ then

$$\lim_{n \to \infty} \frac{1}{n} \sum_{k=1}^{n} X_k \overset{a.s.}{\to} \mu$$

∎

Theorem B.4   Central Limit Theorem

$X_n, n \geq 1$, is a sequence of i.i.d. random variables with finite mean μ and finite variance σ^2, then

$$\frac{1}{\sigma \sqrt{n}} \left(\sum_{k=1}^{n} (X_k - \mu) \right) \overset{dist}{\to} \Phi$$

where Φ is the normal or Gaussian distribution with mean 0 and variance 1 (also called the standard normal distribution).   ∎

B.3.5   Stationarity and Ergodicity

Strict Stationarity

A random process $X_n, n \geq 0$, is said to be *strictly stationary* or just *stationary* if for all $k \geq 1$, and indices $n_1, n_2, \ldots, n_k$, and m,

$$\left(X_{n_1}, X_{n_2}, \ldots, X_{n_k} \right) \overset{dist}{=} \left(X_{n_1+m}, X_{n_2+m}, \ldots, X_{n_k+m} \right)$$

where $\overset{dist}{=}$ denotes equality in (joint) distribution. Thus if we take any subset of random variables of the process and then shift this subset in time by any amount, then the joint distribution of the constituent random variables is unchanged.

Stationary Increments

Consider a Poisson process $A(t), t \geq 0$, where $A(t)$ is the number of points in the interval $(0, t]$. Then obviously, $A(t)$ is nondecreasing, and hence cannot be stationary. However, consider time points $t_1, t_2, \ldots, t_n$, and look at the *increments* of the process $A(t)$, $(A(t_2) - A(t_1), A(t_3) - A(t_2), \ldots, A(t_n) - A(t_{n-1}))$; this is the random vector of the number of arrivals over the intervals $(t_1, t_2], (t_2, t_3], \ldots, (t_{n-1}, t_n]$. Let us now shift all these intervals by some amount τ, yielding the increments $(A(t_2 + \tau) - A(t_1 + \tau), A(t_3 + \tau) - A(t_2 + \tau), \ldots, A(t_n + \tau) - A(t_{n-1} + \tau))$. If these two random vectors have the same distribution for any choice of n, $t_1, t_2, \ldots, t_n$, and τ, then we say that $A(t), t \geq 0$, has stationary increments. The Poisson process has stationary increments. The term can, of course, be applied to discrete time processes as well. In fact, in addition, nonoverlapping increments of a Poisson process are also independent; the random variables $(A(t_2) - A(t_1), A(t_3) - A(t_2), \ldots, A(t_n) - A(t_{n-1}))$

are mutually independent, for any choice of $t_1 \leq t_2 \leq t_3 \leq \cdots \leq t_n$. Thus a Poisson process is said to have *stationary and independent increments*.

Ergodicity and the Ergodic Theorem

Consider a random process $X_n, n \geq 0$. Let us ask a question about this process, whose answer does not depend on whether we ask the question about $X_n, n \geq 0$, or about $X_{n+k}, n \geq 0$, for any $k \geq 1$. Each such question yields an event on which the answer is a "yes" and its complement on which the answer is a "no." For example, consider the question, "Does $X_n, n \geq 0$, converge?" Then, the event $\{\omega : X_n(\omega) \text{ converges}\}$ is the same as the event $\{\omega : X_{n+k}(\omega) \text{ converges}\}$ for any $k \geq 1$, since whether or not a sequence converges does not depend on any finite shift of the sequence (i.e., does not depend on the time origin from which we start looking at the sequence). Such events are called *invariant* events. We say that $X_n, n \geq 0$, is *ergodic* if each such event has probability 0 or 1.

The following is a generalization of the strong law of large numbers to stationary and ergodic processes.

Theorem B.5 Birkhoff's Strong Ergodic Theorem

$X_n, n \geq 0$, is a stationary and ergodic process, and $f(\cdot)$ is a function that maps realizations of the process (i.e., $X_n(\omega), n \geq 0$) to $\mathbb{R}$, such that $\mathsf{E}(|f(X_n, n \geq 0)|) < \infty$. Then, with probability one,

$$\lim_{n \to \infty} \frac{1}{n} \sum_{k=0}^{n-1} f(X_{n+k}, n \geq 0) = \mathsf{E}(f(X_n, n \geq 0)) \qquad \blacksquare$$

Note that the notation $f(X_{n+k}, n \geq 0)$ means "f evaluated at $X_{n+k}(\omega), n \geq 0$, i.e., a left shift of $X_n(\omega), n \geq 0$ by k steps." Thus, for each n the term $\frac{1}{n} \sum_{k=0}^{n-1} f(X_{n+k}, n \geq 0)$ is a random variable. The theorem states that this sequence of random variables converges with probability one to a constant.

B.4 Notes on the Literature

Hoffman's book on real analysis [55] provides the rigorous theory but also a lot of intuition is developed in the discussion. The book [108], by Papoulis, is a popular engineering textbook on probability, random variables, and random processes. Bremaud's book [17] is a rigorous but highly accessible book on probability theory.

APPENDIX C

Convex Optimization

We follow the notation and approach of the text by Bazaraa et al. [6].

C.1 Convexity
Definition C.1

A set $\mathcal{X} \subset \mathbb{R}^n$ is said to be *convex* if for any $x_1, x_2 \in \mathcal{X}$ it holds that $\lambda x_1 + (1 - \lambda)x_2 \in \mathcal{X}$, for every $\lambda \in [0, 1]$. ∎

Thus a set of real vectors is convex if the entire line segment joining any pair of elements of the set lies entirely within the set.

Definition C.2

$\mathcal{X}$ is a convex set in $\mathbb{R}^n$. A function $f : \mathcal{X} \to \mathbb{R}$ is said to be *convex* (resp. *concave*) if for $x_1, x_2 \in \mathcal{X}$, we have $f(\lambda x_1 + (1 - \lambda)x_2) \leq$ (resp. $\geq$) $\lambda f(x_1) + (1 - \lambda)f(x_2)$, for every $\lambda \in [0, 1]$. The function is said to be *strictly convex* or *strictly concave* if the inequality is strict for distinct x_1 and x_2 and $\lambda \in (0, 1)$. ∎

If $f : \mathcal{X} \to \mathbb{R}$ is convex over $\mathcal{X}$ then $-f$ is concave over $\mathcal{X}$.

Theorem C.1

A function $f : \mathbb{R} \to \mathbb{R}$ is convex, if and only if, for every $a < b < c$, the following holds:

$$\frac{f(b) - f(a)}{b - a} \leq \frac{f(c) - f(a)}{c - a}$$

∎

C.2 Local and Global Optima
Definition C.3

For the problem of minimizing a function $f : \mathcal{X} \to \mathbb{R}$, an element $x^* \in \mathcal{X}$ is said to be a *global optimal* solution, or just a *solution*, if $f(x^*) \leq f(x)$, for all $x \in \mathcal{X}$. An element $\hat{x} \in \mathcal{X}$ is said to be locally optimal if for some $\epsilon > 0$, $f(\hat{x}) \leq f(x)$, for all $x \in \{x \in \mathcal{X} : \|x - \hat{x}\| < \epsilon\}$. ∎

Observe that a global optimal solution need not be unique, and must be locally optimal. A corresponding definition, obviously, holds for the maximization of f over $\mathcal{X}$.

Theorem C.2

Given $f : \mathcal{X} \to \mathbb{R}$, such that $\mathcal{X}$ is a convex set, and f is a convex function over $\mathcal{X}$, a local minimum of f over $\mathcal{X}$ is also a global minimum. If f is strictly convex over $\mathcal{X}$ then a local optimum is the unique global optimum.   ■

C.3 The Karush-Kuhn-Tucker Conditions

We now turn to an important *sufficient* condition for the optimality of a point $\mathbf{x}^* \in \mathcal{X} \subset \mathbb{R}^n$ for the following *primal* problem.

Primal Problem

$$\min f(\mathbf{x})$$

subject to

$$g_i(\mathbf{x}) \leq 0, \qquad 1 \leq i \leq m$$

$$\mathbf{x} \geq 0 \tag{C.1}$$

We will limit our presentation here to the special situation in which $f : \mathbb{R}^n \to \mathbb{R}$ and, for $1 \leq i \leq m$, $g_i : \mathbb{R}^n \to \mathbb{R}$, *are all convex and differentiable functions* over $\mathbb{R}^n$. It can then be shown that this is a problem of minimizing a convex function over a convex constraint set.

Theorem C.3   Karush-Kuhn-Tucker

Given $\mathbf{x}^* \in \mathbb{R}^n$, if there exists $\lambda \in \mathbb{R}^m$, with $\lambda \geq 0$, such that

$$\nabla f(\mathbf{x}^*) + \sum_{i=1}^{m} \lambda_i \nabla g_i(\mathbf{x}^*) = 0 \tag{C.2}$$

and

$$\sum_{i=1}^{m} \lambda_i g_i(\mathbf{x}^*) = 0 \tag{C.3}$$

then x^* is a global optimal solution for the Primal Problem.   ■

The sufficient conditions stated in this theorem are called the Karush-Kuhn-Tucker (KKT) conditions. If $\mathbf{x}^*$ satisfies the KKT conditions then it is called a *KKT point*. The elements of the vector λ are called *Lagrange multipliers* or *dual variables*; the latter term will become clear later in this section when we discuss duality. There is one dual variable for every constraint $g_i(\mathbf{x}) \leq 0$. The condition stated in Equation C.3 is called a *complementary slackness* condition; notice that it implies that if the primal constraint $g_i(\mathbf{x}) \leq 0$ is met with a strict inequality at $\mathbf{x}^*$ (i.e., there is a *slack* in the ith constraint) then (since the vector λ is nonnegative) the corresponding $\lambda_i = 0$. If at point $\mathbf{x}^*$ a constraint is met with equality then that constraint is said to be *binding* or *active* at that point. Thus, owing to the complementary slackness condition, we see that in the condition of Equation C.2 only the active constraints will appear.

In general, the KKT conditions are *not necessary*; a point $\mathbf{x} \in \mathbb{R}^n$ can be optimal for the Primal Problem C.1, and yet it may not satisfy the KKT conditions. The following theorem provides two simple conditions under which the KKT conditions are necessary and sufficient for the Primal Problem C.1. Recall that we have limited our discussion to convex and differentiable $f(\mathbf{x})$ and $g_i(\mathbf{x}), 1 \leq i \leq m$.

Theorem C.4   Necessity and Sufficiency of KKT Conditions

a. If $\mathbf{x}^*$ is such that there exists an $\mathbf{x} \in \mathbb{R}^n$ with $g_i(\mathbf{x}) < 0$ if i is active at $\mathbf{x}^*$, then $\mathbf{x}^*$ is optimal if and only if the KKT conditions hold at $\mathbf{x}^*$.

b. *Linear Constraints:* If the constraints are linear, i.e., there are vectors $\mathbf{a}_i \in \mathbb{R}^n$ such that $g_i(\mathbf{x}) = \mathbf{a}_i^T \mathbf{x}$, for $1 \leq i \leq n$, then $\mathbf{x}^*$ is optimal if and only if the KKT conditions hold at $\mathbf{x}^*$. ∎

C.4 Duality

For the general Primal Problem of Equation C.1, define for $\lambda \in \mathbb{R}^m, \lambda \geq 0$,

$$\Theta(\lambda) = \inf\left\{\mathbf{x} \geq 0 : f(\mathbf{x}) + \sum_{i=1}^{m} \lambda_i g_i(\mathbf{x})\right\} \tag{C.4}$$

$\Theta(\lambda)$ is called the *Lagrangian dual function* and is obtained by *relaxing* the constraints $g_i(\mathbf{x}) \leq 0, 1 \leq i \leq m$. The following can now be shown

a. $\Theta(\lambda)$ is a concave function on $\lambda \in \mathbb{R}^m, \lambda \geq 0$.

b. If $\lambda \geq 0$, and $\mathbf{x}$ satisfies the constraints of the Primal Problem, then $\Theta(\lambda) \leq f(\mathbf{x})$.

Now let us consider the following optimization problem.

Dual Problem

$$\max_{\lambda \geq 0} \Theta(\lambda) \tag{C.5}$$

It can be seen, from the observations enumerated above, that the solution to the Dual Problem will lower bound the solution to the primal problem. Under convexity, which we have been assuming throughout this section, more is true, as is stated by the following theorem. Again, we are stating the special case that suffices for our purposes.

Theorem C.5   Strong Duality

Let the Primal and Dual Problems be as defined in Equations C.1 and C.5, respectively. Suppose there exists $x \in \mathbb{R}^n$ such that $g_i(x) < 0, 1 \leq i \leq m$. Then the primal and dual problems have the same optimum values. If the optimum value is finite, and if x^* and λ^* are solutions to the Primal Problem and Dual Problem then $\sum_{i=1}^{m} \lambda_i^* g_i(x^*) = 0$.   ∎

APPENDIX D

Discrete Event Random Processes

D.1 Stability Analysis of Discrete Time Markov Chains (DTMCs)

In analyzing queuing models, one of the first questions one asks is whether the system is "stable." When a queuing system is modeled by a (time homogeneous) Markov chain then one of the criteria for stability is that the Markov chain characterizing the queuing system be positive recurrent. This at least ensures that there is a steady state distribution of the associated processes. One method for determining whether an irreducible DTMC (with transition probability matrix $\mathbf{P}$) is positive recurrent is to look for a positive solution for the system of equations $\pi = \pi \mathbf{P}, \sum_{i \in S} \pi_i = 1$. Although this can be done in many simple cases, in general this approach is intractable. Fortunately, there is another approach based on the technique of drift analysis of a suitable *Lyapunov function*. The following are the main theorems that provide sufficient conditions for an irreducible DTMC to be recurrent, transient, or positive. The state space S is countable and is viewed as $S = \{1, 2, 3, \ldots\}$. In the following three theorems the function $f(\cdot)$ often is called a Lyapunov function.

Theorem D.1
An irreducible DTMC $X_n, n \geq 0$, is **recurrent** if there exists a nonnegative function $f(j), j \in S$, such that $f(j) \to \infty$ as $j \to \infty$, and a *finite* set $A \subset S$, such that, for all $i \notin A$, $\mathsf{E}\big(f(X_{n+1})|X_n = i\big) - f(i) \leq 0$. ∎

Theorem D.2
An irreducible DTMC $X_n, n \geq 0$, is **transient** if there exists a nonnegative function $f(j), j \in S$, and a set $A \subset S$ such that, for all $i \notin A$, $\mathsf{E}\big(f(X_{n+1})|X_n = i\big) - f(i) \leq 0$, and, there exists a $j \notin A$ such that, for all $i \in A$, $f(j) < f(i)$. ∎

Theorem D.3
An irreducible DTMC $X_n, n \geq 0$, is **positive recurrent** if there exists a nonnegative function $f(j), j \in S$, and a *finite* set $A \subset S$, such that, for all $i \notin A$, $\mathsf{E}\big(f(X_{n+1})|X_n = i\big) - f(i) \leq -\epsilon$, for some $\epsilon > 0$, and, for all $i \in A$, $\mathsf{E}\big(f(X_{n+1})|X_n = i\big) < B$, for some finite number B. ∎

It may appear, intuitively, that if $E(X_{n+1}|X_n = i) - i \geq 0$ for all i larger than some finite value then the Markov chain should not be positive recurrent. However, this is not the case and there are counterexamples. Basically, in these examples, even though the mean drift of the process is positive, the chain can return to small values of state *in a single transition* from any state i no matter how large. The following theorem basically eliminates this possibility in order to provide a "converse" to Theorem D.3.

Theorem D.4

An irreducible DTMC $X_n, n \geq 0$, on $i \in \{0, 1, 2, \ldots\}$, is *not* positive recurrent if there exist finite values $K > 0$ and $B > 0$, such that, for all $i \geq 0$,

$$E(X_{n+1} \mid X_n = i) < \infty,$$

for all $i \geq K$,

$$E(X_{n+1} \mid X_n = i) - i \geq 0,$$

and, for all $i \geq K$,

$$E((X_n - X_{n+1})^+ \mid X_n = i) \leq B$$

∎

In the context of Theorems D.1, D.2, and D.3 this theorem is stated for $f(j) = j$. In this theorem, the last requirement states that for large i, the mean *downward* drift must be bounded for states $i \geq K$ (notice that $(X_n - X_{n+1})^+$ is nonzero only if $X_{n+1} < X_n$).

D.2   Continuous Time Markov Chains

We have a continuous time stochastic process $\{X(t), t \geq 0\}$ that takes values in the discrete state space $\mathcal{S}$.

Definition D.1

$\{X(t)\}$ is a continuous time Markov chain (CTMC) on $\mathcal{S}$ if for all $t, s \geq 0$, for all $j \in \mathcal{S}$,

$$Pr(X(t + s) = j|X(u), u \leq t) = Pr(X(t + s) = j|X(t))$$

∎

We will assume time homogeneity and write $p_{i,j}(t) := Pr(X(t + s) = j|X(s) = i)$, which will be the elements of the transition probability matrix over time t, denoted by $P(t)$. For all $t \geq 0$, define $W(t) = \inf\{s > 0 : X(t + s) \neq X(t)\}$; i.e., $W(t)$ is the time after which the process leaves the state it is in at time t. The Markov property itself leads to the following important result.

Theorem D.5

For a CTMC $\{X(t)\}$, for all $i \in S$, and all $t \geq 0$, and $u \geq 0$,

$$\Pr(W(t) > u|X(t) = i) = e^{-a_i u}$$

for some constant $a_i \geq 0$. ■

Thus the time for which the CTMC stays in a state is exponentially distributed with a parameter that depends only on the state.

A state $i \in S$ is called *absorbing* if $a_i = 0$. We will assume that for all states $a_i < \infty$. Under this assumption a CTMC is a *pure jump process*; the process enters a state, spends a positive amount of time in that state, and then moves on to another state (unless the state entered is absorbing). With this picture in mind, define $T_0 = 0, T_1, T_2, \ldots$ to be the successive jump instants of a CTMC, and let $X_n = X(T_n)$. The sequence $T_n, n \geq 0$, is called a sequence of *embedded instants*, and the state sequence $X_n, n \geq 0$, is called the *jump chain*, or an *embedded process*. The following is an important characterization of the jump instants and the jump process.

Theorem D.6

Given a CTMC $\{X(t)\}$, with jump instants $T_n, n \geq 0$, and jump chain $X_n, n \geq 0$, for $i_0, i_1, \ldots, i_{n-1}, i, j \in S, t_0, t_1, \ldots, t_n, u \geq 0$,

$$\Pr\{X_{n+1} = j, T_{n+1} - T_n > u|X_0 = i_0, \ldots, X_{n-1} = i_{n-1},$$

$$X_n = i, T_0 = t_0, \ldots, T_n = t_n\} = p_{i,j}e^{-a_i u},$$

where $p_{i,j} \geq 0, \sum_{j \in S} p_{i,j} = 1$, and if $a_i > 0$, then $p_{i,i} = 0$. ■

This result states that, given the entire state process until the jump T_n (including the state, i, entered at this jump), the time spent in this state (i.e., $T_{n+1} - T_n$) and the state entered at the next jump (i.e., X_{n+1}) are independent, with the time spent in the state (i.e., i) being exponential with the same parameter as that of the unconditional sojourn time in the state.

We can conclude from this result that the embedded process is a DTMC with transition probabilities $p_{i,j}$, and further that the time that the process spends in a state is independent of the past, and is exponentially distributed with a parameter determined only by the state. Define $\mathbf{P}$ to be the transition probability matrix of the embedded DTMC. Thus a CTMC can be constructed as follows. First generate the DTMC using the transition probability matrix $\mathbf{P}$, then generate the sequence of state sojourn times, say $W_0, W_1, W_2, \ldots$ using the parameters a_i. The jump times $T_n, n \geq 0$, are obtained by concatenating the sojourn times; i.e., $T_0 = 0$, and, for $n \geq 1, T_n = \sum_{i=0}^{n} W_i$. Then define $X(t) = X_n$ if $t \in [T_n, T_{n+1})$, for $n \geq 0$. In general, this construction defines the process for all t only if $\sum_{i=0}^{\infty} W_i = \infty$. Such CTMCs are called *regular*. We will assume that this is the case, with probability one, for the

CTMCs with which we are concerned. Hence by this construction we can think of a (regular) CTMC in terms of the jump DTMC and the sequence of state sojourn times. The following are two criteria for recognizing that a CTMC is regular.

Theorem D.7

$\{X(t)\}$ is a CTMC with embedded DTMC $\{X_n\}$. The sojourn time parameters are $a_i, i \in S$.

    a. If there exists v such that $a_i \leq v$ for all i, then the CTMC is regular.

    b. If $\{X_n\}$ is recurrent then $\{X(t)\}$ is regular.

                                                                             ■

It can easily be seen that a CTMC is irreducible if and only if its DTMC is irreducible. For a state $j \in S$, let $X(0) = j$ and define $\tau_{j,j}$ to be the time until the process returns to state j after once leaving it.

Definition D.2

The state j in a CTMC is said to be *recurrent* if $\Pr(\tau_{j,j} < \infty) = 1$; otherwise j is *transient*. A recurrent state j is *positive* if $E(\tau_{j,j}) < \infty$; otherwise it is *null*.   ■

Just as in the case of DTMCs, it can be shown that the states of an irreducible CTMC are either all transient or all positive or all null. Correspondingly we say that an irreducible CTMC is transient, positive, or null.

It can be argued that j is recurrent in the CTMC $\{X(t)\}$ if and only if it is recurrent in its DTMC $\{X_n\}$, and an irreducible CTMC is recurrent if and only if its DTMC is recurrent. A similar result does *not* hold for positivity of states of the CTMC.

Given the transition probabilities (i.e., $p_{i,j}$) of the embedded DTMC $\{X_n\}$, and the state sojourn time parameters $a_i, i \in S$, define the $|S| \times |S|$ matrix Q as follows. For $i, j \in S, i \neq j$, $q_{i,j} = a_i p_{i,j}$, and for $i \in S$, $q_{i,i} = -a_i$. The off-diagonal terms in Q can be interpreted as the rate of leaving the state i to enter the state j conditional on being in the state i. Notice that the sum of each row of Q is 0. The following theorem provides an important criterion for the positivity of an irreducible regular CTMC.

Theorem D.8

An irreducible regular CTMC is positive if and only if there exists a positive probability vector π (i.e., $\pi_i, i \in S$, is a probability distribution on S), such that $\pi Q = 0$. When such a π exists it is unique.   ■

For an irreducible regular CTMC a probability vector π such that $\pi Q = 0$ is also a stationary probability vector; if $\Pr(X(0) = i) = \pi_i$, then $\Pr(X(t) = i) = \pi_i$ for all t. It can also be shown that $\pi_j = \frac{\frac{1}{a_j}}{E(\tau_{j,j})}$, the fraction of time that the process stays in state j. Further, unlike DTMCs, there is no notion of periodicity in CTMCs and the following holds for an irreducible positive recurrent CTMC:

$$\lim_{t\to\infty} p_{i,j}(t) = \pi_j$$

where π is the stationary probability vector.

When π is the stationary probability vector, the set of linear equations $\pi Q = 0$ has an important interpretation. The jth equation is

$$\sum_{i\in S, i\neq j} \pi_i q_{i,j} = \pi_j a_j$$

The right-hand side of this equation is the *unconditional* rate of leaving the state j, and each term in the summation on the left-hand side is the unconditional rate of leaving the state i to enter j (and, hence, the sum is the unconditional rate of entering the state j).

For many simple Markov chain models, examining the solutions of the system of linear equations $\pi Q = 0$, and $\sum_{i\in S} \pi_i = 1$, is the standard way for obtaining a condition for positive recurrence, and the corresponding stationary probability distribution.

Example D.1

Customers arrive to a queue with infinite storage space in a Poisson process with rate λ. The customers' service requirements are i.i.d. and exponentially distributed with mean $\frac{1}{\mu}$. There is a single server that serves at the rate of one unit of work per unit time. This is the M/M/1 queuing model. Let $X(t)$ be the number of customers at time t. Owing to the Poisson arrivals, and the exponential service requirements, it is easily seen that $X(t)$ is a CTMC. When $X(t) = 0$, the next state change occurs on an arrival and hence $a_0 = \lambda$. When $X(t) \geq 1$, the next state change occurs at the earliest of two instants: the current service completion whose residual time is exponentially distributed with mean $\frac{1}{\mu}$, or the next arrival whose residual arrival time is exponentially distributed with mean $\frac{1}{\lambda}$; hence the time until the next event is exponentially distributed with mean $\frac{1}{\mu+\lambda}$. It follows that $a_i = \mu + \lambda$, for $i \geq 1$. Clearly $p_{0,1} = 1$. Further, it can be seen that, for $i \geq 1$, $p_{i,(i-1)} = \frac{\mu}{\mu+\lambda}$, and $p_{i,(i+1)} = \frac{\lambda}{\mu+\lambda}$. Thus we find that the transition rate matrix Q has the following form: $q_{0,0} = -\lambda = -q_{0,1}$; for $i \geq 1$, $q_{i,(i-1)} = \mu$, $q_{i,(i+1)} = \lambda$, and $q_{i,i} = -(\mu + \lambda)$. It can now be checked that the system of equations $\pi Q = 0$ has a positive, summable solution if and only if $\lambda < \mu$ (the arrival rate is less than the maximum rate at which customers can be served), and further, defining $\rho = \frac{\lambda}{\mu}$, $\pi_i = (1 - \rho)\rho^i$, for $i \in \{0, 1, 2, \ldots\}$, is the stationary distribution. The fraction of time that the system has i customers (counting anyone in service) is π_i; in particular, the queue is empty during a fraction $(1 - \rho)$ of the time. If $\Pr(X(0) = i) = (1 - \rho)\rho^i$, then the CTMC is a stationary process.   ∎

D.3   Renewal Processes

There is a sequence of mutually independent random variables $X_k, k \in \{1, 2, 3, \ldots\}$, such that $X_k, k \geq 2$ are i.i.d., and X_1 can have a possibly different distribution from the rest. Think of the $X_k, k \geq 1$, as the *life-times* of a component that fails and is repeatedly replaced (e.g., a light bulb in a particular socket in one's home). The *renewal instants*, $Z_k, k \geq 1$, are defined as

$$Z_k = \sum_{i=1}^{k} X_k$$

The number of renewals in the interval $[0, t]$ is called the *renewal process*, and will be denoted by $M(t)$. Note that $M(t)$ jumps at each instant that a renewal occurs and stays "flat" in between. Since the life-time distributions can have a point mass at 0, there could be multiple jumps at the same instant (e.g., a new light bulb can fail the moment it is first switched on).

D.3.1   Renewal Reward Processes

A useful class of models arises when there are renewal instants, and also there is a reward that accumulates over time. There is a total reward associated with each renewal interval. Consider a renewal process with life-times $X_k, k \geq 1$. Associated with the life-time X_k is a reward R_k, such that the $R_k, k \geq 1$, are mutually independent. However, R_k can depend on X_k. In this renewal reward framework, let us define $C(t)$ to be the total reward accrued until time t. Then we may be interested in the reward rate, i.e., in $\lim_{t \to \infty} \frac{C(t)}{t}$.

Theorem D.9   Renewal Reward Theorem

For $E(|R_k|) < \infty$ and $E(X_k) < \infty$ the following hold:

a.

$$\lim_{t \to \infty} \frac{C(t)}{t} = \frac{E(R_2)}{E(X_2)}$$

where the convergence is in the almost sure sense.

b.

$$\lim_{t \to \infty} \frac{E(C(t))}{t} = \frac{E(R_2)}{E(X_2)}$$

The reason for the subscript 2 appearing in the limit on the right-hand side is that $X_k, k \geq 2$, and $R_k, k \geq 2$, are i.i.d., and the values in the first cycle do not matter in the limit. The second part of the theorem is just a convergence of a deterministic sequence of numbers, and does not follow simply by taking expectation in the first part, since convergence almost surely does not in general imply convergence of expectations.

D.3.2 The Excess Distribution

Given a nonnegative random variable X with distribution $F(x)$, and with finite mean $EX = \int_0^\infty (1 - F(u))du$, define a distribution $F_e(\cdot)$ as follows:

$$F_e(y) := \frac{1}{EX} \int_0^y (1 - F(u))du$$

Clearly, $F_e(\cdot)$ has the properties of a probability distribution function; in particular, it is nondecreasing in its argument, and $\lim_{y\to\infty} F_e(y) = 1$. The term "excess" distribution, or "excess life" distribution comes from the following fact. Consider a renewal process with i.i.d. life-times $X_k, k \geq 1$, with common distribution $F(\cdot)$, and with finite mean life-time. Let $Y(t)$ be the *residual life* or *excess life* at time t; i.e., at t, $Y(t)$ is the time until the first renewal in (t, ∞). Consider

$$\lim_{t\to\infty} \frac{1}{t} \int_0^t I_{\{Y(u)\leq y\}} du$$

or

$$\lim_{t\to\infty} \frac{1}{t} \int_0^t \Pr(Y(u) \leq y)\, du$$

The first expression is the long run fraction of time that the excess life is less than or equal to y, and the second expression is the time average probability of the excess life being less than or equal to y. It can be shown, using Theorem D.9, that in each case these limits exist and are equal to $F_e(y)$. The first expression converges to this limit in the almost sure sense; the second converges as an ordinary limit of real numbers. Thus $F_e(\cdot)$ can be interpreted as the residual life distribution seen by a random observer of the renewal process.

D.3.3 Markov Renewal Processes

$X_n, n \geq 0$, is a random sequence taking values in $\mathcal{S}$, and $T_0 \leq T_1 \leq T_2 \ldots$ is a nondecreasing sequence of random times.

Definition D.3

The random sequence $(X_n, T_n), n \geq 0$, is a Markov Renewal process (MRP) if for $i_0, i_1, \ldots, i_{n-1}, i, j \in \mathcal{S}, t_0 \leq t_1 \leq \ldots \leq t_n, u \geq 0$,

$$\Pr\{X_{n+1} = j, T_{n+1} - T_n \le u | X_0 = i_0, \ldots, X_{n-1} = i_{n-1}, X_n = i, T_0 = t_0, \ldots, T_n = t_n\}$$

$$= \Pr\{X_{n+1} = j, T_{n+1} - T_n \le u | X_n = i\}$$

∎

Thus given X_n, the random vector $(X_{n+1}, T_{n+1} - T_n)$ is independent of anything else in the past. Note that this property holds for the EMC and the jump times of a CTMC (see Theorem D.6). In a CTMC, however, $X_{(n+1)}$ and $(T_{n+1} - T_n)$ are independent given X_n. Hence an MRP is a generalization. Further, in an MRP the state sojourn times $(T_{n+1} - T_n), n \ge 0$, need not be exponential. Define $p_{i,j} = \lim_{u \to \infty} \Pr\{X_{n+1} = j, T_{n+1} - T_n \le u | X_n = i\}$, assuming that the limit does not depend on n. Then $X_n, n \ge 0$, is a DTMC on $\mathcal{S}$ with transition probabilities $p_{i,j}, i, j, \in \mathcal{S}$. Further define, for $i, j, \in \mathcal{S}$,

$$H_{i,j}(u) = \Pr\big((T_{n+1} - T_n) \le u | X_n = i, X_{n+1} = j\big)$$

Thus $H_{i,j}(u)$ is the distribution of the sojourn time in a state given this state and the state entered at the end of this sojourn. The distribution of the sojourn time in state i is given by

$$H_i(u) = \sum_{j \in \mathcal{S}} p_{i,j} H_{i,j}(u)$$

and the mean sojourn time in state i, denoted by σ_i, is given by

$$\sigma_i = \sum_{j \in \mathcal{S}} p_{i,j} \sigma_{i,j}$$

where $\sigma_{i,j}$ is the mean of the distribution $H_{i,j}(u)$.

Consider now a reward $R_k \ge 0$ associated with the interval (T_{k-1}, T_k), for $k \ge 1$, such that R_k is independent of anything else given (X_{k-1}, X_k) and $(T_k - T_{k-1})$ (the states at the endpoints of and the length of the interval (T_{k-1}, T_k)). Let r_j be the expected reward in an interval that begins in the state j. Let $C(t)$ be the cumulative reward until time t. Suppose that $X_k, k \ge 0$, is a positive recurrent DTMC on $\mathcal{S}$, with stationary probability $\pi_j, j \in \mathcal{S}$. Then, as in the renewal reward theorem (Theorem D.9), under the condition that $\sum_{j \in \mathcal{S}} \pi_j \sigma_j < \infty$, it can be shown that

$$\lim_{t \to \infty} \frac{C(t)}{t} = \frac{\sum_{j \in \mathcal{S}} \pi_j r_j}{\sum_{j \in \mathcal{S}} \pi_j \sigma_j}$$

where the convergence is in the almost sure sense.

D.4 Some Topics in Queuing Theory

D.4.1 Little's Theorem

Consider a system into which packets arrive at the instants $a_k, k \geq 1$. Let $A(t)$ denote the cumulative number of arrivals in the interval $[0, t]$. Let d_k denote the instant at which the kth arriving packet leaves the system. Note that packets need not depart in their arrival order. Let $W_k = d_k - a_k$ denote the *sojourn time* of the k-th packet in the system. Associated with packet k is a function $I_k(t), t \geq 0$, such that

$$I_k(t) = \begin{cases} 1 & \text{if packet } k \text{ is present at time } t \\ 0 & \text{otherwise,} \end{cases}$$

i.e., $I_k(t) = 1$ for $a_k \leq t \leq a_k + W_k$, and $I_k(t) = 0$ otherwise. Packet k is in the system at time t if and only if $I_k(t) = 1$. It follows that, for all $t \geq 0$, the number in the system is given by

$$N(t) = \sum_{j=1}^{\infty} I_j(t),$$

since the right hand side of the equation counts all those packets that are in the system at time t.

Little's theorem relates the long run averages of $N(t)$, W_k, and $A(t)$.

Theorem D.10   Little

(i) If, for a sample path (i.e., realization of the process) denoted by ω, $\frac{A(t)}{t} \rightarrow \lambda(\omega)$, and $\frac{1}{n} \sum_{j=1}^{n} W_j \rightarrow W(\omega)$, then $\lim_{t \to \infty} \frac{1}{t} \int_0^t N(u) du = N(\omega)$ exists, and $N(\omega) = \lambda(\omega) W(\omega)$.

(ii) If the limits, for the various sample paths ω, are equal to the constants λ and W with probability 1 then $\lim_{t \to \infty} \frac{1}{t} \int_0^t N(u) du = N(\omega) = \lambda W$ with probability 1; we denote by N the average number in the system, and obtain $N = \lambda W$ with probability 1. ∎

Part (i) of Little's theorem holds for any sample path for which the required time averages exist. The second part (which is the one that is used most often) states that if the packet average system time and the packet arrival rate are constant over a set of sample paths of probability 1, taking the values W and λ on this set, then, with probability 1, the time average number of packets in the system is $N = \lambda W$ with probability 1. In applications, we would know of the *existence* of N, λ, and W, by some other means, and Little's theorem would be used to relate these quantities; for example, to determine one quantity when the other two are known.

D.4.2   Poisson Arrivals See Time Averages (PASTA)

In a discrete event model, we may need to ask a question about a process $X(t), t \geq 0$, if it is observed only at a sequence of random time points $t_k, k \geq 0$. How does the answer of such a question relate to the process observed over all time? For example, consider a stable D/D/1 queue. In such a system, customers arrive periodically at intervals of length a, and require a service time of $b < a$. Let $X(t)$ be the number of customers in the system at time t. Note that $X(t) \in \{0, 1\}$. Clearly, the following holds:

$$\lim_{t \to \infty} \frac{1}{t} \int_0^t X(u)du = \frac{b}{a}$$

The average number of customers *over all time* is $\frac{b}{a}$. Let $t_k = ka$, for $k \geq 0$, and let us look at

$$\lim_{n \to \infty} \frac{1}{n} \sum_{k=0}^{n-1} X(t_{k-})$$

where $X(t_{k-})$ means $X(\cdot)$ observed just before t_k. Thus this is the average number of customers *seen by arrivals*. This is clearly 0, since arrivals always find the system empty (the previous arrival always leaves before the next arrival comes in a stable D/D/1 queue). However, when the instants at which the process is observed form a Poisson process, and under an additional independence assumption, the time averages observed at these Poisson points are the same as the averages over all time. We now state this formally.

Let $X(t), t \geq 0$, be a random process. Let B denote a set in the state space of $X(t)$. $A(t)$ is a Poisson process of rate λ; $A(t)$ denotes the number of Poisson arrivals in $(0, t]$, and $t_k, k \geq 1$, denote the points of the Poisson process. Define

$$V^B(t) = \frac{1}{t} \int_0^t I_{\{X(u) \in B\}} du$$

$V^B(t)$ is the fraction of time over $[0, t]$ that the process $X(\cdot)$ is in the set B. Also, define

$$V_A^B(t) = \frac{1}{A(t)} \sum_{k=1}^{A(t)} I_{\{X(t_{k-}) \in B\}}$$

$V_A^B(t)$ is the fraction of arrivals over $(0, t]$ that see the process in the set B (not counting the arriving customer). We are interested in relating the limit of $V_A^B(t)$, the fraction of customers that find $X(\cdot)$ in the set B, to the limit of $V^B(t)$, i.e., the fraction of time that the process $X(\cdot)$ spends in the set B. An independence assumption is required:

Lack of Anticipation Assumption: For all $t \geq 0$, $A(t + u) - A(t), u \geq 0$, is independent of $X(s), 0 \leq s \leq t$; i.e., for all $t \geq 0$, the future arrivals (more precisely, the future increments of the arrival process) are independent of the past of the process $X(\cdot)$.

Remark D.1

Where a system receives external arrivals as independent Poisson processes, this independence assumption clearly holds. This follows since the Poisson process has independent increments. $X(s), 0 \leq s \leq t$, depends on the arrival process up to t, which is independent of the increments of the arrival process after t (see Appendix B, Section B.3.5).

Theorem D.11   PASTA

Under the Lack of Anticipation Assumption, $V_B(t) \overset{w.p.\ 1}{\to} \bar{V}_B$ if and only if $V_A^B(t) \overset{w.p.\ 1}{\to} \bar{V}_B$; i.e., the time average and the arrival average converge with probability 1, and to the same values.   ∎

As an application, consider the M/G/c/c model (see Section D.5.1) that is used for modeling telephone calls using a trunk group of c trunks. The arrival rate is λ, the mean holding time is $\frac{1}{\mu}$, and $\rho := \lambda\mu$. If $N(t)$ is the number of trunks occupied at time t, then Markov chain analysis (see Section D.5.1) yields, for $0 \leq n \leq c$,

$$\frac{1}{t}\int_0^t I_{\{N(u)=n\}}\,du \overset{w.p.\ 1}{\to} \frac{\frac{\rho^n}{n!}}{\sum_{j=0}^c \frac{\rho^j}{j!}} =: \pi_n$$

Since the arrival process is Poisson and independent of the future evolution of the $N(u)$ process, we can conclude from PASTA (i.e., Theorem D.11) that

$$\frac{1}{A(t)}\sum_{k=1}^{A(t)} I_{\{N(t_{k-})=c\}} \overset{w.p.\ 1}{\to} \pi_c$$

The left-hand side is the number of arrivals that find all c trunks occupied, and hence is the probability of call blocking.

D.5   Some Important Queuing Models

D.5.1   The M/G/c/c Queue

Consider a queuing system in which there are c servers, each of which serves at unit rate. Customers arrive in a Poisson process of rate $\lambda, 0 < \lambda < \infty$. The service requirements are i.i.d. and are generally distributed, with distribution $F(\cdot)$, with finite mean $\frac{1}{\mu}, 0 < \mu < \infty$. Each arriving customer is assigned to a free server if one exists, otherwise the arriving customer is denied admission and it goes away never to be heard from again. Since a customer "holds" a dedicated server for the entire duration of its service, the service requirements are also called *holding times*. Let $X(t)$ denote the number of customers in the queue at time t. Define $\rho = \frac{\lambda}{\mu}$. Observe that ρ is the average number of new arrivals during the holding time of a customer. This is a measure of *load* or *traffic intensity* and is given the units of *Erlangs*. Now, it can be seen that if the holding time distributions are exponential, then $X(t)$ is a positive recurrent CTMC on the state space $\{0, 1, \ldots, c\}$. The stationary distribution is given by

$$\pi_n = \frac{\frac{\rho^n}{n!}}{\sum_{j=0}^{c} \frac{\rho^j}{j!}} \tag{D.1}$$

The performance measure that is commonly of interest in the M/G/c/c model is the probability of blocking. In Section D.4.2 we have shown how PASTA is used to establish that the probability of blocking when the load is ρ Erlangs, is given by

$$B(\rho, c) = \pi_c \tag{D.2}$$

When the holding times are not exponentially distributed, then $X(t)$ is not a Markov chain. When $X(t) = n$, let $Y_i(t), 1 \le i \le n$, denote the residual service requirements of the customers in the system. It is then easily seen that the process $(X(t), Y_1(t), \ldots, Y_{X(t)}(t))$ is a Markov process. To see this we observe that, given this state at time t, no past information is required to evolve the process; the arrivals come in an independent Poisson process, and each of the residual service times reduces at the rate 1. It can be shown that the stationary joint distribution of the process is given by

$$\Pr\big(X(t) = n, Y_1 \le y_1, \ldots, Y_n \le y_n\big) = \pi_n \prod_{i=1}^{n} F_e(y_i) \tag{D.3}$$

where π_n is as displayed in (D.1), and $F_e(\cdot)$ is the excess distribution of the holding time distribution (see Section D.3.2).

D.5.2 The Processor Sharing Queue

Consider a queue with an infinite amount of waiting space. Customers arrive in a Poisson process with rate λ. The service requirements are i.i.d. and are generally distributed with common distribution $F(\cdot)$, with finite mean. As long as there is work to be done, the server reduces the total amount of unfinished work at the rate of 1 unit per second. When there are n customers in the system then the unfinished work on the ith customer reduces at rate $\frac{1}{n}$. This is called the M/G/1 *processor sharing (PS)* queuing model; the G in the notation refers to the generally distributed service requirements. Let $X(t)$ denote the number of customers at time t. If the service requirements are exponentially distributed with mean $\frac{1}{\mu}$, then $X(t)$ is a CTMC, which is positive recurrent if and only if $\frac{\lambda}{\mu} < 1$. In that case, defining $\rho := \frac{\lambda}{\mu}$, the stationary distribution of $X(t)$ is given by

$$\pi_n = (1 - \rho)\rho^n$$

In general, when the service requirements are not exponentially distributed, $X(t)$ is not a Markov chain. When $X(t) = n$, let $Y_i(t), 1 \leq i \leq n$, denote the residual service requirements of the customers in the system. It is then easily seen that the process $(X(t), Y_1(t), \ldots, Y_{X(t)}(t))$ is a Markov process. To see this observe that, given this state at time t, no past information is required to evolve the process; the arrivals come in an independent Poisson process, and each of the residual service times reduces at the rate $\frac{1}{n}$. Further, if $\rho < 1$, it can be shown that the stationary joint distribution of the process is given by

$$\mathsf{Pr}\big(X(t) = n, Y_1 \leq y_1, \ldots, Y_n \leq y_n\big) = (1 - \rho)\rho^n \prod_{i=1}^{n} F_e(y_i) \qquad \text{(D.4)}$$

where $F_e(\cdot)$ is the excess distribution of the service time distribution (see Section D.3.2).

From the form of the stationary distribution displayed, we conclude that the residual service times are independent conditional on the number of customers in the system. Further, the stationary residual service time distribution is just the excess life distribution of the service time distribution. Finally, a very important conclusion from Equation D.4 is that the stationary distribution of the number of customers in an M/G/1 PS queue (i.e., the marginal distribution of $X(t)$) is the same as that in an M/M/1 queue, and, hence, is *insensitive* to the distribution of the service time (except through its mean).

Since the stationary distribution of the number of customers in an M/G/1 PS queue is $\pi_n = (1 - \rho)\rho^n$, the time average number of customers is $\mathsf{E}(X) = \frac{\rho}{1-\rho}$, hence by Little's Theorem (Theorem D.10) the mean sojourn time is

$$\mathsf{E}(W) = \frac{\mathsf{E}(S)}{1 - \rho}$$

where S is the service time random variable. Again, we see that the mean sojourn time in the M/G/1 PS queue is insensitive to the actual distribution of the service time (except through its mean). Further, an important result is that the mean conditional sojourn time of a customer with service requirement s is given by

$$\mathsf{E}(W|S = s) = \frac{s}{1 - \rho}$$

D.6 Notes on the Literature

In this chapter we have brought together some classical results in Markov chains, renewal processes, and queuing theory, as a ready reference when reading the main chapters of the book. There are many excellent books on these topics. Our treatment of Markov chains and renewal theory is based on the textbook by Wolff [136]. A first course on probabilistic modeling and queuing theory is provided by

Mitrani [102]. A more sophisticated treatment is available in the books by Fayolle et al. [34] and by Bremaud [18]. The Lyapunov drift criteria for the stability analysis of Markov chains are available in these two books. The two volumes by Kleinrock ([77] and [78]) provide a compendium of results on a vast variety of single station queuing models.

Bibliography

[1] N. Abramson. The ALOHA system—another alternative for computer communications. In *Proceedings of Fall Joint Computing Conference*, pages 281–285, 1970.

[2] R. Agrawal, R. Berry, J. Huang, and V. Subramanian. Optimal scheduling for OFDMA systems. In *Asilomar Conference on Signals, Systems, and Computers*, pages 1347–1351, 2006.

[3] F. Anjum and L. Tassiulas. On the behavior of different TCP algorithms over a wireless channel with correlated packet losses. In *Proceedings of ACM SIGMETRICS*. ACM, 1999.

[4] N. Bambos. Toward power-sensitive network architectures in wireless communications: Concepts, issues, and design approaches. *IEEE Personal Communications*, pages 50–59, June 1998.

[5] N. Bansal and Z. Liu. Capacity, delay and mobility in wireless ad-hoc networks. In *Proc. of IEEE INFOCOM*, 2003.

[6] M. S. Bazaraa, H. D. Sherali, and C. M. Shetty. *Nonlinear Programming—Theory and Algorithms*. Series in Discrete Mathematics and Optimization. John Wiley, 2nd edition, 1993.

[7] Y. Bejerano, S.-J. Han, and Li (Erran) Li. Fairness and load balancing in wireless LANs using association control. In *Proc. of ACM MobiCom*, pages 315–329, Philadelphia PA, USA, 2004.

[8] P. Bender, P. Black, M. Grob, R. Padovani, N. Sindhushayana, and A. Viterbi. CDMA/HDR: A bandwidth-efficient high-speed wireless data service for nomadic users. *IEEE Communications Magazine*, pages 70–77, July 2000.

[9] R. A. Berry and R. G. Gallager. Communication over fading channels with delay-constraints. *IEEE Transactions on Information Theory*, 48(5):1135–1149, May 2002.

[10] D. Bertsekas and R. G. Gallager. *Data Networks*. Prentice Hall, 2nd edition, 1992.

[11] V. Bharghavan, A. Demers, S. Shenker, and L. Zhang. MACAW: A media access protocol for wireless LANs. In *Proceedings of ACM SIGCOMM*, pages 212–225, 1994.

[12] G. Bianchi. Performance analysis of the IEEE 802.11 distributed coordination function. *IEEE Journal on Selected Areas in Communications*, 18(3):535–547, March 2000.

[13] J. A. C. Bingham. Multicarrier modulation for data transmission: An idea whose time has come. *IEEE Communications Magazine*, pages 5–14, May 1990.

[14] B. Bollobas. *Random Graphs*. Academic Press, 1985.

[15] T. Bonald and A. Proutiere. Wireless downlink data channels: User performance and cell dimensioning. In *Proc. ACM MobiCom*, 2003.

[16] S. Boyd, L. El Ghaoui, E. Feron, and V. Balakrishnan. *Linear Matrix Inequalities in System and Control Theory*. SIAM, 1994.

[17] P. Bremaud. *An Introduction to Probabilistic Modelling*. Springer-Verlag, 1988.

[18] P. Bremaud. *Markov Chains: Gibbs Fields, Monte Carlo Simulation, and Queues*. Springer, 1999.

[19] F. Cali, M. Conti, and E. Gregori. IEEE 802.11 wireless LAN: Capacity analysis and protocol enhancement. In *Proc. of IEEE INFOCOM*, 1998.

[20] F. Cali, M. Conti, and E. Gregori. IEEE 802.11 protocol: Design and performance evaluation of an adaptive backoff mechanism. *IEEE Journal on Selected Areas in Communications*, 18(9):1774–1780, September 2000.

[21] M. M. Carvalho and J. J. Garcia-Luna-Aceves. Delay analysis of the IEEE-802.11 in single-hop networks. In *Proc. of the 11th IEEE International Conference on Network Protocols (ICNP'03)*, Atlanta GA, USA, November 4–7, 2003.

[22] Y. C. Cheng and T. G. Robertazzi. Critical connectivity phenomena in multihop radio models. *IEEE Transactions on Communication*, 37(7):770–777, July 1989.

[23] H. S. Chhaya and S. Gupta. Performance modeling of asynchronous data transfer methods of IEEE 802.11 MAC protocol. *Wireless Networks*, 3:217–234, 1997.

[24] M. de Berg, M. van Kreveld, M. Overmars, and O. Schwarzkopf. *Computational Geometry: Algorithms and Applications*. Springer-Verlag, 1st edition, 1997.

[25] N. G. De Bruijn. *Asymptotic Methods in Analysis*. Dover Publications, New York, 1981.

[26] M. P. Desai and D. Manjunath. On the connectivity of finite ad hoc networks. *IEEE Communications Letters*, 10(6):437–490, 2002.

[27] L. Doherty, K. Pister, and L. El Ghaoui. Convex position estimation in wireless sensor networks. In *Proc. of IEEE INFOCOM*, pages 1665–1663, 2001.

[28] X. J. Dong, M. Ergen, P. Varaiya, and A. Puri. Improving the aggregate throughput of access points in IEEE 802.11 wireless LANs. In *Proc. of the 28th Annual Local Computer Networks Conference*, 2003.

[29] O. Dousse, F. Baccelli, and P. Thiran. Impact of interference on connectivity of ad hoc networks. *IEEE Transactions on Networking*, 13(2):425–436, April 2005.

[30] O. Dousse and P. Thiran. Connectivity *vs* capacity in dense ad hoc networks. In *Proc. of IEEE INFOCOM*, 2004.

[31] M. Ergen, S. Coleri, and P. Varaiya. QoS aware adaptive resource allocation techniques for fair scheduling in OFDMA based broadband wireless access systems. *IEEE Transactions on Broadcasting*, 49(4):362–370, December 2003.

[32] J. S. Evans and D. Everitt. Effective bandwidth-based admission control for multiservice CDMA cellular networks. *IEEE Transactions on Vehicular Technology*, 48(1):36–46, January 1999.

[33] G. Fayolle, E. Gelenbe, and J. Labetoulle. Stability and optimal control of the packet switching broadcast channel. *Journal of the ACM*, 24(3):375–386, 1977.

[34] G. Fayolle, V. A. Malyshev, and M. V. Menshikov. *Topics in the Constructive Theory of Countable Markov Chains*. Cambridge University Press, 1995.

[35] G. Foschini and Z. Miljanic. A simple distributed autonomous power control and its convergence. *IEEE Transactions on Vehicular Technology*, 42(4):641–646, November 1993.

[36] M. Franceschetti, O. Dousse, D. N. C. Tse, and P. Thiran. Closing the gap in the capacity of wireless networks via percolation theory. *IEEE Transactions on Information Theory*, 53(3):1009–1018, March 2007.

[37] A. El Gamal, J. P. Mammen, B. Prabhakar, and D. Shah. Optimal throughput-delay scaling in wireless networks: Part I: The fluid model. *IEEE/ACM Transactions on Networking*, 14(SI):2568–2592, 2006.

[38] A. Gamst. Homogeneous distribution of frequencies in a regular hexagonal cell system. *IEEE Transactions on Vehicular Technology*, 31(3):132–144, 1982.

[39] V. K. Garg and J. E. Wilkes. *Wireless and Personal Communications Systems*. Prentice Hall PTR, 1996.

[40] L. Georgiadis, M. J. Neely, and L. Tassiulas. Resource allocation and cross layer control in wireless networks. *Foundations and Trends in Networking*, 1(1):1–144, 2006.

[41] A. Giridhar and P. R. Kumar. Computing and communicating functions over sensor networks. *IEEE Journal on Selected Areas in Communications*, 23(4):755–764, April 2005.

[42] E. Godehardt and J. Jaworski. On the connectivity of a random interval graph. *Random Structures and Algorithms*, 9:137–161, 1996.

[43] A. Goldsmith. *Wireless Communications*. Cambridge University Press, New York, 2005.

[44] A. J. Goldsmith and P. P. Varaiya. Capacity of fading channels with channel side information. *IEEE Transaction on Information Theory*, 43(6):1986–1992, November 1997.

[45] M. Goyal, A. Kumar, and V. Sharma. Power constrained and delay optimal policies for scheduling transmission over a fading channel. In *Proceedings of IEEE INFOCOM*. IEEE, April 2003.

[46] R. L. Graham, D. E. Knuth, and O. Patashnik. *Concrete Mathematics*. Addison-Wesley, 2nd edition, 1998.

[47] M. Grossglauser and D. Tse. Mobility increases the capacity of wireless ad hoc networks. In *Proc. of IEEE INFOCOM*, pages 1360–1369, 2001.

[48] B. G. Gupta, S. K. Iyer, and D. Manjunath. On the topological properties of one dimensional exponential random geometric graphs. *Revision submitted*, 2006.

[49] P. Gupta and P. R. Kumar. Critical power for asymptotic connectivity in wireless networks. In W. M. McEneaney, G. Yin, and Q. Zhang, editors, *Stochastic Analysis, Control, Optimization and Applications: A Volume in Honor of W. H. Fleming*. Birkhauser, Boston, 1998.

[50] P. Gupta and P. R. Kumar. The capacity of wireless networks. *IEEE Transactions on Information Theory*, 46(2):388–404, 2000.

[51] B. Hajek and G. Sasaki. Link scheduling in polynomial time. *IEEE Transactions on Information Theory*, 34(5, Part 1):910–917, September 1988.

[52] S. V. Hanly. An algorithm for combined cell-site selection and power control to maximise cellular spread spectrum capacity. *IEEE Journal on Selected Areas in Communications*, 13(7):1332–1340, September 1995.

[53] S. Harsha, A. Kumar, and V. Sharma. An analytical model for the capacity estimation of combined VoIP and TCP file transfers over EDCA in an IEEE 802.11e WLAN. In *Proc. of the 14th IEEE International Workshop on Quality of Service (IWQoS)*, Yale University, New Haven CT, USA, June 2006.

[54] M. Hassan, A. Nayandoro, and M. Atiquzzaman. Internet telephony: Service, technical challenges, and products. *IEEE Communications Magazine*, 38(4):96–103, April 2000.

[55] K. Hoffman. *Analysis in Euclidean Space*. Prentice Hall, 1975.

[56] D. Hole and F. A. Tobagi. Capacity of an IEEE 802.11b wireless LAN supporting VoIP. In *Proc. of IEEE ICC*, 2004.

[57] T. Holliday, A. Goldsmith, P. Glynn, and N. Bambos. Distributed power and admission control for time varying wireless channels. In *Proc. IEEE Globecom*, pages 768–774, 2004.

[58] H. Holma and A. Toskala. *WCDMA for UMTS: Radio Access for Third Generation Mobile Communications*. John Wiley and Sons, 2002.

[59] J. Huang, V. Subramanian, R. Agrawal, and R. Berry. Downlink scheduling and resource allocation for OFDM systems. In *Conference on Information Sciences and Systems (CISS)*, 2006.

[60] J. Y. Hui. Resource allocation in broadband networks. *IEEE Journal on Selected Areas in Communications*, 6:1598–1608, 1988.

[61] IEEE. *IEEE std. 802.15.4, "Part 15.4: Wireless MAC and PHY Specifications for Low-Rate Wireless Personal Area Networks."* IEEE, New York, May 2003.

[62] C. Intanagonwiwat, R. Govindan, D. Estrin, and J. Heidemann. Directed diffusion for wireless sensor networking. *IEEE/ACM Transactions on Networking*, 11(1):2–16, February 2003.

[63] S. K. Iyer and D. Manjunath. Topological properties of random wireless networks. *Sadhana: Proceedings of the Indian Academy of Sciences*, 2006.

[64] J. W. Jaromczyk and G. T. Toussaint. Relative neighbourhood graphs and their relatives. *Proceedings of the IEEE*, 80(9):1502–1517, September 1992.

[65] I. Kalet. The multitone channel. *IEEE Transactions on Communications*, 37(2): 119–124, February 1989.

[66] K. Kar, S. Sarkar, and L. Tassiulas. Achieving proportional fairness using local information in aloha networks. *IEEE Transactions on Automatic Control*, 49(10):1858–1863, October 2004.

[67] N. Karamchandani, D. Manjunath, and S. K. Iyer. On the clustering properties of exponential random networks. In *Proc. of the IEEE Intnl. Symp. World of Wireless, Mobile and Multimedia Networks (WoWMoM)*, Taormina, Italy, June 14–16, 2005.

[68] P. Karn. A new channel access method for packet radio. In *Proceedings of Ninth Computer Networking Conference*, pages 134–140, 1990.

[69] G. Kasbekar, J. Kuri, and P. Nuggehalli. Online association policies in IEEE 802.11 WLANs. In *Proc. of the Fourth International Symposium on Modeling and Optimization in Mobile, Ad Hoc, and Wireless Networks (WiOpt)*, Boston MA, April 2006.

[70] G. Kasbekar, J. Kuri, and P. Nuggehalli. Online client-AP association policies WLANs. In *Proc. of the Workshop on Resource Allocation in Wireless Networks (RAWNET)*, Boston MA, April 2006.

[71] M. Kaufmann, S. Rajasekaran, and J. F. Sibeyn. Matching the bisection bound for routing and sorting in a mesh. In *Proc. of ACM Symposium on Parallel Algorithms and Architectures*, pages 31–40, San Diego CA, USA, 1992.

[72] V. Kawadia and P. R. Kumar. A cautionary perspective on cross layer design. *IEEE Wireless Communications Magazine*, 12(1):3–11, February 2005.

[73] F. P. Kelly. Stochastic models of computer communication systems. *Journal of the Royal Statistical Society: Series B*, 47(3):379–395, 1985.

[74] F. P. Kelly. Effective bandwidths at multiclass queues. *Queueing Systems*, 9:5–16, 1991.

[75] F. P. Kelly, A. Maulloo, and D. Tan. Rate control for communication networks: Shadow price proportional fairness and stability. *Journal of the Operations Research Society*, 49:237–252, 1998.

[76] S. Kittipiyakul and T. Javidi. A-fresh-look at optimal subcarrier allocation in OFDMA systems. In *43rd IEEE Conference on Decision and Control*, pages 3289–3294, December 2004.

[77] L. Kleinrock. *Queueing Systems*, volume 1. John Wiley, 1975.

[78] L. Kleinrock. *Queueing Systems*, volume 2. John Wiley, 1976.

[79] L. Kleinrock and F. Tobagi. Packet switching in radio channels: Part I: Carrier sense multiple access and their throughput delay characteristics. *IEEE Transactions on Communications*, 23(12):1400–1412, 1975.

[80] M. Kodialam and T. Nandagopal. Characterizing achievable rates in multi-hop wireless mesh networks with orthogonal channels. *IEEE/ACM Transactions on Networking*, 13(4):868–880, August 2005.

[81] T. J. Kostas, M. S. Borella, I. Sidhu, G. M. Schuster, J. Grabiec, and J. Mahler. Real-time voice over packet-switched networks. *IEEE Network Magazine*, pages 18–27, 1998.

[82] E. Kranakis, H. Singh, and J. Urrutia. Compass routing on geometric networks. In *Proceedings of the 11th Canadian Conference on Computational Geometry*, pages 51–54, August 1999.

[83] F. Kuhn, R. Wattenhofer, Y. Zhang, and A. Zollinger. Geometric ad-hoc routing: Of theory and practice. In *Proceedings of the ACM Symposium on the Principles of Distributed Computing (PODC)*, pages 63–72, 2003.

[84] S. R. Kulkarni and P. Viswanath. A deterministic approach to throughput scaling in wireless networks. *IEEE Transactions on Information Theory*, 50(6):1041–1049, June 2004.

[85] A. Kumar. Comparative performance analysis of versions of TCP in a local area network with a lossy link. *IEEE Transactions on Networking*, 6(4):485–498, August 1998.

[86] A. Kumar, E. Altman, D. Miorandi, and M. Goyal. New insights from a fixed point analysis of single cell IEEE 802.11 WLANs. In *Proc. of IEEE INFOCOM*, 2005.

[87] A. Kumar and J. M. Holtzman. Performance analysis of versions of TCP in a local network with a mobile radio link. *SADHANA: Indian Academy of Sciences Proceedings in Engineering Sciences*, 23(1):113–129, February 1998. Also a WINLAB Technical Report, Rutgers University, 1996.

[88] A. Kumar and V. Kumar. Optimal association of stations and aps in an IEEE 802.11 WLAN. In *Proc. of the National Conference on Communications (NCC)*, IIT Kharagpur, February 2005.

[89] A. Kumar, D. Manjunath, and J. Kuri. *Communication Networking: An Analytical Approach*. Morgan-Kaufmann (an imprint of Elsevier), San Francisco, May 2004.

[90] G. Kuriakose, Sri Harsha, A. Kumar, and V. Sharma. Analytical models for capacity estimation of IEEE 802.11 WLANs using DCF for internet applications. *Wireless Networks*, accepted, to appear.

[91] S. S. Lam. A carrier sense multiple access protocol for local networks. *Computer Networks*, 4(1):21–32, 1980.

[92] E. A. Lee and D. G. Messerschmitt. *Digital Communication*. Kluwer Academic Publishers, February 1988.

[93] B. Li and R. Battiti. Supporting service differentiation with enhancements of the IEEE 802.11 MAC protocol: Models and analysis. Technical Report DIT-03-024, Dept. of Information and Communication Technology, University of Trento, May 2003.

[94] X. Lin, N. B. Shroff, and R. Srikant. A tutorial on cross-layer optimization in wireless networks. *IEEE Journal on Selected Areas in Communications*, 24(8):1452–1463, August 2006.

[95] X. J. Lin and N. B. Shroff. Joint rate control and scheduling in multihop wireless networks. In *Proc. of IEEE Conference on Decision and Control*, pages 1484–1489, Paradise Island, Bahamas, December 2004.

[96] V. H. MacDonald. Advanced mobile phone service: The cellular concept. *The Bell System Technical Journal*, 58:15–41, 1979.

[97] S. Mangold, S. Choi, P. May, O. Klein, G. Hiertz, and L. Stibor. IEEE 802.11e wireless LAN for quality of service. In *Proceedings of European Wireless*, February 2002.

[98] N. McKeown, A. Mekkittikul, V. Anantharam, and J. Walrand. Achieving 100% throughput in an input queued switch. *IEEE Transactions on Communications*, 47(8):1260–1267, 1999.

[99] K. Medepalli, P. Gopalakrishnan, D. Famolari, and T. Kodama. Voice capacity of IEEE 802.11b, 802.11a and 802.11g wireless LANs. In *Proc. of IEEE GLOBECOM*, 2004.

[100] D. Miorandi, A. A. Kherani, and E. Altman. A queueing model for HTTP traffic over IEEE 802.11 WLANs. *Computer Networks*, 50, 2006.

[101] D. Mitra and J. A. Morrison. A novel distributed power control algorithm for classes of service in cellular CDMA networks. In J. Holtzman and M. Zorzi, editors, *Advances in Wireless Communications,* pages 187–202. Kluwer Academic Publishers, 1998.

[102] I. Mitrani. *Probabilistic Modelling*. Cambridge University Press, 1997.

[103] M. L. Molle, K. Sohraby, and A. N. Venetsanopoulos. Space-time models for asynchronous CSMA protocols for local area networks. *IEEE Journal on Selected Areas in Communications*, 5(6):956–968, 1987.

[104] M. Mouly and M.-B. Pautet. *The GSM System Mobile Communications*. By the Authors, 1992.

[105] J. Mussachio and J. Walrand. WiFi access point pricing as a dynamic game. *IEEE Transactions on Networking*, 14(2), April 2006.

[106] R. Nagpal, H. Shrobe, and J. Bachrach. Organizing a global and coordinate system from local information on an ad hoc sensor network. In *Proc. of IPSN*, Lecture Notes in Computer Science 2634, April 2003.

[107] M. J. Neely, E. Modiano, and C. E. Rohrs. Dynamic power allocation and routing for time varying wireless networks. In *Proc. of IEEE INFOCOM*, San Francisco CA, USA, 2003.

[108] A. Papoulis. *Probability, Random Variables, and Stochastic Processes*. McGraw Hill, 1984.

[109] M. D. Penrose. *Random Geometric Graphs*. Oxford University Press, 2003.

[110] T. K. Philips, S. S. Panwar, and A. N. Tantawi. Connectivity properties of a packet radio network model. *IEEE Transactions on Information Theory*, 35(5):1044–1047, September 1989.

[111] P. Piret. On the connectivity of radio networks. *IEEE Transactions on Information Theory*, 37(5):1490–1492, September 1991.

[112] F. Lo Presti. Joint congestion control: Routing and media access control optimization via dual decomposition for ad hoc wireless networks. In *Proc. of the 8th ACM international Symposium on Modeling, Analysis and Simulation of Wireless and Mobile Systems (WiOpt)*, 2005.

[113] J. G. Proakis. *Digital Communications*. Electrical Engineering. McGraw Hill International Editions, 1995.

[114] V. Ramaiyan, A. Kumar, and E. Altman. Fixed point analysis of single cell IEEE 802.11e WLANs: Uniqueness, multistability and throughput differentiation. In *Proc. ACM Sigmetrics*, 2005.

[115] B. Raman and K. Chebrolu. Experiences in using WiFi for rural internet in India. *IEEE Communications Magazine*, pages 104–110, January 2007.

[116] T. S. Rappaport. *Wireless Communications: Principles and Practice*. Prentice Hall PTR, 1996.

[117] R. Rom and M. Sidi. *Multiple Access Protocols: Performance and Analysis*. Springer-Verlag, 1990.

[118] L. Romdhani, Q. Ni, and T. Turletti. Adaptive EDCF: Enhanced service differentiation for IEEE 802.11 wireless ad-hoc networks. In *Proc. of Wireless Communications and Networking Conference (WCNC)*, New Orleans LA, USA, March 16–20, 2003.

[119] M. Schwartz. *Broadband Integrated Networks*. Prentice Hall PTR, 1996.

[120] E. Seneta. *Nonnegative Matrices and Markov Chains*. Springer-Verlag, 2nd edition, 1980.

[121] G. Sharma, A. Ganesh, and P. Key. Performance analysis of contention based medium access control protocols. In *Proc. of IEEE INFOCOM*, Barcelona, Spain, 2006.

[122] G. Sharma, R. Mazumdar, and N. Shroff. Delay and capacity trade-offs in mobile ad hoc networks: A global perspective. In *Proc. of IEEE INFOCOM*, Barcelona, Spain, 2006.

[123] G. L. Stuber. *Principles of Mobile Communication*. Kluwer Academic Publishers, 2nd edition, 2001.

[124] H. Takagi and L. Kleinrock. Optimal Transmission Ranges for Randomly Distributed Packet Radio Terminals. *IEEE Transactions on Communications*, 32(3):246–257, March 1984.

[125] G. Tan and J. Guttag. Capacity allocation in wireless LANs. Technical Report 973, MIT CSAIL, Cambridge MA, UK, November 2004.

[126] L. Tassiulas and A. Ephremides. Jointly optimal routing and scheduling in packet ratio networks. *IEEE Transactions on Information Theory*, 38(1):165–168, January 1992.

[127] L. Tassiulas and A. Ephremides. Stability properties of constrained queueing systems and scheduling policies for maximum throughput in multihop radio networks. *IEEE Transactions on Automatic Control*, 37(12):1936–1948, 1992.

[128] L. Tassiulas and A. Ephremides. Dynamic server allocation to parallel queues with randomly varying connectivity. *IEEE Transactions on Information Theory*, 39(2):466–478, March 1993.

[129] Y. C. Tay and K. C. Chua. A capacity analysis for the IEEE 802.11 MAC protocol. *Wireless Networks*, 7:159–171, 2001.

[130] O. Tickoo and B. Sikdar. Queueing analysis and delay mitigation in IEEE 802.11 random access MAC based wireless networks. In *Proc. of IEEE INFOCOM*, Hong Kong, China, March 2004.

[131] D. Tse and P. Viswanath. *Fundamentals of Wireless Communication*. Cambridge University Press, New York, 2005.

[132] A. J. Viterbi. *CDMA Principles of Spread Spectrum Communication*. Wireless Communications. Addison-Wesley, 1995.

[133] J. Walrand and P. Varaiya. *High-Performance Communication Networks*. Morgan Kaufman, 1996.

[134] X. Wang, G. Xing, Y. Zhang, C. Lu, R. Pless, and C. Gill. Integrated coverage and connectivity configuration in wireless sensor networks. In *Proc. of ACM SENSYS*, pages 28–39, November 2003.

[135] D. B. West. *Introduction to Graph Theory*. Pearson Education, Prentice Hall, 2nd edition, 2006.

[136] R. W. Wolff. *Stochastic Modelling and the Theory of Queues*. Prentice Hall, Englewood Cliffs, New Jersey, 1989.

[137] F. Xue and P. R. Kumar. Scaling laws for ad hoc wireless networks: An information theoretic approach. *Foundations and Trends in Networking*, 1(2):145–270, 2006.

[138] R. D. Yates. A framework for uplink power control in cellular radio systems. *IEEE Journal on Selected Areas in Communications*, 13(7):1341–1347, September 1995.

[139] W. Ye, J. Heidemann, and D. Estrin. Medium access control with coordinated adaptive sleeping for wireless sensor networks. *IEEE/ACM Transactions on Networking*, 12(3):493–506, June 2004.

[140] E. Yoon, D. Tujkovic, and A. Paulraj. Exploiting channel statistics to improve the average sum rate in OFDMA systems. In *IEEE Vehicular Technology Conference, VTC 2005*, volume 2, pages 1053–1057, 2005.

[141] E. Yoon, D. Tujkovic, and A. Paulraj. Subcarrier and power allocation for an OFDMA uplink based on tap correlation information. In *ICC 2005*, pages 2744–2748, 2005.

[142] J. Zander. Performance of optimum transmitter power control in cellular radio systems. *IEEE Transactions on Vehicular Technology*, 41(1):57–62, February 1992.

[143] H. Zhang and J. Hou. On deriving the upper bound of α-lifetime for large sensor networks. In *Proc. of ACM MobiHoc*, pages 121–132, 2004.

[144] M. Zorzi, A. Chockalingam, and R. R. Rao. Throughput analysis of TCP on channels with memory. *IEEE Journal on Selected Areas in Communications*, 18(7):1289–1300, July 2000.

[145] M. Zorzi and R. R. Rao. Effect of correlated errors on TCP. In *Proceedings of the Conference on Information Sciences and Systems (CISS)*, March 1997.

[146] M. Zorzi, R. R. Rao, and L. B. Milstein. On the accuracy of a first-order Markov model for data transmission on fading channels. In *Proceedings of ICUPC*, November 1995.

Index

A

ABR. *See* Available bit rate
AC. *See* Access classes
Access classes (AC), 223, 224f
Access networks, 3–4
Access point (AP), 10, 205
  STA and, 226–227, 226f,
    234–235
ACK. *See* Acknowledgment
Acknowledgment (ACK), 71,
  73, 77, 226
Additive white Gaussian noise
  (AWGN), 18, 20, 25, 32,
  34, 44
Ad hoc networks, 3, 4f, 8, 10,
  205, 206f
Ad hoc wireless sensor
  networks
  applications of, 337
  communication coverage,
    339–341
  design and analysis of, 337
  distributed computing
    algorithms, 338
  function computation, 359
  intrusion detection, 338
  localization, 348
  packet scheduling
    algorithms, 338
  random deployment, 338,
    363f
  routing, 353
  scheduling, 368–369
  sensing coverage, 341–348
Advanced mobile phone service
  (AMPS), 120
AIFS. *See* Arbitration
  interframe space
Aloha protocol, 7, 190. *See
  also* Slotted Aloha
  collision, 191–192, 193f
  instability of, 195–197

packet transmission attempts
  in, 191–192, 193f
performance of, 190
propagation delay, 190
stability of, 197–198
transmission and reception
  in, 190, 191f
AMPS. *See* Advanced mobile
  phone service
Anchors, 339, 350–351
  sensor nodes and, 349–350,
    353f
  transmit power of, 348
AP. *See* Access point
Arbitrary networks
  exclusion disk, 319–320
  protocol model, 318–321
  transport capacity of,
    318–322
Arbitration interframe space
  (AIFS), 223–225
ARQ protocol, 75
Asynchronous transfer mode
  (ATM), 70
ATM. *See* Asynchronous
  transfer mode
Attribute-based routing,
  357–359
Available bit rate (ABR), 70
Average cost Markov decision
  problem, 176
AWGN. *See* Additive white
  Gaussian noise

B

Bandpass white Gaussian noise,
  18, 22
Bandwidth delay product
  (BDP), 72
Bandwidth sharing
  dynamic control of, 69–70

explicit feedback, 70
implicit feedback, 70
in wide area Internet, 71
in wireless mesh networks,
  274
Baseband pulse, 17–19
Base station (BS)
  CDMA cellular systems, 126
  FDM-TDMA systems, 82,
    85–86
  WiMAX, 56
Base station controller (BSC),
  6, 117
Base station subsystem, 117
Base transceiver station (BTS),
  117
Basic service set (BSS),
  205, 208
BDP. *See* Bandwidth delay
  product
Beacon transmissions, 208,
  235, 372
BER. *See* Bit error rate
Bernoulli distribution, 139
Binary modulation, 17–20,
  31
Birkhoff's strong ergodic
  theorem, 387
Bisection bound, 325
Bit carrier infrastructure, 2
Bit error rate (BER), 20, 23,
  128
  of binary modulation
    scheme, 31
  with fading, 30–32
  on wireless link, 73, 75
Blocking probability
  for FCA and MPA, 108–109
  for FDM-TDMA systems, 85
  for fixed channel allocation,
    108, 110
  for handovers, 115–117

Printed and bound by CPI Group (UK) Ltd, Croydon, CR0 4YY

03/10/2024

01040314-0005